Härdle • Klinke • Müller
XploRe – Learning Guide

Springer
Berlin
Heidelberg
New York
Barcelona
Hong Kong
London
Milan
Paris
Singapore
Tokyo

W. Härdle • S. Klinke • M. Müller

XploRe –
Learning Guide

Springer

Wolfgang Härdle
Sigbert Klinke
Marlene Müller
Humboldt-Universität zu Berlin
Institut für Statistik und Ökonometrie
Spandauer Straße 1
10178 Berlin, Germany
e-mail: haerdle/sigbert/marlene@wiwi.hu-berlin.de

Cover art: Detail of "Portrait of the Franciscan monk Luca Pacioli, humanist and mathematician (1445-1517)" by Jacopo de Barbari, Museo di Capodimonte, Naples, Italy

CIP data applied for
Die Deutsche Bibliothek – CIP-Einheitsaufnahme

Härdle, Wolfgang:
XploRe - learning guide / W. Härdle; S. Klinke; M. Müller.- Berlin; Heidelberg; New York;
Barcelona; Hong Kong; London; Milan; Paris; Singapore; Tokyo: Springer, 2000

ISBN-13: 978-3-540-66207-5   e-ISBN-13: 978-3-642-60232-0
DOI: 10.1007/978-3-642-60232-0

ISBN 978-3-540-66207-5 Springer-Verlag Berlin Heidelberg New York

Mathematics Subject Classification (1991): 62-04, 62-00

Typesetting: Camera-ready copy by authors
SPIN 10733011   40/3143CK-5 4 3 2 1 0 – Printed on acid-free paper

Contents

2 Descriptive Statistics   43

Marlene Müller

3 Graphics   69

Sigbert Klinke

Part II: Statistical Libraries   167

8   Neural Networks   **229**

Wolfgang Härdle and Heiko Lehmann

14 Wavelets   **375**

Yuri Golubev, Wolfgang Härdle, Zdeněk Hlávka, Sigbert Klinke,
Michael H. Neumann and Stefan Sperlich

Part III: Programming   409

15 Reading and Writing Data   411

Sigbert Klinke, Jürgen Symanzik and Marlene Müller

16 Matrix Handling   429

Yasemin Boztug and Marlene Müller

17 Quantlets and Quantlibs   465

Wolfgang Härdle, Zdeněk Hlávka and Sigbert Klinke

Appendix   503

Preface

It is generally accepted that training in statistics must include some exposure to the mechanics of computational statistics. This learning guide is intended for beginners in computer-aided statistical data analysis. The prerequisites for XploRe — the statistical computing environment — are an introductory course in statistics or mathematics. The reader of this book should be familiar with basic elements of matrix algebra and the use of HTML browsers. This guide is designed to help students to XploRe their data, to learn (via data interaction) about statistical methods and to disseminate their findings via the HTML outlet. The XploRe APSS (Auto Pilot Support System) is a powerful tool for finding the appropriate statistical technique (quantlet) for the data under analysis. Homogeneous quantlets are combined in XploRe into quantlibs.

The XploRe language is intuitive and users with prior experience of other statistical programs will find it easy to reproduce the examples explained in this guide. The quantlets in this guide are available on the CD-ROM as well as on the Internet. The statistical operations that the student is guided into range from basic one-dimensional data analysis to more complicated tasks such as time series analysis, multivariate graphics construction, microeconometrics, panel data analysis, etc.

The guide starts with a simple data analysis of pullover sales data, then introduces graphics. The graphics are interactive and cover a wide range of displays of statistical data. The regression chapter guides the user first to linear scatter-plot regression and then to multiple regression and nonlinear models. The teachware quantlets comprise a basic set of interactive, illustrative examples in introductory statistics. For the student, they provide an opportunity to understand some important basic concepts in statistics through trial and error. For the teacher, they can aid the instruction by allowing the students to independently work on exploratory examples at their own pace. Additionally, with a modicum of understanding of the XploRe programming language, the teacher can modify these examples to fit his/her own preferences.

In the second part, nonparametric smoothing methods are introduced. They provide a flexible analysis tool, often based on interactive graphical data representation. The focus there is on function estimation by kernel smoothing. More refined tools are the neural networks. They consist of many simple processing units (neurons) that are connected by communication channels. This guide shows how to adapt such networks to real data. The time series quantlets cover the standard analysis techniques and offer nonlinear methods for modern time series data analysis. The Kalman filtering module is a set of recursive quantlets based on the construction of an estimate from the previous periods and observations available at the time. The `finance` quantlib considers the statistical analysis of financial markets. This quantlib offers functions to predict, to simulate and to estimate finance data series such as, for example, stock returns, to determine option prices and to evaluate different scenarios (e.g. for portfolio strategies).

The microeconometrics section introduces the tools available in XploRe for analyzing micro data, i.e. data sets consisting of observations on individual units, such as persons, households or firms. The quantlib `metrics` provides the statistical tools such as panel analysis for analyzing observed individual behavior. The analysis of risk is one of the primary objectives of insurance firms and the profit and loss analysis on financial markets. It is also applied to flood discharges and high concentration of air pollutants. The 'value at risk' and the extreme (upper or lower) parts of a sample are estimated and analyzed in this module. Wavelets comprise a powerful tool which can be used for a wide range of applications, namely describing a signal for parsimonious representation, for data compression, for smoothing and image denoising and for jump detection. This guide introduces the wavelet quantlib `twave`.

XploRe offers a large variety of commands and tools for creating and manipulating multidimensional objects called matrices and lists. The last part of this guide presents the basic instructions for matrix handling and illustrates further topics in matrix algebra and list handling with XploRe. The quantlet construction technique and the generation of a user defined APSS are the topics of the final chapter on quantlets and quantlibs.

XploRe and this learning guide have benefited at several stages from cooperation with many colleagues and students. We want to mention in particular: Gökhan Aydinli, Silke Baars, Alexander Benkwitz, Marco Bianchi, Rong Chen, Vila Co, Stefan Daske, Sven Denkert, Ulrich Dorazelski, Michaela Dranganska, Dorit Feldmann, Jörg Feuerhake, Frank Geppert, Birgit Grund, Janet Grassmann, Christian Hafner, Susanne Hannappel, Nicholas Hengartner, Holger

Jakob, Christopher Kath, Jussi Klemelä, Petra Korndörfer, Gunnar Krause, Katharina Kröl, Stephan Lauer, Pascal Lavergne, Michel Lejeune, Torsten Kleinow, Thomas Kötter, Thomas Kühn, Hua Liang, Steve Marron, Danilo Mercurio, Hans-Joachim Mucha, Fabian Nötzel, Isabel Proenca, Andreas Ruppin, Jerome Saracco, Swetlana Schmelzer, Levon Shahbaghyan, Hizir Sofyan, Hans Gerhard Strohe, Wolfgang Stockmeyer, Elke Tscharf, Berwin Turlach, Rodrigo Witzel, Lijian Yang, Kerstin Zanter.

We owe special thanks to M. Feith, Springer Verlag, for professional editorship and continuous encouragement at several critical junctions. We gratefully acknowledge the support of the Deutsche Forschungsgemeinschaft, Sonderforschungsbereich 373 "Quantifikation und Simulation Ökonomischer Prozesse".

Berlin, May 1999

Contents of the CD-ROM

```
Acrobat_Reader
  ar32e301.exe        Setup file for Acrobat Reader 3.01

e-books
  mva.pdf             Multivariate Analysis textbook in PDF format
  xplbook.pdf         This Learning Guide in PDF format
  xplbook.html        This Learning Guide in HTML format

Netscape
  cc32e46.exe         Setup file for Netscape Communicator 4.6

Readme_Installation
                      Installation instructions and Registration Form

XploRe
  Setup.exe           Setup file for XploRe

XploRe-Client
  xplinst.exe         XploRe Java client for worldwide Quantlet Service
```

Part I: First Steps

1 Getting Started

Here we show how to get started with XploRe and support you in performing your first statistical analysis. We explain how you can get help and how you can do your own quantlet programming.

1.1 Using XploRe

1.1.1 Input and Output Windows

Once you have clicked on the XploRe icon, three windows open up on the screen.

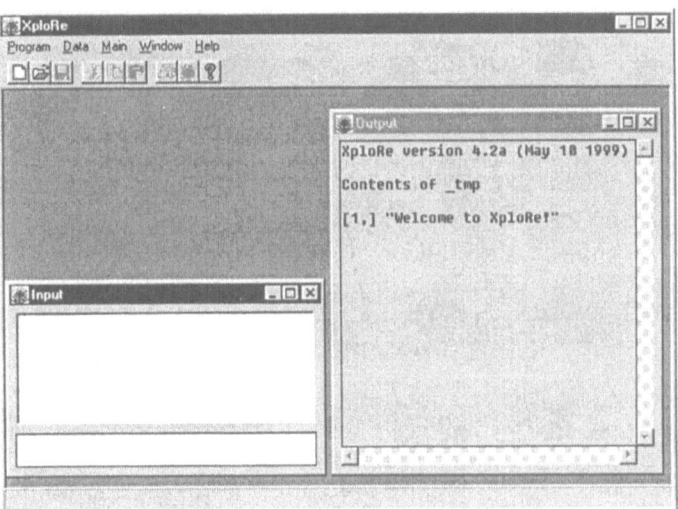

These windows are

- the main window, with its **menu bar**,

- the **input window**, in which you can enter instructions interactively,

- the **output window**, which displays the numerical results and text information.

1.1.2 Simple Computations

Let's carry out some elementary calculations to get familiar with XploRe. Assume that you wish to calculate the sum of 2 numbers, e.g. 1 and 2. Then you have to follow these steps:

1: point the cursor to the command line, i.e. the lower part of the input window,

2: enter 1+2,

3: press the <Return> key.

The screen changes as follows:

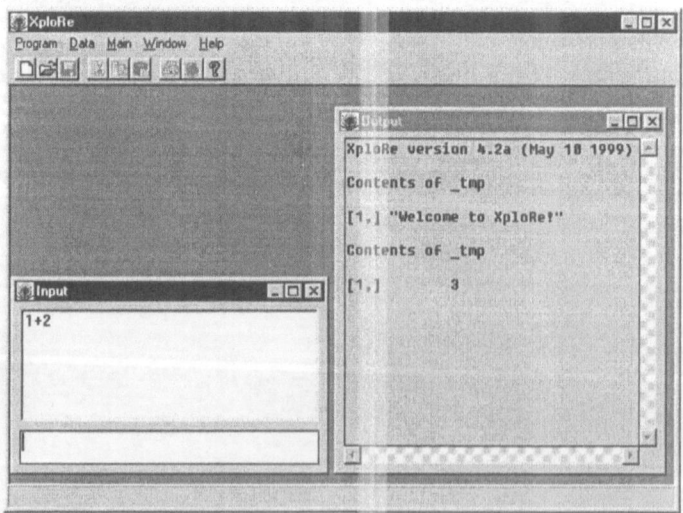

The outcomes of the above sequence of operations are

- the upper part of the input window now contains the command entered, the command line is empty and ready to receive new commands,

- the output window contains the result, i.e. 3.

1.1.3 First Data Analysis

The typical basic steps for a statistical data analysis in XploRe are

1: read the data,

2: apply a statistical method.

Let's load our first data set. The ASCII file `pullover.dat` contains data on pullovers sales in 10 time periods. The four columns of `pullover.dat` correspond to four variables. These are the number of pullovers sold, the price (in DM), costs for advertising (in DM), and the presence of a shop assistant (in hours per period), see Appendix B.7 for more information.

We read the data file `pullover.dat` into XploRe by entering

```
x=read("pullover")
```

at the command line. With this instruction, we have assigned the contents of the data file `pullover.dat` to the XploRe variable `x`. We can print the contents of `x` by issuing

```
x
```

at the command line. This shows

```
Contents of x
[ 1,]     230     125     200     109
[ 2,]     181      99      55     107
[ 3,]     165      97     105      98
[ 4,]     150     115      85      71
```

[5,]	97	120	0	82
[6,]	192	100	150	103
[7,]	181	80	85	111
[8,]	189	90	120	93
[9,]	172	95	110	86
[10,]	170	125	130	78

As an example of a statistical analysis of the data, let's compute the mean function (the average of the columns) here:

```
mean(x)
```

returns

```
Contents of mean
[1,]     172.7     104.6        104        93.8
```

in the output window. This shows that during the 10 considered periods, 172.7 pullovers have been sold on average per period, the average price was 104.6 DM, the average advertising costs were 104 DM and the shop assistant was present for 93.8 hours on average.

1.1.4   Exploring Data

In the previous example, we applied the XploRe built-in function mean which provides the sample mean of the data. Apart from the built-in functions, XploRe offers libraries (= quantlibs) of functions (= quantlets) that must be loaded before usage.

In general, the statistical analysis of data comprises the following steps:

1: read the data,

2: select the interesting variables from the data,

3: load the necessary library,

4: explore the data,

5: find a statistical model and apply a statistical method,

6: display the results.

We continue our analysis with the `pullover` data. The first column of this data set contains measurements on the sales of "classic blue" pullovers in different shops whereas the second column contains the corresponding prices. Let's say we are interested in the relation between prices and sales.

We read the data again and now select the price and sales columns (second and first columns) only:

```
x=read("pullover")
x=x[,2|1]
```

One of the strengths of XploRe is the graphical exploration of data. A scatter plot of the data should give us a first impression on the relation of both variables. We will show the scatter plot by means of the function `plot`. Since this function is part of the quantlib `plot`, we must load this library first:

```
library("plot")
plot(x)
```

The last instruction creates a display, i.e. a new graphics window, which contains the scatter plot:

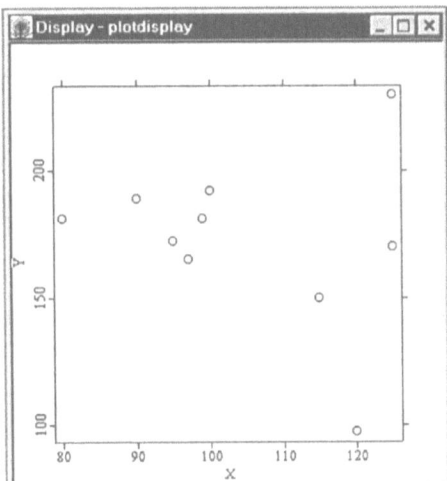

Looking at this scatter plot, it is difficult to find a clear tendency in the relation between price and sales. It is the task of **regression** analysis — discussed in Chapter 4 — to determine the appropriate functional relation between variables. We will now use one of the regression methods introduced there:

```
regx=grlinreg(x)
plot(x,regx)
```

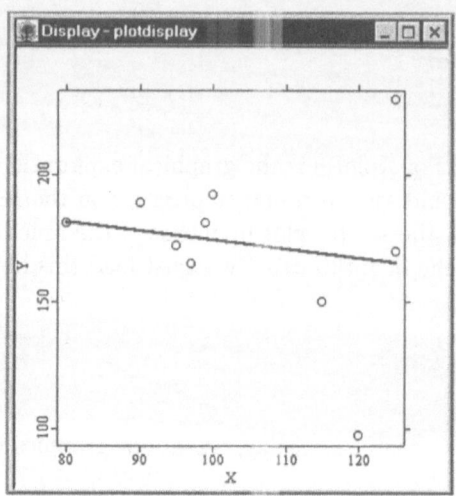

Figure 1.1. Regression line and data.

The resulting plot in Figure 1.1 shows the regression line **regx** and the data **x** as circles. The regression line has a negative slope. We can conclude that (on average) the number of sold pullovers decreases if the price of the pullover increases. However, this result may be influenced by the two extreme observations in the upper right and lower right of the figure. XploRe can easily identify such "outliers". For example, the instruction

```
x=paf(x,(100<x[,2])&&(x[,2]<200))
```

would only keep those lines of x where the sales observation is above 100 and below 200. You could now redo the previous regression in order to see how the regression line changes.

1.1.5 Printing Graphics

XploRe offers several ways to produce quality graphics for publication. You can modify the objects in a plot (point and line style, title and axes labels) and finally save the graphical display in different file formats.

Let's continue with the regression example from the previous subsection. We can improve the graphic by several graphic tools. For example,

```
x=setmask(x,"large","blue","cross")
plot(x)
setgopt(plotdisplay,1,1,"xlabel","price","ylabel","sales")
```

will show the data points as blue crosses and the axes labels with the appropriate names of the variables. We can set a title for the display in the same way:

```
setgopt(plotdisplay,1,1,"title","Pullover Data")
```

The final plot is shown in Figure 1.2.

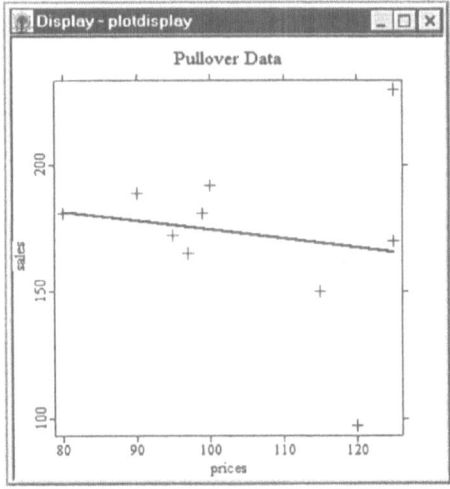

Figure 1.2. Final regression plot.

Graphical displays can be printed or saved to a file. If you click on the display `plotdisplay`, the Print menu will appear in the menu bar of XploRe. This

menu offers three choices: to Printer prints the display directly on your default printer, to Bitmap file ... saves the display to a Windows Bitmap file, to
PostScript file ... saves the display to a PostScript file. The two latter menu
items open a file manager box, where you can enter the file name. Here you
see the resulting PostScript plot:

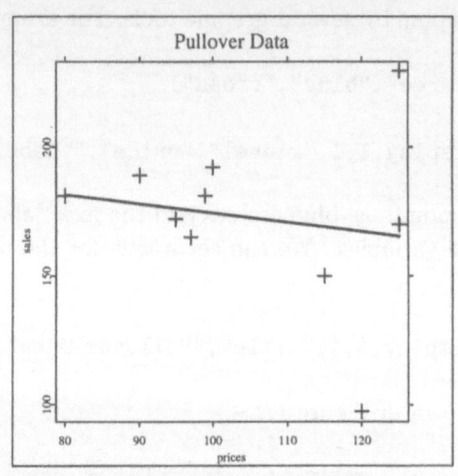

PostScript files can also be printed by an XploRe instruction:

```
print(plotdisplay,"Plot1.ps")
```

will save the display `plotdisplay` into the file `Plot1.ps`.

1.2   Quantlet Examples

Throughout the book we use XploRe example code which can be directly typed into the command line. We have seen this in the previous sections when calculating the mean and creating the scatter plot for the `pullover` data.

Since it can be cumbersome to type many lines of code, we summarized instructions that belong together into **quantlet files**. These quantlet files are either displayed within the text, as e.g. ☞ `start01.xpl` or printed at the end of a sequence of instructions:

```
x=read("pullover")   ; reads data
x=x[,2|1]
library("plot")
plot(x)              ; shows scatter plot
```
☞ start02.xpl

You can find all example quantlets in the **examples** directory of your XploRe installation. This means, if XploRe is installed in `C:\XploRe`, you can find the examples in `C:\XploRe\examples`.

To load an example quantlet, use the Open item from the Program menu. Change the directory to **examples** and double-click on the appropriate file. This opens an **editor window** containing the example code. We execute the quantlet by clicking the Execute item in the menu bar or by entering <Alt E>.

Quantlets contain one XploRe instruction in each line. Additionally, it is possible to add comments. For example, in

```
x = 1   ;  assign the value 1 to x
y = 2   // assign the value 2 to y
```

everything beyond the ; and // is a comment. You can also use /* and */ to mark the beginning and the end of several lines of comments.

The following subsections intend to give you an impression of the capabilities of XploRe. We will present some of the statistical methods that are explained in detail later on.

1.2.1   Summary Statistics

XploRe has several functions to print summary statistics of all columns of a data file. The following example uses the function **summarize** from the **stats** library:

```
library("stats")
x=read("pullover")
library("stats")
setenv("outputstringformat","%s")
summarize(x, "sales"|"price"|"ads"|"assistance")
```

Q start03.xpl

This returns the following table in the output window:

```
Contents of summ
[1,]
[2,]              Minimum   Maximum    Mean   Median   Std.Error
[3,]          ------------------------------------------------------
[4,] sales          97       230    172.7      172      33.948
[5,] price          80       125    104.6       99      15.629
[6,] ads             0       200      104      105      53.996
[7,] assistance     71       111     93.8       93      14.038
[8,]
```

You will learn more about summary statistics in Chapter 2.

1.2.2   Histograms

A histogram visualizes univariate data. Hence, we produce separate histograms for the variables (columns) of a data matrix. The following example shows the histogram for the sixth column (diagonal length) of the Swiss bank notes data stored in the file bank2.dat (see Appendix B.7).

```
library("plot")
y=read("bank2")
y=y[,6]
library("plot")
setsize(640,480)      ; sets display size
```

```
plothist(y)          ; plots histogram
```

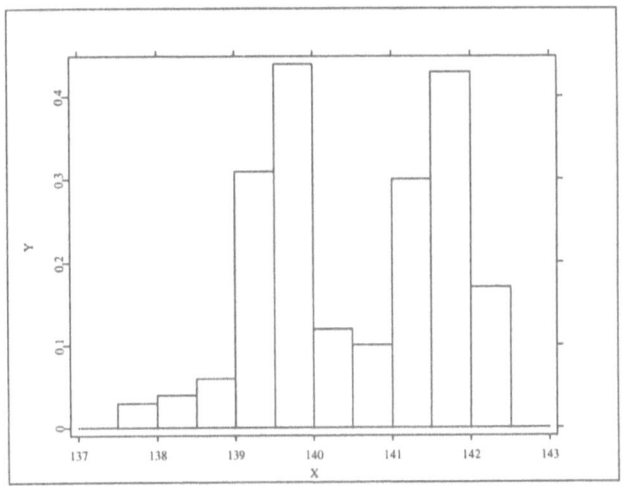 start04.xpl

Figure 1.3 shows the resulting histogram. We recognize two modes in the distribution of the data. This indicates two clusters.

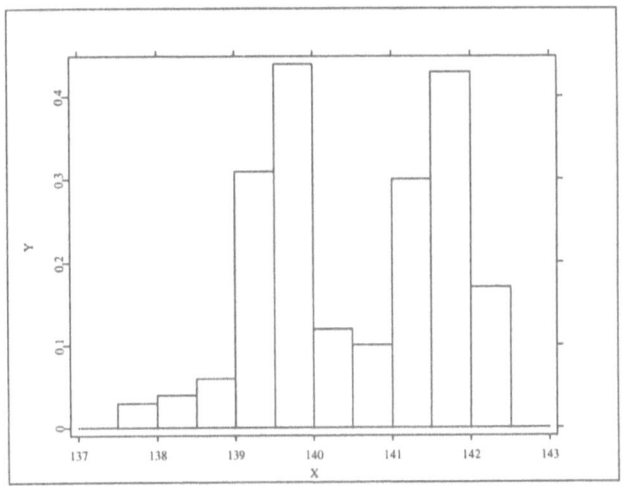

Figure 1.3. Histogram.

1.2.3   2D Density Estimation

The method of kernel density estimation allows us to estimate the distribution of data in a very flexible way. We introduce this topic in Chapter 6. Here we compute the density estimate for the joint density of the upper frame length and the diagonal length (fifth and sixth variable) of the **bank2** data (see Appendix B.7).

```
library("smoother")
library("plot")
x = read("bank2")
x=x[,5:6]
library("smoother")
```

```
library("plot")
fh = denxestp(x)                ; density estimation
fh = setmask(fh,"surface","blue")
setsize(640,480)
axesoff()
cu = grcube(fh)                 ; box
plot(cu.box,cu.x,cu.y, fh)   ; plot box and fh
setgopt(plotdisplay,1,1,"title","2D Density Estimate")
axeson()
```

Q start05.xpl

This example computes the density estimate on a grid of points. It returns a graphical display containing the estimated function. To obtain the three-dimensional surface in Figure 1.4, you need to rotate the estimate by using the cursor keys. The vertical axis shows the estimated density while the horizontal axes show the upper frame (to the right) and the diagonal length (to the left), respectively.

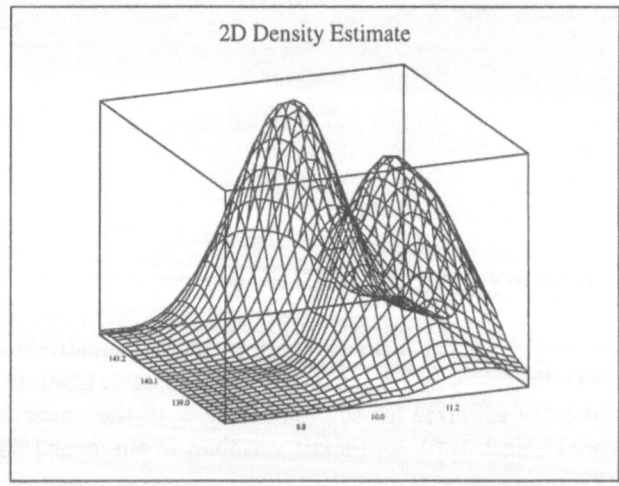

Figure 1.4. Two-dimensional density estimate.

The density estimate confirms our impression from Subsection 1.2.2 that the data set features two clusters.

1.2.4   Interactive Kernel Regression

XploRe is particularly suited for interactive use. The following quantlet computes a nonparametric regression smoother and asks interactively for the bandwidth.

```
proc()=interactive(x)
  error(cols(x)!=2,"x has more than 2 columns!")
  x=setmask(x,"red","cross")
  h=(max(x[,1])-min(x[,1]))/5
  stop=0
  while (stop!=1)
    mh=setmask(regxest(x,h),"line")
    plot(x,mh)
    hnew=readvalue("Bandwidth h (0 to stop)",h)
    while (hnew<0)
      hnew=readvalue("Bandwidth h (0 to stop)",h)
    endo
    if (hnew!=0)
      h=hnew
    else
      stop=1
    endif
  endo
endp

library("plot")
library("smoother")
x=read("nicfoo")
interactive(x)
```

Q start06.xpl

When the quantlet is executed, it first shows a graphical display with the nonparametric regression curve (solid line) and the data (red crosses). Next, a dialog box opens which allows us to modify the currently used bandwidth parameter. The quantlet stops, if 0 (zero) has been entered.

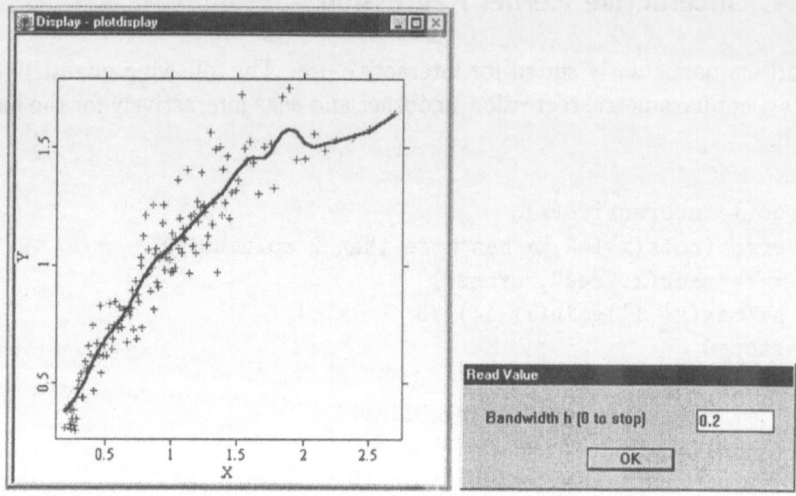

Nonparametric smoothing is introduced in Chapter 6. If you want to program interactive quantlets yourself, we recommend that you study Chapter 17.

1.3   Getting Help

XploRe has an extensive help system, the Auto Pilot Support System (APSS). To start the APSS, select the Help item from the menu bar of the XploRe main window and select APSS Help. This will open your default World Wide Web browser and display the start page as shown in Figure 1.5.

The APSS provides several entry points for searching help using and programming in XploRe:

- **Function Groups** lists all functions by topic,

- **Alphabetical Index** lists all available functions alphabetically,

- **Keywords** lists all functions by keyword,

- **Tutorials** links to tutorials on a broad range of topics.

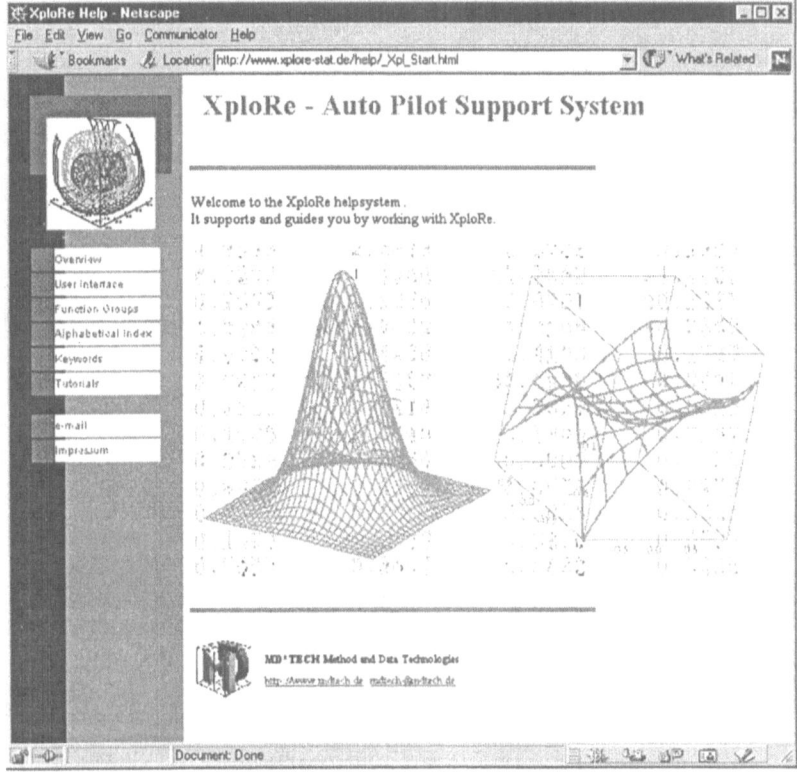

Figure 1.5. APSS start page.

Figure 1.6 shows — as an example — the help page for the function **abs**
which computes the absolute value. As an advanced user of XploRe, you will
be able to generate your help pages for your self-written functions. You are
able to add your own functions to the APSS such that you can create an
XploRe environment according to your own needs. Consult Chapter 17 for
more information.

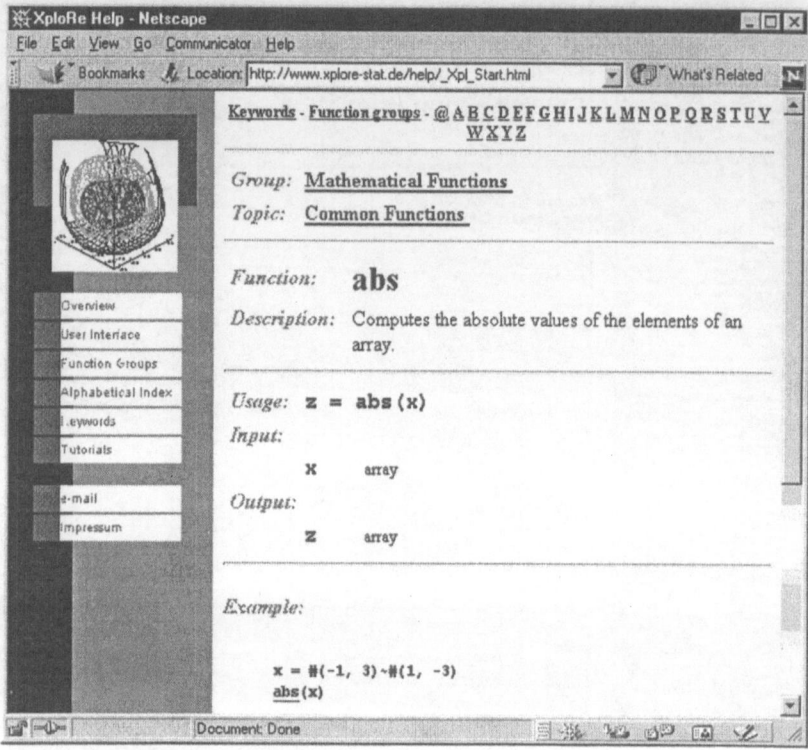

Figure 1.6. APSS help file for `abs`.

1.4   Basic XploRe Syntax

1.4.1   Operators

In the introductory example in Section 1.1, we have only considered two numbers 1 and 2, and the addition operator +. Other standard operators are available:

- the symbol – stands for subtraction,

- the symbol * stands for multiplication,

- the symbol / stands for division,

- the caret symbol ∧ stands for exponentiation.

These five basic operators are ordered by **precedence rules**, which are the same as in pencil and paper calculus: entering in the input line the arithmetic expression

    2*2+3

gives in the output window

    7

since the operator * has precedence on the operator +. Thus, the computation 2*2 is carried out first and gives 4 as result, which is added to 3 to yield 7.

The exponent operator ∧ has the highest level of precedence. Next, the operators * and / have the same order of precedence, and have precedence over the operators + and -. As in standard paper and pencil calculus, you can force these precedence rules by using parentheses: the command line

    (3*4)^2

gives

    144

as the expression (3*4) is enclosed in parentheses, and is then evaluated independently from the other operations. The outcome of this operation, 12, is squared to obtain 144.

As in pencil and paper calculus, parentheses could be nested, although left and right parentheses must match. Assume that you enter the instruction:

    (1+(3*4)^2

which is incomplete since the two left parentheses are matched by only one right parenthesis. This incomplete instruction ends in a **error message**, i.e. the syntax analyzer of XploRe, or **parser**, has detected that something went wrong and gives the cause of the mistake.

1.4.2 Variables

In the previous subsection, we have considered numeric expressions which — once evaluated — are "lost" since they are not stored anywhere. If you need to keep the result of a numeric expression for further calculation, you can store it in a precise location by **assigning** it to a **variable**.

The assignment is done with the **assignment operator** denoted by the symbol = which means "equal". For example,

```
a = 2*3
```

assigns the value 6 to the variable **a**.

We assign a variable by either storing in it a number or the content of another variable. The following command

```
b = a
```

assigns into the variable **b** the content of the variable **a**. We can verify this by displaying the content of **b** by typing its name:

```
b
```

prints

```
Contents of b
 [1,]          6
```

in the output window.

1.4.3 Variable Names

The name of a variable should be a string of alphabetic or numeric characters: a, b, product, result are possible variable names. Not allowed for the use in variable names are spaces (blanks) and underscore symbols (_).

The XploRe parser is **case sensitive**, i.e. the variables a, A, as well as result, Result, ResulT and RESULT are considered as distinct. Thus, assigning the value 5 to the variable A

```
A = 5
```

does not affect the content of the variable a which still contains the number 6.
We display the content of both A and a by typing

```
A
a
```

which gives in the output window

```
Contents of A
[1,]        5
Contents of a
[1,]        6
```

Lastly, some labels which represent standard constants in XploRe are **protected** and cannot be used as variable names. These are pi for the constant $\pi = 3.1415926\ldots$ and eh for the constant $e = 2.7182818\ldots$.

1.4.4 Functions

The environment XploRe provides several hundreds of functions for data handling, statistical data analysis, graphics and user interaction. There are two types of functions:

- **commands** = built-in functions,

- **quantlets** = functions programmed in the XploRe language.

Among the commands, you find all important mathematical functions, such as

log, log10, exp	logarithm and exponential functions
sin, cos, tan	trigonometric functions
abs	absolute value
sqrt	square root
floor, ceil, rint	rounding functions

Since XploRe is a **matrix-oriented** language, applying a mathematical function to a data matrix means that the function is applied to all elements of this matrix. For example,

```
x=#(1,2,3)  ; vector of 1, 2, and 3
log(x)
```

returns

```
Contents of log
[1,]        0
[2,]  0.69315
[3,]    1.0986
```

i.e. the log has been applied to all elements of the vector x. You find more information on available functions for matrix handling and data analysis in Chapters 2 and 16.

Quantlets are used in the same way as the built-in commands, except for the fact that a quantlib has to be loaded first. For example, the function mean is a built-in command while the function median is a quantlet from the xplore library. If you want to compute the median of a data vector, you first have to load the xplore library:

```
library("xplore")
x=#(1,2,3)  ; vector of 1, 2, and 3
median(x)
```

To find out which library a function belongs to, consult the APSS help file of this function.

1.4.5   Quantlet files

Any file which consists of XploRe instructions is called a **quantlet**. Quantlets are saved as text files using the file extension .xpl.

We use the term quantlet for two types of .xpl files:

- Files of XploRe instructions that contain the definition of a **procedure**. This means that the quantlet adds a new function to the XploRe environment. An example of this type of quantlet is the above-mentioned median function which is part of the xplore library.

- Files of XploRe instructions that we **execute** line by line.
  Examples of this type of quantlet file are the example quantlets which come with your XploRe installation, see Section 1.2.

All quantlet files can be loaded by using the Open item from the Program menu. This opens an **editor window** containing the quantlet code. We execute the quantlet by clicking the Execute item in the menu bar or by entering <Alt E>.

If the quantlet contains a code to be executed line by line, the execution is immediately started. If the quantlet defines a procedure, an additional instruction to carry out the procedure is required. We recommend to study Chapter 17 for using such procedure quantlets.

2 Descriptive Statistics

Marlene Müller

A descriptive analysis of a given data set is typically the first part of statistical modeling and evaluation of data. In the following, we will show how the functions that come with XploRe can be used for this purpose.

All routines for descriptive analysis that we present are part of the libraries xplore (basic routines) and stats (basic statistical methods). We will not use graphical methods here. Tools for the graphical exploration of data are extensively discussed in Chapter 3.

Before proceeding to the next section, please type

```
library("xplore")
library("stats")
```

to load the libraries xplore and stats. Note also that we use

```
setenv("outputstringformat","%s")
```

throughout this section. This command suppresses the quotation marks in text output.

2.1 Data Matrices

In XploRe, data can be stored in matrices ($n \times p$) or arrays ($n \times p \times \dots$). Here, we will concentrate on data matrices. Small data matrices can be created directly from the command line or within an XploRe quantlet. Large data matrices are typically read from data files.

The following subsections provide a short introduction on matrix and data handling. Consult Chapter 15 to learn more about loading data files into XploRe. More details on data and matrix manipulation can be found in Chapter 16.

2.1.1   Creating Data Matrices

```
z = # (x1, x2, ..., xn)
    creates a column vector z from scalar numbers x1, x2, ... , xn

z = x | y
    concatenates two arrays x and y rowwise

z = x ~ y
    concatenates two arrays x and y columnwise
```

Small data matrices can be directly given at the command line or within an XploRe program. The following XploRe codes are all available from the quantlet ◫ desc01.xpl. As a first example, consider the data matrix

$$\begin{pmatrix} 1 & 2.0 & 3.4 \\ 5 & 6.0 & 7.8 \\ 9 & 0.0 & 1.44 \\ 8 & 7.0 & 10.432 \end{pmatrix}$$

which has dimension 4×3, i.e. four rows and three columns. To construct this matrix in XploRe, we create each column vector separately and then concatenate these column vectors. A column vector can be created by means of the # or the | operator. The following two lines are equivalent:

```
col1=#(1,5,9,8)
col1=1|5|9|8
```

Both create the column vector

$$\begin{pmatrix} 1 \\ 5 \\ 9 \\ 8 \end{pmatrix}.$$

We can check the contents of col1 by issuing just

```
col1
```

at the command line, which results in

```
Contents of col1
[1,]         1
[2,]         5
[3,]         9
[4,]         8
```

in the output window. In the same way as for col1, we build the second and third columns:

```
col2=#(2.0,6.0,0.0,7.0)
col3=#(3.4,7.8,1.44,10.432)
```

and group all three vectors together by means of the $\sim$ operator:

```
mat=col1~col2~col3
```

When we check the contents of mat we see

```
Contents of mat
[1,]         1        2       3.4
[2,]         5        6       7.8
[3,]         9        0       1.44
[4,]         8        7      10.432
```

Note that we could have created mat within a single step

```
mat= #(1,5,9,8) ~ #(2.0,6.0,0.0,7.0) ~ #(3.4,7.8,1.44,10.432)
```

Let us also remark that XploRe does not distinguish between integer and float values. Therefore, the first two columns of the matrix mat appear in the same format.

It is also possible to create text matrices. For example

```
textmat= #("aa","c") ~ #("b","d2")
```

creates the text matrix

$$\begin{pmatrix} \text{aa} & \text{b} \\ \text{c} & \text{d2} \end{pmatrix}$$

Note that text and numeric values need to be stored in different matrices.

2.1.2   Loading Data Files

```
x = read ("file")
     reads numeric data from file.dat

x = readm ("file")
     reads mixed text and numeric data from file.dat
```

Large data sets are usually stored in data files. XploRe can read data from ASCII files, consisting of both numeric and text data. In the following we will use two data sets: cps85 and uscomp2 (see Appendix B.2).

The file cps85.dat consists of a subsample of the 1985 U.S. current population survey. The file contains only numeric data. We will assign columns 1 (years of education), 2 (= 1 if living in south), 5 (= 1 if female) 8 (years of labor market experience), 10 (= 1 if working on a union job), 11 (natural logarithm of average hourly earnings) and 12 (age in years) to the XploRe variable earn:

```
earn=read("cps85")
earn=earn[,1|2|5|8|10|11|12]
```

                                                             ⌀ desc02.xpl

The file uscomp2.dat contains information on 79 U.S. companies. The data set has 8 columns, two of them text (columns 1,8) and six numeric (columns 2 to 7). We will only use column 8 (branch, text) and columns 3 and 5 (sales and profits, both numeric) and assign them to the XploRe variables branch and salpro:

```
uscomp=readm("uscomp2")
```

```
branch=uscomp.text[,2]
salpro=uscomp.double[,2|4]
```

desc02.xpl

Since text and numeric data are stored in different XploRe objects, we find column 8 of uscomp2 as the second text column and columns 3 and 5 as the second and fourth numeric columns, respectively. readm is a function written in the XploRe language, which can be used for reading mixed text and numeric data.

2.1.3 Matrix Operations

```
d = dim(x)
    shows the dimension of an array x

n = rows(x)
    shows the number of rows of an array x

p = cols(x)
    shows the number of columns of an array x

y = x[i,j] or y = x[i,] or y = x[,j]
    extracts element i,j or row i or column j from x
```

The first step in data analysis is to find out information on the dimension of the data. This can be done generally by using the function dim. We apply this function now to the data matrices mat, earn, branch, and salpro that we specified in Subsections 2.1.1 and 2.1.2. The codes for this section are available from the quantlet desc02.xpl.

```
dim(mat)
dim(earn)
dim(branch)
dim(salpro)
```

yields

```
Contents of dim
[1,]        4
[2,]        3
Contents of dim
[1,]        534
[2,]        7
Contents of dim
[1,]        79
Contents of dim
[1,]        79
[2,]        2
```

and tells us that mat is a 4 × 3 matrix, earn is 534 × 7, branch is a 79 × 1 vector and salpro is 79 × 2. If we are just interested in the number of rows or columns, we can use the commands rows and cols. For example,

```
rows(earn)
cols(earn)
```

gives

```
Contents of rows
[1,]        534
Contents of cols
[1,]        7
```

To extract elements or submatrices of a matrix, we can use the subarray operator []. The following three lines extract the first row, the second column and (4, 3)-element (fourth row, third column), for example:

```
mat[1,]
mat[,2]
mat[4,3]
```

This operator can also be used for extracting several rows and columns at once. The statement mat[1:3,1|3] extracts the elements which are in the 1st to 3rd rows of mat and in the 1st and 3rd columns. The operator : is used to specify a range of subsequent integers.

2.2 Computing Statistical Characteristics

Given a data matrix and basic information about the values in it, the next analysis step is to compute some characteristic numbers from the data.

We will start our first steps in statistical analysis by investigating the data matrix `salpro`, which contains the variables sales and profits from the U.S. companies database. If you want to check the contents of this matrix, just issue `salpro` at the command line. Recall that this is a small data set of only 79 rows, containing realizations of two metric variables. For all observations, we have information about the branch in which the company is working. This information is contained in the text vector `branch`. To give you a first impression of the data, Figure 2.1 shows a scatter plot of sales against profits.

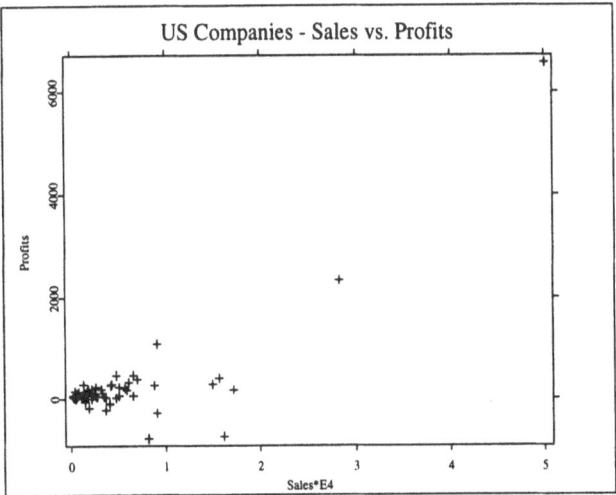

Figure 2.1. Scatter plot of sales vs. profits for U.S. companies data.
Q `desc03.xpl`

Later in this section, we will continue our course by studying the data matrix `earn`. `earn` contains data of a different structure. Some of the variables are continuous but most of the variables are discrete (i.e. can take on only a few values) or are in fact nonmetric.

In the following subsections, we will compute different characteristics of the data matrices `salpro` and `earn`. All XploRe codes can be found in the quantlets

ⓠ desc04.xpl and ⓠ desc05.xpl, respectively.

2.2.1   Minimum and Maximum

```
mx = min(x {,d})
    computes the minimum of an array x, optionally with respect to
    dimension d

mx = max(x {,d})
    computes the maximum of an array x, optionally with respect to
    dimension d
```

A basic information of a data set are the **minimum** and **maximum** values of the variables. For the variables sales and profits of the U.S. companies data set we obtain these values from

```
min(salpro)
max(salpro)
```

which computes

```
Contents of min
[1,]      176   -771.5
Contents of max
[1,]    50056     6555
```

In XploRe, the minima and maxima can be computed for all rows or columns of a matrix in one step. By default, the columnwise minima and maxima are computed. This means for the above output, 176 is the minimum of sales and -771.5 the minimum of profits. The maximum of sales is 50056 and the maximal profit is 6555.

For rowwise computation of minima and maxima, a second parameter needs to be given to min and max. For example, min(salpro,2) would compute the rowwise minimum of salpro. Let us remark that rowwise computations make of course no sense in the case of salpro, since the variables (sales and profits) are stored columnwise in this data matrix.

2.2.2   Mean, Variance and Other Moments

---

mx = mean(x {,d})
    computes the mean of an array x, optionally with respect to di-
    mension d

vx = var(x {,d})
    computes the variance of an array x, optionally with respect to
    dimension d

kx = kurtosis(x)
    computes the (columnwise) kurtosis of an array x

sx = skewness(x)
    computes the (columnwise) skewness of an array x

To describe numeric data, it is useful to study the empirical values for the
moments of the underlying distribution. Suppose we have a data vector x
which contains n observations $x_1, \ldots, x_n$. The **mean** $\overline{x}$ and the **variance** v^2

$$\overline{x} = \frac{1}{n} \sum_{i=1}^{n} x_i, \quad v^2 = \frac{1}{n-1} \sum_{i=1}^{n} (x_i - \overline{x})^2$$

are the average (arithmetic mean) of all observations and average quadratic
deviation from the mean, respectively. Analogous to minimum and maximum,
mean and variance can be computed for all rows or columns of a matrix in one
step. By default, columnwise means and variances are computed. For example,
the commands

```
mean(salpro)
var(salpro)
```

give the means and variances for the variables sales and profits of the U.S.
companies data set:

```
Contents of mean
[1,]    4178.3    209.84
Contents of var
[1,]    4.9163e+07  6.3517e+05
```

The standard deviations of sales and profits can be obtained by taking the square root from the variance: `sqrt(var(salpro))`.

The **skewness** s and the **kurtosis** k measure the skewness and the departure from the normal distribution, respectively:

$$s = \frac{1}{n} \sum_{i=1}^{n} \frac{(x_i - \overline{x})^3}{v^3}, \quad k = \left\{ \frac{1}{n} \sum_{i=1}^{n} \frac{(x_i - \overline{x})^4}{v^4} \right\}$$

The skewness should be close to 0 for a distribution that is symmetric around $\overline{x}$. The kurtosis should be close to 3 for a distribution resembling the normal. For the variables sales and profits from the U.S. companies data we find

```
skewness(salpro)
kurtosis(salpro)
```

resulting in

```
Contents of s
[1,]    4.2659    6.5908
Contents of k
[1,]    25.426    51.577
```

which are all far away from 0 or 3, respectively. We can conclude that we have a skewed distribution. See Figure 2.1 for a scatter plot of both variables. Of course, this skewness also implies the observed non-normality.

2.2.3   Median and Quantiles

```
mx = median(x)
     computes the (columnwise) median of an array x

qx = quantile(x, alpha)
     computes the (columnwise) quantile of an array x at level alpha
```

An alternative way to characterize numeric data are the **median** and **quantiles** of the data. The empirical median of a data set is given by the value that separates the data into the 50% smallest and 50% highest values. Formally, the empirical median is calculated as

$$
x_{\text{med}} = \begin{cases} x_{[(n+1)/2]} & \text{if } n \text{ is odd,} \\ \frac{1}{2}\left(x_{[n/2]} + x_{[(n/2)+1]}\right) & \text{if } n \text{ is even,} \end{cases}
$$

where $x_{[(n+1)/2]}$ and $x_{[(n/2)+1]}$ are the elements at position $(n+1)/2$ and $(n/2)+1$ of the ranked data. In simpler words, the median gives the central value of the data.

If the distribution of the data were symmetric, arithmetic mean and median should roughly coincide. Let us check this for sales and profits of the U.S. companies:

```
mean(salpro)
median(salpro)
```

result in the following output

```
Contents of mean
[1,]   4178.3   209.84
Contents of med
[1,]    1754     70.5
```

We observe a mean of 4178.3 and a median of 1754 for the sales as well as a mean of 209.84 and a median of 70.5 for the profits. Obviously, the mean $\bar{x}$ and median x_{med} for both variables are quite different. This confirms the conclusions we had already obtained from the computation of the skewness for both variables.

While the median reflects the central value which partitions the data into two 50% portions, we might also be interested in partitioning into fractions of other size. The corresponding values from the data are then empirical values for the quantiles of the distribution.

Theoretically, the α-quantile of a random variable X is the value q_α, such that

$$
q_\alpha = F^{-1}(\alpha) \quad \text{or} \quad F(q_\alpha) = \alpha \quad \text{or} \quad \int_{-\infty}^{q_\alpha} f(x)\,dx = \alpha.
$$

Here, $F(\bullet)$ and $f(\bullet)$ denote the cumulative distribution function (cdf) and the probability density function (pdf) of X, respectively. If $\alpha = 0.5 = 50\%$,

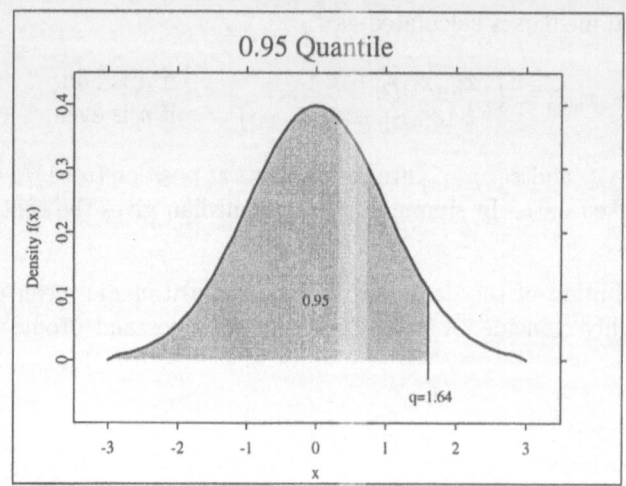

Figure 2.2. The $q_{0.95}$ quantile of a Gaussian distribution.

then $q_{0.5}$ is called the (theoretical) median of the distribution. For illustration, Figure 2.2 visualizes the 0.95 quantile $q_{0.95} \cong 1.64$ of a Gaussian distribution.

The empirical quantiles of a data set can be found by sorting the data and then finding the sample value that partitions the data into portions of size α and $1 - \alpha$. Together with the median (0.5 quantile of the data), the **quartiles** (0.25 and 0.75 quantiles) are often computed for a data set.

```
quantile(salpro,0.25)
median(salpro)
quantile(salpro,0.75)
```

computes both quartiles and medians for our running example and gives

```
Contents of q
[1,]      749      37.8
Contents of med
[1,]     1754      70.5
Contents of q
[1,]     4781     195.3
```

As you can easily calculate, the absolute difference between median and the lower quartile (25% quantile) is less than the absolute difference between median and the upper quartile (75% quantile):

```
abs(median(salpro)-quantile(salpro,0.25))
abs(median(salpro)-quantile(salpro,0.75))
```

yields

```
Contents of abs
[1,]     1005      32.7
Contents of abs
[1,]      3027     124.8
```

This indicates that the densities of both variables are more flat at the right tail and more steep at the left tail. Recall also the scatter plot of both variables in Figure 2.1.

2.2.4   Covariance and Correlation

cx = cov(x)
> computes the covariance matrix of a data matrix **x** (the covariance matrices of an array **x**, respectively)

rx = corr(x)
> computes the correlation matrix of a data matrix **x** (the covariance matrices of an array **x**, respectively)

Recall the scatter plot of the data from the matrix **salpro** which was given in Figure 2.1. The figure shows that both variables sales and profits are highly dependent. The dependence of metric continuous variables (to be exact, the linear dependence) can be measured by the covariance or correlation between the variables.

Suppose we have realizations x_{1i} and x_{2i} $(i = 1, \ldots, n)$ for two variables X_1 and X_2. The empirical covariance between X_1 and X_2 is then defined as

$$\text{cov}(X_1, X_2) = \frac{1}{n-1} \sum_{i=1}^{n} (x_{1i} - \bar{x}_1)(x_{2i} - \bar{x}_2).$$

For a data matrix consisting of several variables, often the **covariance matrix** is given. This matrix contains the variances in the diagonal and the covariances $\text{cov}(X_\ell, X_k)$ in the off-diagonal elements. Note that $\text{cov}(X_\ell, X_k) = \text{cov}(X_k, X_\ell)$, so that covariance matrices are always symmetric matrices.

To compute the covariance matrix of a data matrix, XploRe provides the function cov. For example, from

```
cov(salpro)
```

we obtain the covariance matrix

```
Contents of s
[1,]   4.9163e+07   4.5475e+06
[2,]   4.5475e+06   6.3517e+05
```

On the diagonal of the 2×2 matrix we discover the variances of the variables sales and profits. The covariance between sales and profits is the off-diagonal element 4.5475e+06, which we can access by

```
covmat=cov(salpro)
covmat[1,2]
```

As mentioned above, the covariance measures the (linear) dependence between two variables. If many or all terms $(x_{1i} - \overline{x}_1)$ have the same sign as $(x_{2i} - \overline{x}_2)$, we observe positive dependence (X_1 increases if X_2 increases and vice versa). If the signs are typically different, we observe negative dependence (X_1 increases if X_2 decreases and vice versa). However, it is difficult to assess the magnitude of dependence from the covariance alone. For measuring the strength of dependence, the covariance is considered in relative value to the variances of X_1 and X_2. The resulting coefficient is the **correlation coefficient** which is defined as

$$\rho(X_1, X_2) = \frac{\text{cov}(X_1, X_2)}{v_1 \cdot v_2}$$

with v_1 and v_2 denoting the standard deviations of X_1 and X_2, respectively. For the correlation coefficient, it always holds that $-1 \leq \rho(X_1, X_2) \leq 1$. The values -1 or 1 indicate perfect negative or positive correlation, respectively.

Similar to the covariance matrix, the correlation coefficients for a data matrix can be stored in a **correlation matrix**. This matrix has values 1 on the diagonal and the correlation coefficients $\rho(X_\ell, X_k)$ on the off-diagonal elements. Of course, correlation matrices are symmetric matrices too.

XploRe provides the function `corr` to compute the correlation matrix:

```
corr(salpro)
```

results in

```
Contents of r
[1,]         1  0.81378
[2,]   0.81378       1
```

which means that there is correlation of 0.81378 between sales and profits. As we expected, both variables are highly correlated.

2.2.5   Categorical Data

```
{xr, r} = discrete(x {,y})
        reduces a matrix to its distinct rows and gives the number of repli-
        cations of each row in the original data set; optionally a second
        matrix y can be given, which is summed up accordingly
```

Up to now we have dealt mainly with continuous (metric) variables. In applied sciences, however, dichotomous or categorical variables often play an important role. The first question in descriptive analysis of such variables is which **categories** can be found with which **frequencies**. This information is available from the function `discrete`. Let us consider the third column of the matrix **earn** which is 1 for females and 0 for males:

```
{cat,freq}=discrete(earn[,3])
cat
freq
```

Q desc06.xpl

gives

```
Contents of cat
[1,]          0
[2,]          1
Contents of freq
[1,]        289
[2,]        245
```

As can be seen, the data contain 245 observations of females and 289 observations of males.

The function discrete can be used for text matrices as well. Let us consider the observations in the branch vector from the U.S. companies:

```
{cat,freq}=discrete(branch)
cat
freq
```

◯ desc06.xpl

results in

```
Contents of cat
[1,] Communication
[2,] Energy
[3,] Finance
[4,] HiTech
[5,] Manufacturing
[6,] Medical
[7,] Other
[8,] Retail
[9,] Transportation
Contents of freq
[1,]          2
[2,]         15
[3,]         17
[4,]          8
[5,]         10
[6,]          4
[7,]          7
[8,]         10
[9,]          6
```

We will come back to categorical variables in Subsection 2.3.2. This subsection will present a more convenient way to summarize categories and frequencies in a frequency table. We will also see how to tabulate two variables and to study the dependence between them.

2.2.6 Missing Values and Infinite Values

---

nx = countNaN(x)
    counts missing values in an array x

nx = countNotNumber(x)
    counts missing and infinite values in an array x

ix = isNaN(x)
    determines whether the elements of an array x are missing values

ix = isInf(x)
    determines whether the elements of an array x are infinite values

ix = isNumber(x)
    determines whether the elements of an array x are regular numeric values

y = paf(x, i)
    deletes all rows of x for which the corresponding element of i equals 0

y = replace(x, w, b)
    replaces all elements of x which equal w by the value b

Numeric data sometimes contain **missing** values. Also, operations on the data may lead to missing or **infinite** values. For a subsequent statistical analysis, it is important to identify these values, to delete them or replace them by other values. XploRe encodes missing values by NaN, positive infinite values by Inf and negative infinite values by -Inf.

By means of countNaN and countNotNumber, the existence of values that are
NaN or not numbers (NaN, Inf, -Inf) can be investigated. Here, let us check
our data matrix earn:

```
countNaN(earn)
countNotNumber(earn)
```

give both the result 0, which tells us that the matrix earn contains neither
missing values nor infinite values.

Consider now a small matrix that contains such values. The XploRe code for
the following examples can be found in the quantlet ◘ desc07.xpl.

```
matnan=#(NaN, 1, 2) ~ #(Inf, -Inf, 3) ~ #(4, 5, 6)
matnan
```

generates the following 3 × 3 matrix

```
Contents of matnan
[1,]        NaN        Inf         4
[2,]          1       -Inf         5
[3,]          2          3         6
```

that has one missing and two infinite values. The lines

```
countNaN(matnan)
countNotNumber(matnan)
```

result consequently in

```
Contents of n
[1,]          1
Contents of n
[1,]          3
```

To identify the location of the problematic values, we can use the functions
isNaN, isInf, and isNumber. isNaN produces a matrix of the same size as the
input matrix, indicating NaN values by 1 and all other values by 0:

```
isNaN(matnan)
```

shows

```
Contents of y
[1,]       1        0        0
[2,]       0        0        0
[3,]       0        0        0
```

Note that `isNaN(x)` is equivalent to `x==NaN`. `isInf` searches in the same way for `Inf` and `-Inf` values. `isNumber` indicates numbers by 1 and missing and infinite values by 0.

We can now use the results from `isNaN`, `isInf`, and `isNumber` to delete rows of our data matrix that contain these problematic values. For example, to extract all rows without missing or infinite numbers, type

```
inum=prod(isNumber(matnan),2)
inum
matnew=paf(matnan,inum)
matnew
```

The first line computes the rowwise product of `isNumber(matnan)` by means of the function `prod`. The result `inum` indicates all rows with missing or infinite numbers with the value 0, otherwise the value 1 is produced. The third line of code uses `paf` to extract all rows from `matnan` for which the corresponding element of `inum` is 1. Hence, `paf` extracts only the last line of `matnan`:

```
Contents of inum
[1,]       0
[2,]       0
[3,]       1
Contents of matnew
[1,]       2        3        6
```

Note that `paf` only operates on the rows of a matrix. To extract columns of a matrix, `paf` needs to be applied to the transposed matrix. See Chapter 16 for more information on matrix operations.

Another possibility of handling missing and infinite values is to replace them by some other values. For example, we want to replace all `NaN`, `Inf` and `-Inf` in `matnan` by 0. This is done by the function `replace`:

```
matgood=replace(matnan,#(NaN,Inf,-Inf),0)
matgood
```

gives

```
Contents of matgood
[1,]        0          0          4
[2,]        1          0          5
[3,]        2          3          6
```

2.3 Summarizing Statistical Information

All functions introduced in Section 2.2 are intended to directly compute statistical characteristics from data. These direct computations may be quite cumbersome for everyday work. XploRe offers additional functions which summarize the statistical characteristics of data sets in an efficient way.

2.3.1 Summarizing Metric Data

s = summarize(x {,xvars})
>   computes a short summary of descriptive statistics for each column of a matrix x; optionally a vector of variable names xvars can be given

s = fivenum(x {,xvars})
>   computes the five number summary for each column of a matrix x; optionally a vector of variable names xvars can be given

s = descriptive(x {,xvars})
>   computes detailed descriptive statistics for each column of a matrix x; optionally a vector of variable names xvars can be given

summarize is a tool to obtain a fast overview on a (metric) data matrix. It gives the most important statistical characteristics in the form of a short table. The only required input is the data matrix itself. Optionally, variable names can be provided as a text vector.

The following codes for the data matrix earn are available form the quantlet Q desc08.xpl. For example

```
vearn="educ"|"south"|"female"|"exp"|"union"|"lnwage"|"age"
summarize(earn, vearn)
```

shows the summary of the earn data together with their variable names:

```
[ 1,]
[ 2,]          Minimum  Maximum      Mean   Median   Std.Error
[ 3,]          -------------------------------------------------
[ 4,] educ           2       18    13.019       12      2.6154
[ 5,] south          0        1   0.29213        0     0.45517
[ 6,] female         0        1    0.4588        0     0.49877
[ 7,] exp            0       55    17.822       15       12.38
[ 8,] union          0        1   0.17978        0     0.38436
[ 9,] lnwage         0   3.7955    2.0592   2.0513     0.52773
[10,] age           18       64    36.833       35      11.727
[11,]
```

An alternative to summarize is fivenum which reports the **five number summary** of all columns of a data set. These five numbers are minimum, maximum, median, 25% and 75% quartile of the data. As with summarize, the required input is the data matrix itself and variable names can be provided optionally. For the sake of brevity we show fivenum applied only to column seven of the earn matrix:

```
fivenum(earn[,7],vearn[7])
```

reports

```
Contents of five
[ 1,]
[ 2,] ========================================================
[ 3,]   Five number summary: age
[ 4,]   ------------------------------------------------------
[ 5,]     Minimum                  18
[ 6,]     25% Quartile             28
[ 7,]     Median                   35
[ 8,]     75% Quartile             44
[ 9,]     Maximum                  64
[10,] ========================================================
[11,]
```

The function descriptive provides detailed information about the statistical characteristics of all columns of a data matrix. As in the previous tools, the input for descriptive is the data matrix and optionally variable names:

```
descriptive(earn[,7],vearn[7])
```

produces

```
Contents of desc
[ 1,]
[ 2,] ============================================================
[ 3,]  Variable age
[ 4,] ============================================================
[ 5,]
[ 6,]  Mean               36.8333
[ 7,]  Std.Error          11.7266     Variance            137.513
[ 8,]
[ 9,]  Minimum                 18     Maximum                  64
[10,]  Range                   46
[11,]
[12,]  Lowest cases                   Highest cases
[13,]           350:           18               368:           64
[14,]            94:           18               212:           64
[15,]            48:           18               223:           64
[16,]            78:           18               125:           64
[17,]           298:           19               501:           64
[18,]
[19,]  Median                  35
[20,]  25% Quartile            28     75% Quartile             44
[21,]
[22,]  Skewness          0.545221     Kurtosis          -0.595615
[23,]
[24,]  Observations                   534
[25,]  Distinct observations          47
[26,]
[27,]  Total number of {-Inf,Inf,NaN}   0
[28,]
[29,] ============================================================
[30,]
```

2.3.2 Summarizing Categorical Data

---

s = frequency(x {, xvars {, outwidth}})
    computes a frequency table for each column of a matrix x; op-
    tionally a vector variable names xvars and maximal string length
    for categories can be given

s = crosstable(x{,xvars})
    computes pairwise cross tables from all columns of a data matrix
    x and computes the result of a χ^2 independence test; optionally
    a vector variable names xvars can be given

The functions frequency and crosstable can be applied to numeric as well
as to text matrices. The XploRe codes for this section are available from the
quantlet ℚ desc09.xpl.

frequency produces a text matrix containing the categories and frequencies as
well as cumulative frequencies for all columns of a data matrix. We apply this
function to the first and third columns of the earn data.

```
frequency(earn[,1|3], vearn[1|3])
```

frequency is a different way of reporting information about categories and
frequencies than the function discrete used in Section 2.2.5:

```
Contents of freq
[ 1,]
[ 2,] ======================================================
[ 3,]   Variable educ
[ 4,] ======================================================
[ 5,]              | Frequency  Percent  Cumulative
[ 6,] ------------------------------------------------------
[ 7,]          2 |         1    0.002      0.002
[ 8,]          3 |         1    0.002      0.004
[ 9,]          4 |         1    0.002     · 0.006
[10,]          5 |         1    0.002      0.007
[11,]          6 |         3    0.006      0.013
[12,]          7 |         5    0.009      0.022
[13,]          8 |        15    0.028      0.051
```

[14,]	9 |	12	0.022	0.073
[15,]	10 |	17	0.032	0.105
[16,]	11 |	27	0.051	0.155
[17,]	12 |	219	0.410	0.566
[18,]	13 |	37	0.069	0.635
[19,]	14 |	56	0.105	0.740
[20,]	15 |	13	0.024	0.764
[21,]	16 |	71	0.133	0.897
[22,]	17 |	24	0.045	0.942
[23,]	18 |	31	0.058	1.000
[24,]	------	------	------	------
[25,]	|	534	1.000	
[26,]	==			
[27,]				
[28,]	==			
[29,]	Variable female			
[30,]	==			
[31,]	| Frequency	Percent	Cumulative	
[32,]	------	------	------	
[33,]	0 |	289	0.541	0.541
[34,]	1 |	245	0.459	1.000
[35,]	------	------	------	
[36,]	|	534	1.000	
[37,]	==			
[38,]				

To study the dependence of two categorical variables, one typically analyzes the **contingency table** (or cross table). crosstable provides the cross tables of all columns of a data matrix and additionally computes the χ^2 statistic for testing independence and contingency coefficients. For example,

```
crosstable(earn[,1|3], vearn[1|3])
```

gives

```
Contents of cross
[ 1,]
[ 2,]
[ 3,] Crosstable for variables educ, female
[ 4,]
```

```
[ 5,]                    |      0.0000  1.0000  |
[ 6,] ----------|---------------------|---------
[ 7,]   2.0000  |        1       0     |      1
[ 8,]   3.0000  |        1       0     |      1
[ 9,]   4.0000  |        1       0     |      1
[10,]   5.0000  |        1       0     |      1
[11,]   6.0000  |        1       2     |      3
[12,]   7.0000  |        4       1     |      5
[13,]   8.0000  |        6       9     |     15
[14,]   9.0000  |        7       5     |     12
[15,]  10.0000  |       12       5     |     17
[16,]  11.0000  |       16      11     |     27
[17,]  12.0000  |      109     110     |    219
[18,]  13.0000  |       21      16     |     37
[19,]  14.0000  |       33      23     |     56
[20,]  15.0000  |        7       6     |     13
[21,]  16.0000  |       37      34     |     71
[22,]  17.0000  |       10      14     |     24
[23,]  18.0000  |       22       9     |     31
[24,] ----------|---------------------|---------
[25,]           |      289     245     |    534
[26,]
[27,] Chi^2 test of independence
[28,]
[29,]   chi^2 statistic:                  16.15
[30,]   degrees of freedom:                  16
[31,]   significance level for rejection: 0.4427
[32,]
[33,]   contingency coefficient:           0.17
[34,]   corrected contingency coefficient: 0.24
[35,]
```

The significance value for the χ^2-test of independence between both variables is 0.4427 here, which means that independence cannot be rejected (at the usual 5% or 10% levels).

3 Graphics

Sigbert Klinke

The XploRe graphics support has different levels which are described here. The levels correspond to the ease of use for different types of users.

- the **graphical tools**
  These routines are high-end-level tools for the beginner and possibly inexperienced user. The aim of the graphic toolbox of XploRe is to allow automatic creation of highly illustrative graphics over a wide range of data types. Examples are `plotandrews` for plotting Andrews curves, `plotstar` for star plots. Most plot tools are automatic or the user will be asked interactively for the parameters.

- the **graphical primitives**
  Graphical primitives are the typical graphics used in data analysis. Examples are boxplots, scatter plots, Andrews curves. This includes some routines for manipulation. The aim of these primitives is to ease the construction of user designed graphics. Graphical primitives may be changed by the user. Boxplots for example exist in large variety: some boxplot species mark outliers in a specific way, others do not.

- the **graphical commands**
  These commands are the building blocks of highly complex graphics. For an overview of construction and manipulation of simple graphics see the following sections. Some basic commands are described here.

The first two levels are reflected in the names of the quantlets. Quantlets starting with `gr`... are routines that are part of the **graphic** library. The aim is to generate and manipulate graphical primitives. Quantlets starting with `plot`... are routines from the graphical toolbox and part of the `plot` library.

3.1   Basic Plotting

```
plot (x1 {, x2 {, ...  {x5}}})
    plots the data sets x1, ..., x5

line (x1 {, x2 {, ...  {x5}}})
    plots the lines sets x1, ..., x5

y = setmask (x, opt1 {, opt2 {, ...  {opt9}}})
    modifies a data set for plotting

disp = createdisplay (r,c)
    creates a display disp

show (disp, i, j, x1 {, x2 {, ...  {, xn}}})
    plots the data sets x1, ..., xn in the display disp d
```

3.1.1   Plotting a Data Set

The quantlet plot plots up to five bivariate data sets in a scatter plot. Let us load a data set, the Boston Housing data file **bostonh.dat** (see Appendix B.4). It consists of 14 variables and we select the last two (percentage of lower status people, LSTAT, and median house price for owner occupied homes, MEDV) which are the 13th and 14th columns of the data set. Each observation represents one school district in the Boston metropolitan area.

```
library ("plot")            ; loads library plot
data = read ("bostonh")     ; reads Boston Housing data
x = data[,13:14]            ; selects columns 13 and 14
plot(x)                     ; plots data set
```

                                                                Q graph21.xpl

The plot in Figure 3.1 indicates a decreasing relationship between both variables.

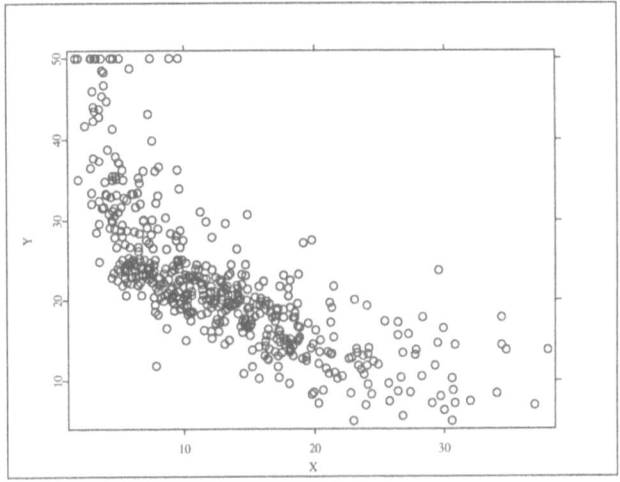

Figure 3.1. Plot of the 13th (LSTAT) and 14th variable (MEDV) of
the Boston Housing data. ⃝ `graph21.xpl`

3.1.2 Plotting a Function

We can plot mathematical functions, e.g. the sine. First we create an equidistant grid from 0 to 2π with 100 grid points.

```
library("plot")            ; loads library plot
xmin = 0                   ; grid minimum
xmax = 2*pi                ; grid maximum
n    = 100                 ; number of grid points
x = xmin + (xmax-xmin)/(n-1) .* (0:n-1)   ; generates grid
y = sin(x)                 ; computes sin(x)
plot(x~y)                  ; plots data set
```

⃝ `graph22.xpl`

Note that both x and y are vectors. The input for plot, x~y, is a matrix
composed of two vectors. The object x in the first example is a 506×2 matrix.

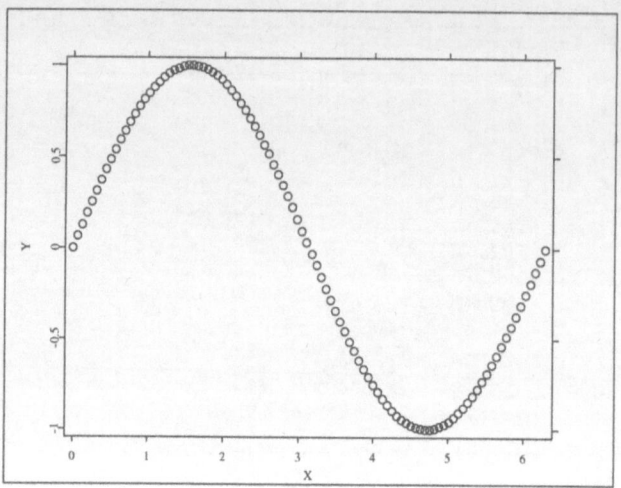

Figure 3.2. Plot of the sine curve in $[0, 2\pi]$. 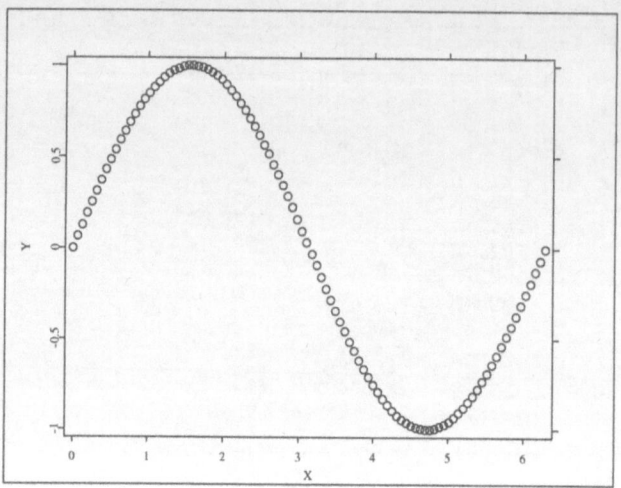graph22.xpl

3.1.3 Plotting Several Functions

We continue with plotting three functions $\sin(x)$, $\sin(3x)$ and $\sin(6x)$. Our last program has to be modified to

```
library("plot")              ; loads library plot
xmin = 0                     ; grid minimum
xmax = 2*pi                  ; grid maximum
n    = 100                   ; number of grid points
x   = xmin + (xmax-xmin)/(n-1) .* (0:n-1)  ; generates grid
y1 = sin(x)                  ; computes sin(x)
y2 = sin(3.*x)               ; computes sin(3x)
y3 = sin(6.*x)               ; computes sin(6x)
plot(x~y1, x~y2, x~y3)       ; plots data sets
```

graph23.xpl

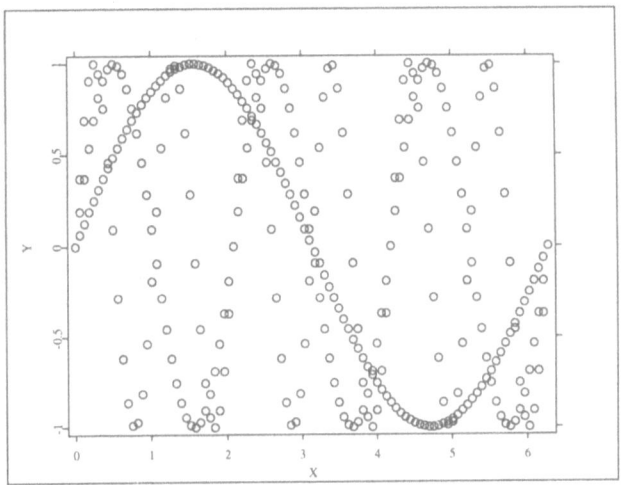

Figure 3.3. Plot of three sine curves in $[0, 2\pi]$. ⌕ graph23.xpl

3.1.4 Coloring Data Sets

Since by default the points are shown as circles it is difficult to distinguish the three functions. We may easily color them with the quantlet setmask.

```
library("plot")                      ; loads library plot
xmin = 0                             ; grid minimum
xmax = 2*pi                          ; grid maximum
n    = 100                           ; number of grid points
x    = xmin + (xmax-xmin)/(n-1) .* (0:n-1)   ; generates grid
y1 = sin(x)                          ; computes sin(x)
y2 = sin(3.*x)                       ; computes sin(3x)
y3 = sin(6.*x)                       ; computes sin(6x)
z1 = setmask(x~y1, "red")            ; colors sin(x) red
z2 = setmask(x~y2, "green")          ; colors sin(x) green
z3 = setmask(x~y3, "blue")           ; colors sin(x) blue
plot(z1, z2, z3)                     ; plots data sets
```

⌕ graph24.xpl

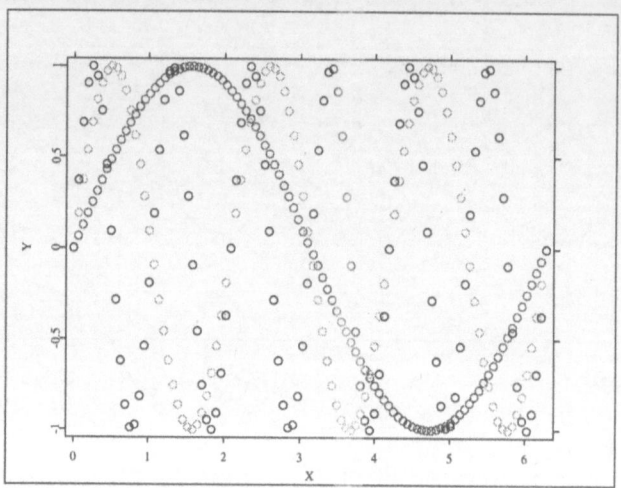

Figure 3.4. Plot of three colored sine curves in $[0, 2\pi]$. graph24.xpl

3.1.5 Plotting Lines from Data Sets

Even now, it is not easy to recognize the underlying functions. We can connect
the data points with the quantlet line instead of coloring the data points.

```
library("plot")                        ; loads library plot
xmin = 0                               ; grid minimum
xmax = 2*pi                            ; grid maximum
n    = 100                             ; number of grid points
x   = xmin + (xmax-xmin)/(n-1) .* (0:n-1)   ; generates grid
y1 = sin(x)                            ; computes sin(x)
y2 = sin(3.*x)                         ; computes sin(3x)
y3 = sin(6.*x)                         ; computes sin(6x)
line(x~y1, x~y2, x~y3)                 ; connects data points
```

graph25.xpl

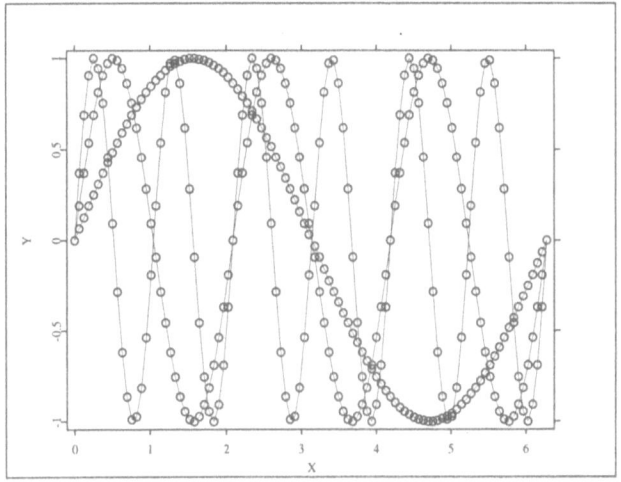

Figure 3.5. Plot of three sine curves as lines in $[0, 2\pi]$. ⊙ graph25.xpl

Let us now color the lines to distinguish them better. We again use the quantlets `setmask` and `plot`.

```
library("plot")              ; loads library plot
xmin = 0                     ; grid minimum
xmax = 2*pi                  ; grid maximum
n    = 100                   ; number of grid points
x   = xmin + (xmax-xmin)/(n-1) .* (0:n-1)  ; generates grid
y1 = sin(x)                  ; computes sin(x)
y2 = sin(3.*x)               ; computes sin(3x)
y3 = sin(6.*x)               ; computes sin(6x)
plot(x~y1, x~y2, x~y3)       ; plots the data sets
z1 = setmask(x~y1, "line", "red")    ; red line for sin(x)
z2 = setmask(x~y2, "line", "green")  ; green line for sin(3x)
z3 = setmask(x~y3, "line", "blue")   ; blue line for sin(6x)
plot(z1, z2, z3)             ; plots lines
```

⊙ graph26.xpl

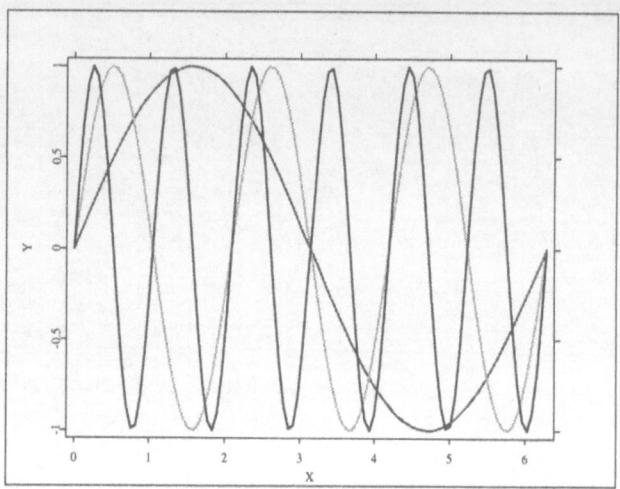

Figure 3.6.   Plot of three colored sine curves as lines in $[0, 2\pi]$.
Ⓠ graph26.xpl

3.1.6   Several Plots

Sometimes we want to compare several plots of data sets with each other;
e.g. imagine a scatter-plot matrix. In most graphical user interfaces, we have
windows available. However, we do not want to compose a scatter-plot matrix
from a set of single windows. Since we are allowed to move the windows around,
we would be able to hide one plot of a scatter-plot matrix. Thus we allow
subwindows in a window. In our terminology, a window is called a **display**
and the nonoverlapping subwindows are called **graphics ports** or **plots**.

In one of our last examples we plotted the points for three different sine curves.
Let us now modify our example such that we plot each of the curves in a single
port.

First we need to create a display which consists of several windows. The command **createdisplay** in the example below creates a display with 2 plots
vertically and 3 plots horizontally.

(1,1)	(1,2)	(1,3)
(2,1)	(2,2)	(2,3)

To identify our display among several displays we need a unique name. The assignment of the result of `createdisplay(2,3)` to `disp` identifies our display uniquely.

Then we create six sine curves with different frequencies. With the command `show`, we show the sine curves in the different plots of the display `disp`. Since we may have more than one display, the first parameter of `show` is the display name `disp`, the next two parameters describe which plot of the display is used for plotting, and the following parameter(s) are the data sets which will be plotted in the plot.

```
disp = createdisplay(2,3)      ; display with 6 windows
xmin = 0                       ; grid minimum
xmax = 2*pi                    ; grid maximum
n    = 100                     ; number of grid points
x  = xmin + (xmax-xmin)/(n-1) .* (0:n-1)   ; generates grid
y1 = sin(x)                    ; computes sin(x)
y2 = sin(2.*x)                 ; computes sin(2x)
y3 = sin(3.*x)                 ; computes sin(3x)
y4 = sin(4.*x)                 ; computes sin(4x)
y5 = sin(5.*x)                 ; computes sin(5x)
y6 = sin(6.*x)                 ; computes sin(6x)
show (disp, 1, 1, x~y1)        ; shows sine curve x~sin(x)
show (disp, 1, 2, x~y2)        ; shows sine curve x~sin(2x)
show (disp, 1, 3, x~y3)        ; shows sine curve x~sin(3x)
show (disp, 2, 1, x~y4)        ; shows sine curve x~sin(4x)
show (disp, 2, 2, x~y5)        ; shows sine curve x~sin(5x)
show (disp, 2, 3, x~y6)        ; shows sine curve x~sin(6x)
```

graph27.xpl

It seems that we used a completely different technique from before to generate our plots. However, you may already guess that `plot(x)` which we have used before consists mainly of

```
d = createdisplay(1,1)
show(d,1,1,x)
```

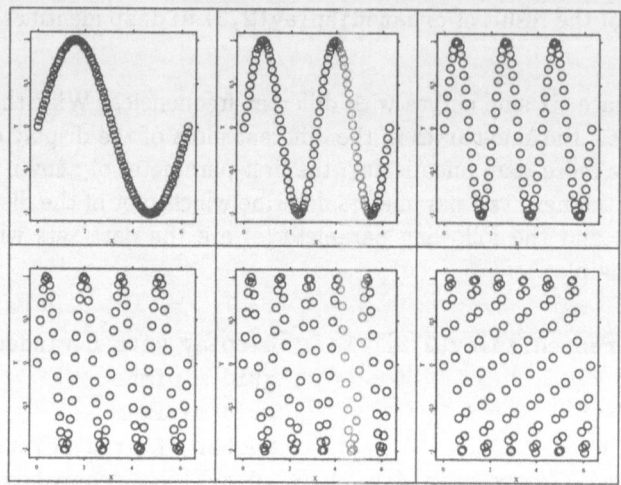

Figure 3.7. Plot of six sine curves, each in its own plot, in $[0, 2\pi]$.
◙ graph27.xpl

Note that in contrast to all other variables in XploRe, displays are always global variables. This is because displays should survive the execution end of a quantlet.

3.2 Univariate Graphics

```
gr = grbox (x {,col})
     generates a boxplot of the data set x

gr = grdot (x {,col})
     generates a dotplot of the data set x

gr = grbar (x {,col})
     generates a bar chart of the data set x

gr = grqq (y, x {,col})
     generates a QQ-plot from the data sets y and x

gr = grqqn (x {,col})
     generates a QQ-plot from the data set x and a normal distribution

gr = grqqu (x {,col})
     generates a QQ-plot from the data set x and a uniform distribu-
     tion

gr = grhist (x {, h {, o {,col}}})
     generates a histogram of the data set x

gr = grash (x {, h {, k {,col}}})
     generates an averaged shifted histogram of the data set x
```

The optional parameter col allows us to produce a graphical object in another color other than black. For details, see Subsection 3.4.3. The other optional parameters will be explained when we introduce **grhist** and **grash**.

In the following examples, we use a mix of graphical primitives which are part of the library **graphic** and high-level routines which are part of the library **plot**. Since a call of library **plot** also loads the library **graphic**, we do not need to call the library **graphic** explicitly.

3.2.1   Boxplots

Let us now examine some variables of the Boston Housing data with statistical
graphics. Since the aim of the data exploration is to predict the median house
price from the variables, let us make a boxplot with the quantlet grbox.

```
library("plot")              ; loads library plot
data = read ("bostonh")      ; reads Boston Housing data
gr = grbox(data[,14])        ; generates a graphical object
plot(gr)                     ; plots graphical object
```

Q graph31.xpl

Note that we generate the boxplot in two steps. First we generate the **graphical object gr** and then we plot it.

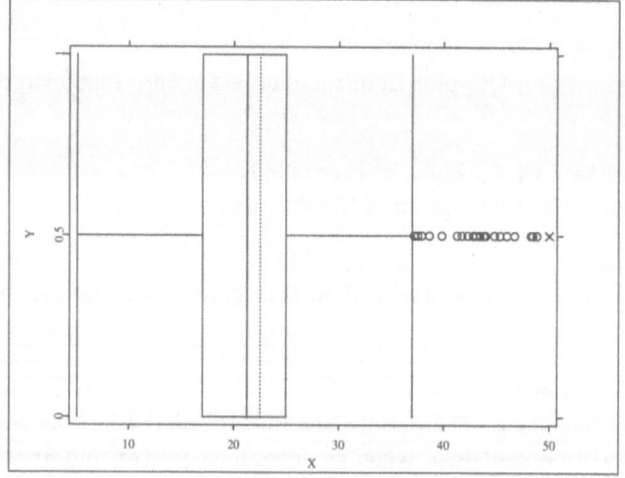

Figure 3.8. Boxplot of the 14th variable (MEDV) of the Boston Housing
data. Q graph31.xpl

We might not be satisfied with the boxplot, since the window size is chosen
such that all the data are visible. Let us now apply an often helpful trick to
get a better plot.

```
library("plot")                  ; loads library plot
```

```
data = read ("bostonh")          ; reads Boston Housing data
x = data[,14]                    ; selects the 14th column
gr = grbox(x)                    ; generates graphical object
scale = #(min(x),max(x))~#(-1, 2) ; generates scaling data set
scale = setmask(scale,"white")   ; makes scaling data "invisible"
plot(gr, scale)                  ; plots boxplot and scaling data
```
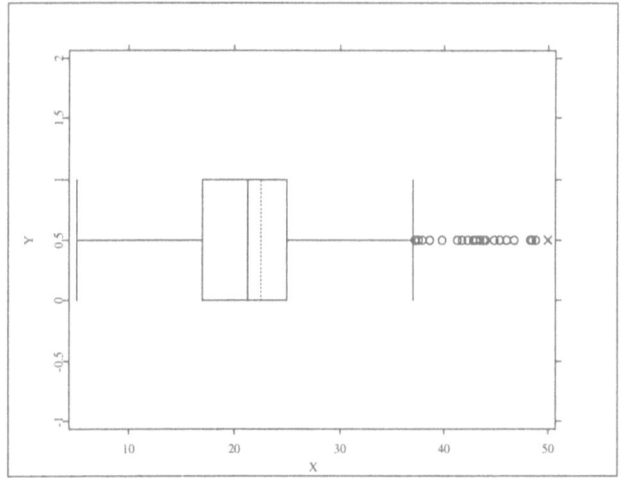 graph32.xpl

Figure 3.9. Rescaled boxplot of the 14th variable (MEDV) of the
Boston Housing data. ⍥ graph32.xpl

We have generated an invisible data set which helps us to scale the boxplot
better in the window.

We learn from the boxplot that the variable MEDV contains several large
outliers. The mean (broken line) and the median (solid line in the box) differ.
Moreover we see on the right outliers marked with circles and crosses. Since the
box borders (25%- and 75%-quantile) and the whiskers ($\geq$ 25%-quantile $-$ 1.5
interquartile range and $\leq$ 75%-quantile $+$ 1.5 interquartile range) have more
or less the same distance from the median, we may consider that the variable
has a symmetrical distribution.

3.2.2   Dotplots

Let us now examine the median house price a little bit more in detail. We use
the quantlet **grdot** to generate a dotplot. In the horizontal direction, a dotplot
takes the value of the observations, in the vertical direction it takes a generated
uniformly distributed random number.

```
library("plot")                   ; loads library plot
data = read ("bostonh")           ; reads Boston Housing data
x = data[,14]                     ; selects 14th column
gr = grdot(x)                     ; generates dotplot
scale = #(min(x),max(x))~#(-1, 2) ; generates scaling data set
scale = setmask(scale,"white")    ; makes scaling data "invisible"
plot(gr, scale)                   ; plots dotplot and scaling data
```
🔍 graph33.xpl

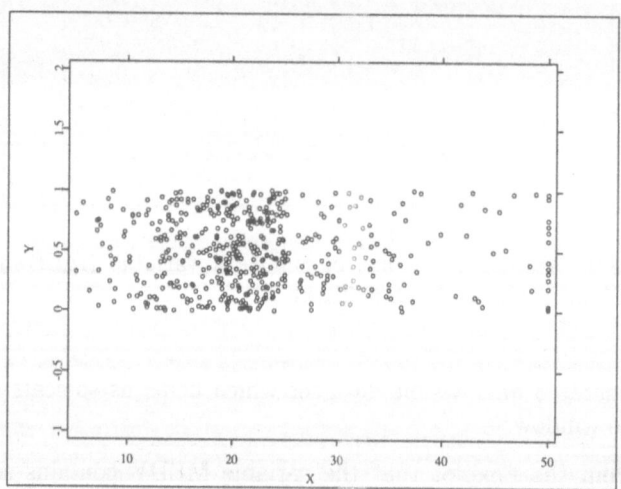

Figure 3.10. Rescaled dotplot of the 14th variable (MEDV) of the
Boston Housing data. 🔍 graph33.xpl

After having rescaled the display, we can detect patterns within the variable
MEDV. It seems we have rather sparse area of data until $x \approx 12$, then a denser

area of data until $x \approx 18$ with a sharp break at $x \approx 25$ and finally another break at $x \approx 36$. We also see that behind the cross in the dotplot, there is more than one observation.

3.2.3   Bar Charts

If we want to plot discrete variables, it does not make sense to use a box- or dotplot. For this purpose we can use bar charts. We generate a bar chart with the quantlet grbar and use the fourth variable (CHAS) which is an indicator variable as to whether the Charles river is part of the school district.

```
library("plot")                        ; loads library plot
data = read ("bostonh")                ; reads Boston Housing data
x = data[,4]                           ; selects 4th column
gr = grbar(x)                          ; generates a bar chart
gr = setmask(gr, "line", "medium") ; changes line thickness
plot(gr)                               ; plots bar chart
```
         Q graph34.xpl

We see in the bar chart a large bar representing zeros (school district does not include Charles river) and a small bar representing ones (school district does include Charles river).

Although most gr... quantlets generate already a useful graphic, they aimed to be building blocks of high-level routines. If the Charles river index variable would be coded by the numbers –1 and 0, we would not be able to tell which bar chart represents the –1 and which represents the 0. The left bar would still start at 0 and the right bar at 1.

Fortunately, there is the more sophisticated quantlet plotbar available which generates a much better bar chart.

```
library("plot")                        ; loads library plot
data = read ("bostonh")                ; reads Boston Housing data
x = data[,4]                           ; selects 4th column
plotbar(x)                             ; plots the bar chart
```
         Q graph35.xpl

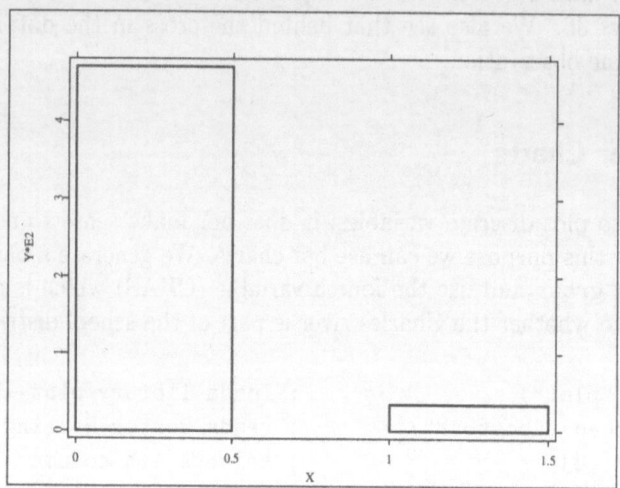

Figure 3.11. Barchart of the 4th variable (CHAS) of the Boston Housing data generated by a graphic primitive routine. ⍾graph34.xpl

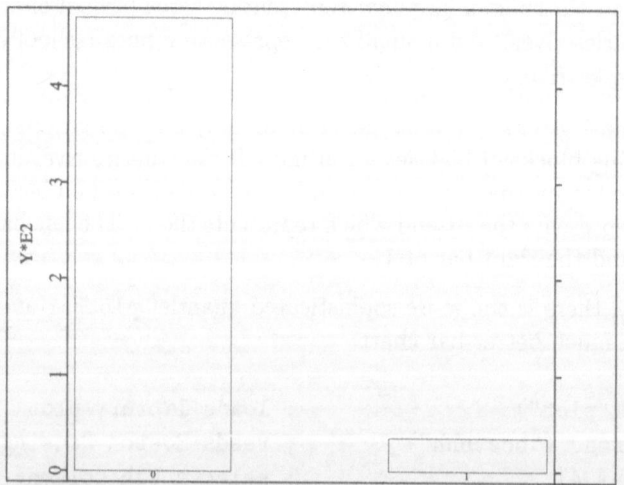

Figure 3.12. Barchart of the 4th variable (CHAS) of the Boston Housing data generated by a graphic high-level routine. ⍾graph35.xpl

3.2.4   Quantile-Quantile Plots

Quantile-Quantile plots are used to compare distributions of two variables
(**grqq**) or to compare one variable with a given distribution (**grqqu** uniform,
grqqn normal).

Let us compare the percentage of lower status people with the appropriate
normal distribution.

```
library("plot")              ; loads library plot
data = read ("bostonh")      ; reads Boston Housing data
x = data[,13]                ; selects 13th column
gr = grqqn(x)                ; generates a qq plot
plot(gr)                     ; plots the qq plot
```
⊙ graph36.xpl

Apparently we have a clear deviation from the line which indicates that the
13th variable is not normally distributed. Since the data points cross the 45
degree line twice, we can say that the distribution is steeper in the center and
thicker in the tails.

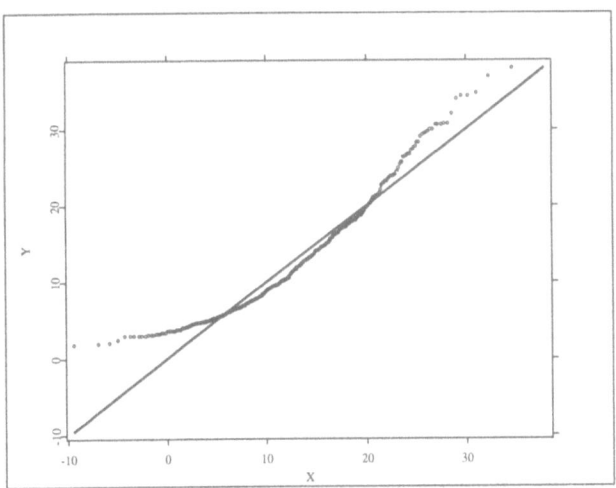

Figure 3.13. QQ-plot of the 13th variable (LSTAT) of the Boston Housing data. ⊙ graph36.xpl

3.2.5   Histograms

The most often used statistical graphics tools to visualize continuous data is
the histogram. Let's now generate a histogram from the median house prices.

```
library("plot")         ; loads library plot
data = read ("bostonh") ; reads Boston Housing data
x = data[,14]           ; selects 14th column
gr = grhist(x)          ; generates histogram
plot(gr)                ; plots histogram
```

graph37.xpl

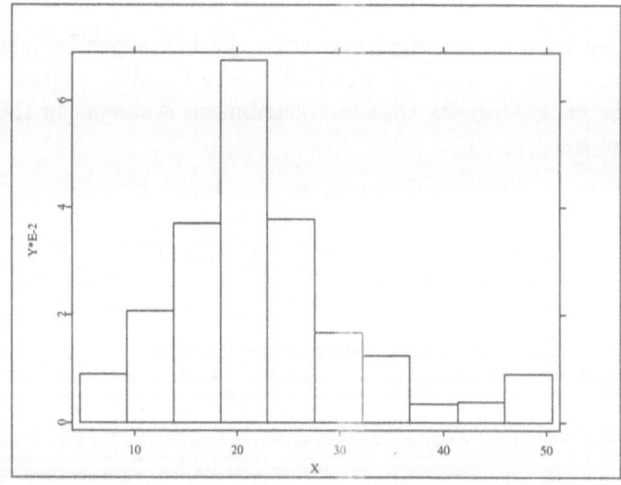

Figure 3.14. Histogram of the 14th variable (MEDV) of the Boston
Housing data. ◙ graph37.xpl

We already noticed some characteristics of this variable when we generated
boxplots. Here we find some of them again, e.g. the central dense region of
data and the outliers at the right border. In contrast to all other univariate
graphics, **grhist** has two optional parameters: h the binwidth and o the origin
of the histogram. The change of the binwidth as well as the origin might reveal
some more patterns within the data. Let us first change the binwidth to 1.

```
library("plot")              ; loads library plot
data = read ("bostonh")      ; reads Boston Housing data
x = data[,14]                ; selects 14th column
gr = grhist(x,1)             ; generates histogram with
                             ;     binwidth 1
plot(gr)                     ; plots histogram
```
Q graph38.xpl

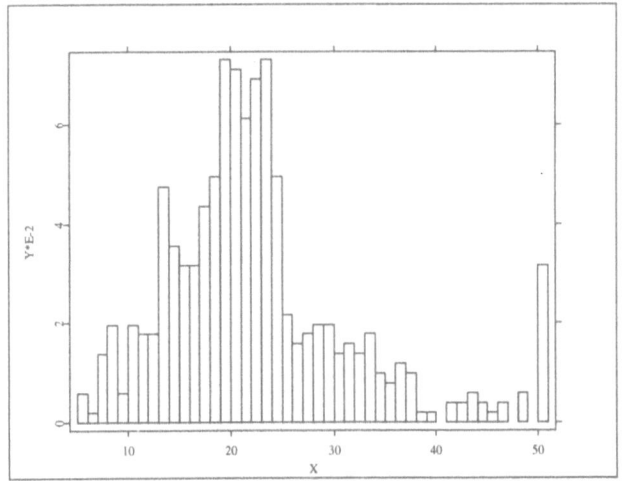

Figure 3.15. Histogram of the 14th variable (MEDV) of the Boston
Housing data with a different binwidth. Q graph38.xpl

Now change the origin to 0.5.

```
library("plot")              ; loads library plot
data = read ("bostonh")      ; reads Boston Housing data
x = data[,14]                ; selects 14th column
gr = grhist(x,1,0.5)         ; generates histogram with
                             ;     binwidth 1 and origin 0.5
plot(gr)                     ; plots histogram
```
Q graph39.xpl

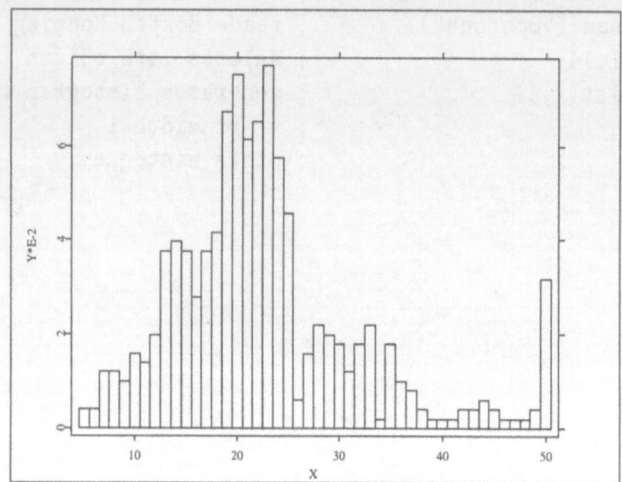

Figure 3.16. Histogram of the 14th variable (MEDV) of the Boston
Housing data with a different origin of the bin. ◻ graph39.xpl

Well-known problems with histograms are the choices of the origin and the binwidth. To overcome these problems, the concept of average shifted histograms
has been developed. In principle, we generate a set of histograms with different
origins, instead of one histogram, and then we average these histograms. We
apply average shifted histograms to our last example with binwidth 1 but 10
different origins.

```
library("plot")          ; loads library plot
data = read ("bostonh")  ; reads Boston Housing data
x = data[,14]            ; selects 14th column
gr = grash(x,1,10)       ; generates average shifted histogram
plot(gr)                 ; plots average shifted histogram
```

                                                              ◻ graph3A.xpl

In both histograms, Figures 3.15 and 3.16, we can speculate about multimodal-
ity of the data, since they show different histograms. However, the average
shifted histogram suggests the existence of three modes between 10 and 25.

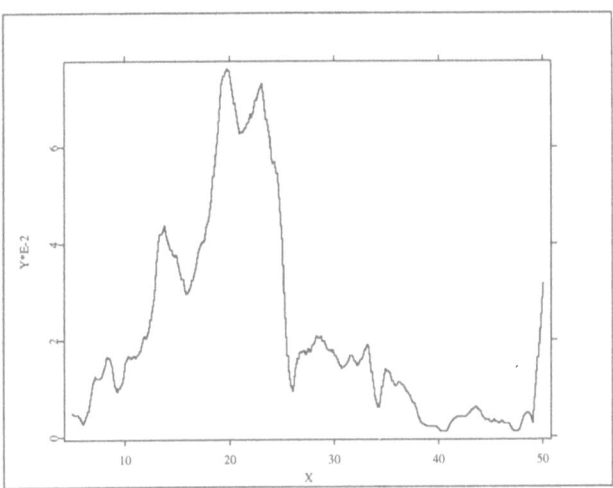

Figure 3.17. Average shifted histogram of the 14th variable (MEDV) of the Boston Housing data. ◻ **graph3A.xpl**

3.3   Multivariate Graphics

---

plot (x1 {, x2 {, ... {x5}}})
     plots the three-dimensional data sets x1, ..., x5

gr = grsurface (x {, col})
     generates surface from the function $f(x, y)$

gr = grcontour2 (x {, c {,col}})
     generates the contour lines $f(x, y) = c$

gr = grcontour3 (x {, c {,col}})
     generates the contour lines $f(x, y, z) = c$

gr = grsunflower (x {, d {, o {,col}}})
     generates a sunflower plot

gr = grlinreg (x {,col})
     generates the linear regression line

gr = grlinreg2 (x {, n {,col}})
     generates the linear regression plane

plot2 (x {, prep {,col}})
     plots two variables

plotstar (x {, prep {,col}})
     plots a star diagram

plotscml (x {, varnames})
     plots a scatter-plot matrix

plotandrews (x {, prep {,col}})
     plots Andrews curves

plotpcp (x {, prep {,col}})
     plots parallel coordinates

The optional parameter col allows us to produce a graphical object in another
color other than black. For details, see Subsection 3.4.3.

3.3.1 Three-Dimensional Plots

Up to now, we have just generated graphical objects or data sets and plotted them in a two-dimensional plot. Sometimes the analysis of data is easier if we can show them as three-dimensional data. Rotating the data point cloud will give us more insight into the data. We can use the `plot` quantlet for this.

```
library("plot")           ; loads library plot
data = read ("bostonh")   ; reads Boston Housing data
x = data[,6|13|14]        ; selects columns 6, 13 and 14
plot(x)                   ; plots the data set
```

Q graph41.xpl

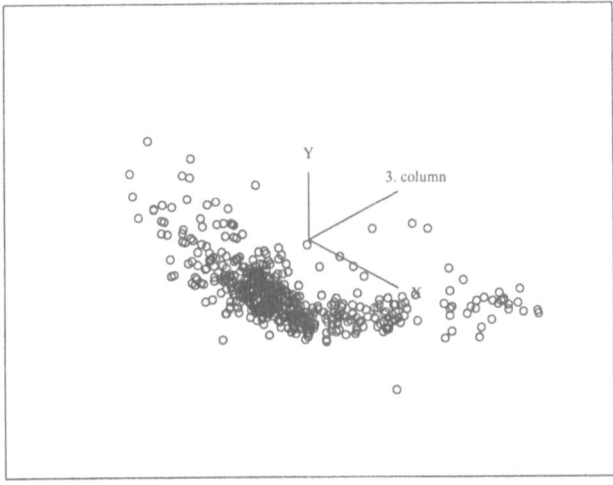

Figure 3.18. Variables 6 (RM), 13 (LSTAT) and 14 (MEDV) of the Boston Housing data plotted in a 3D scatter plot. Q `graph41.xpl`

Now click with the mouse in the window and use the cursor keys to rotate the data set. Note that the only change we made for three-dimensional plotting is that the input matrix we use in `plot` consists of three vectors. We can apply here the quantlet `setmask` to color the data points.

We see in the plot of RM, LSTAT and MEDV a nonlinear relationship which may allow us to estimate the median house prices (MEDV) as a parametric function of percentage of lower status people (LSTAT) and average number of rooms (RM).

3.3.2   Surface Plots

Surfaces are the three-dimensional analogs of curves in two dimensions. Since we have already plotted three-dimensional data, we can imagine what we have to do: generate a data set, generate some lines, and plot them. The quantlet grsurface does this for data on a rectangular mesh.

```
library ("plot")              ; loads library plot
x0 = #(-3, -3)
h  = #(0.2, 0.2)
n  = #(31, 31)
x  = grid(x0, h, n)           ; generates a bivariate grid
f  = exp(-(x[,1]^2+x[,2]^2)/1.5)/(1.5*pi)
                              ; computes density of bivariate
                              ;   normal with correlation 0.5
gr = grsurface(x~f)           ; generates surface
plot(gr)                      ; plots the surface
```
🔍 graph42.xpl

Most of the upper program is used to generate the underlying grid and the density function. We may plot the data set itself by plot(g~f). The surface quantlet grsurface needs three parameters: the x- and the y-coordinates of the grid and the function values at $f(x_i, y_j)$.

3.3.3   Contour Plots

If we view surfaces, then understanding them might be difficult, even if we can rotate them. We can produce contour plots with contour lines $f(x, y) =$ constant.

```
library ("plot")          ; loads the library plot
x0 = #(-3, -3)
h  = #(0.2, 0.2)
```

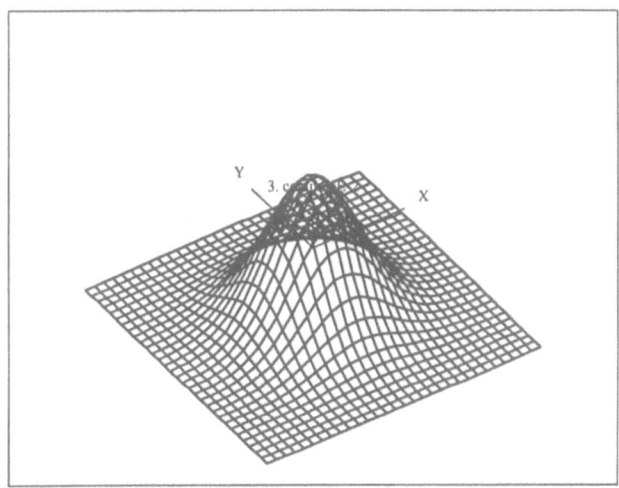

Figure 3.19. Surface plot of the density of the bivariate standard normal
distribution. ⊙graph42.xpl

```
n  = #(31, 31)
x  = grid(x0, h, n)       ; generates a bivariate grid
f  = exp(-(x[,1]^2+x[,2]^2)/1.5)/(1.5*pi)
                          ; computes density of bivariate
                          ;   normal with correlation 0.5
c  = 0.2*(1:4).*max(f)    ; selects contour lines as 10%,...,90%
                          ;   times the maximum density
gr = grcontour2(x~f, c)   ; generates contours
plot(gr)                  ; plots the contours
```

⊙ graph43.xpl

The quantlet grcontour3 can be used to compute contour lines for a function $f(x, y, z)$ = constant. This gives a two-dimensional surface in a three-
dimensional space. An example is the XploRe logo.

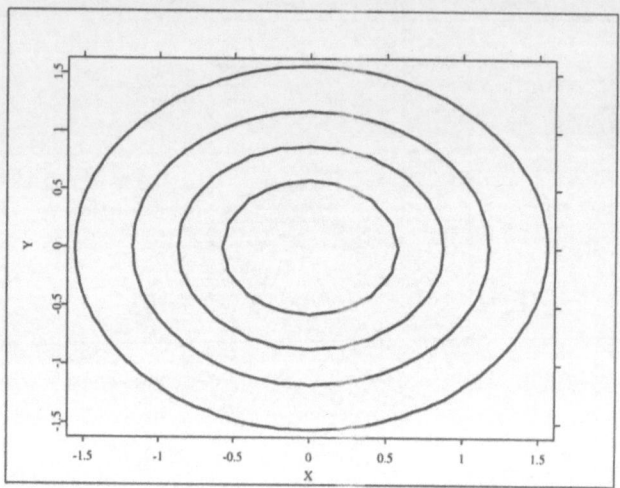

Figure 3.20. Contour plot of the density of the bivariate standard normal distribution. 🔍 graph43.xpl

3.3.4 Sunflower Plots

Sunflower plots avoid the overplotting if we have many data points. We can consider a sunflower plot as a combination of two-dimensional histograms with a contour plot. As with histograms, we must define a two-dimensional binwidth d and a two-dimensional origin o.

Let's compare a bivariate normal distribution with 1000 data points with the equivalent sunflower plot.

```
library ("graphic")        ; loads library graphic
x  = normal(1000, 2)       ; generates bivariate normal data
d  = createdisplay(2,1)    ; creates a display with two plots
show (d, 1, 1, x)          ; plots the original data
gr = grsunflower(x)        ; generates sunflower plot
show (d, 2, 1, gr)         ; plots the sunflower plot
```

🔍graph44.xpl

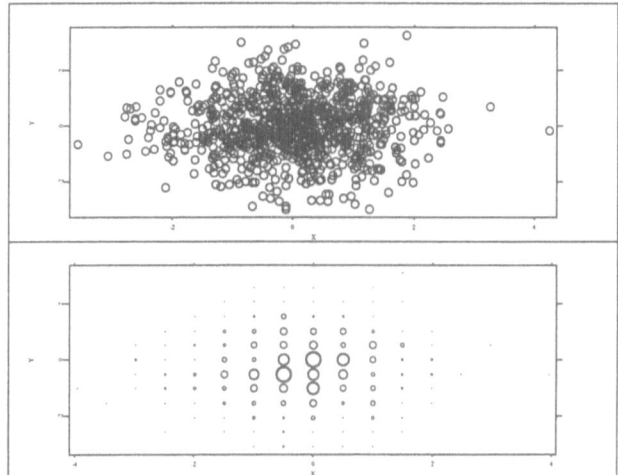

Figure 3.21. Standard 2D plot and sunflower plot of a large random sample of a bivariate standard normal distribution. ⌐graph44.xpl

3.3.5 Linear Regression

An important statistical task is to find a relationship between two variables. The most frequently used technique to quantify the relationship is the least squares linear regression:

$$\sum_{i=1}^{n}(y_i - b_0 - b_1 x_i)^2 \rightarrow \text{ minimal.}$$

To understand (graphically) how well the linear regression picks up the true relationship between two variables, we show the data points and the regression line in one plot:

```
library("plot")             ; loads library plot
data = read ("bostonh")     ; reads Boston Housing data
x0 = data[,13:14]           ; selects columns 13 and 14
l0 = grlinreg(x0)           ; generates regression line
x1 = log(data[,13:14])      ; logarithm of columns 13 and 14
l1 = grlinreg(x1)           ; generates regression line
d  = createdisplay(1,2)     ; creates display with two plots
```

```
show (d, 1, 1, x0, 10)              ; plots data and regression 10
show (d, 1, 2, x1, 11)              ; plots data and regression 11
```

Q graph45.xpl

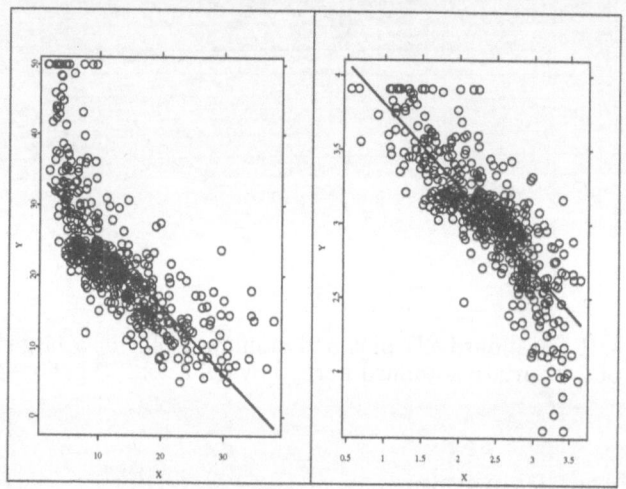

Figure 3.22. Linear regressions of the 14th variable (MEDV) by the 13th variable (LSTAT), left: original variables, right: both variables transformed by logarithm. Q graph45.xpl

We see that the regression line (median house price $= b_0 + b_1$ percentage of lower status people) does not fit well the data. Either a transformation of the data or a nonlinear regression technique seems to be useful. In our example we have taken logarithms of x and y, thus our model is

$$\text{median house price} = \exp(b_0) \text{ percentage lower status people}^{b_1}$$

The transformation of the house price turns out to be especially useful, since it avoids having to achieve negative values for the house prices in the model.

Obviously we can choose other explanatory variables, e.g. nitric oxygen in parts per ten million (NOXSQ) and average numbers of rooms (RM). This results in the following program which draws the three-dimensional data set and the regression plane $(b_0 + b_1 x_1 + b_2 x_2)$ with a 4×4 mesh.

```
library("plot")          ; loads library plot
data = read ("bostonh")  ; reads Boston Housing data
x   = data[,5|6|14]      ; selects columns 5, 6 and 14
p   = grlinreg2(x, 5|5)  ; generates regression plane
                         ;     with 4x4 mesh
plot(x, p)               ; plots data set and regression plane
```

Qgraph46.xpl

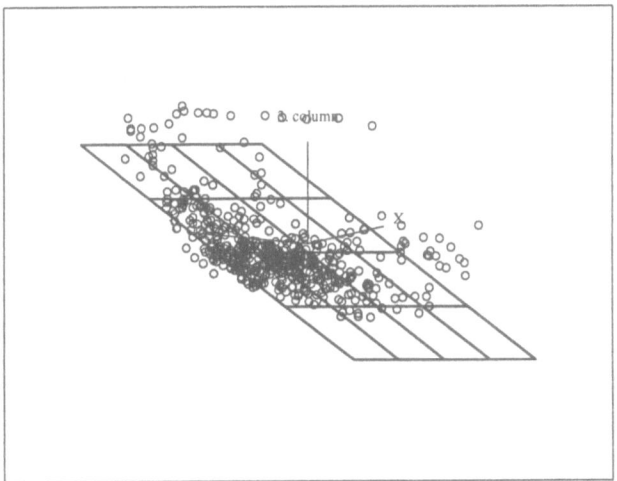

Figure 3.23. Bivariate linear regressions of the 14th variable (MEDV) by the 5th (NOXSQ) and 6th variables (RM). Qgraph46.xpl

3.3.6 Bivariate Plots

We have already plotted two variables, the percentage of lower status people against the median house price. The quantlet plot2 provides a much more powerful way of plotting multivariate data sets. The following program shows the first two principal components (based on the correlation matrix) of the Boston Housing data set. Additionally, if the median house price is less than the mean of the median house price, then we color the observation green, otherwise we color the observation blue.

```
library("plot")              ; loads library plot
data = read ("bostonh")      ; reads Boston Housing data
col  = grc.col.green-grc.col.blue
col  = grc.col.blue+col*(data[,14]<mean(data[,14]))
                             ; colors observations blue and green
plot2 (data, grc.prep.pcacorr, col)
                             ; plots two first principal axes
```

Q graph47.xpl

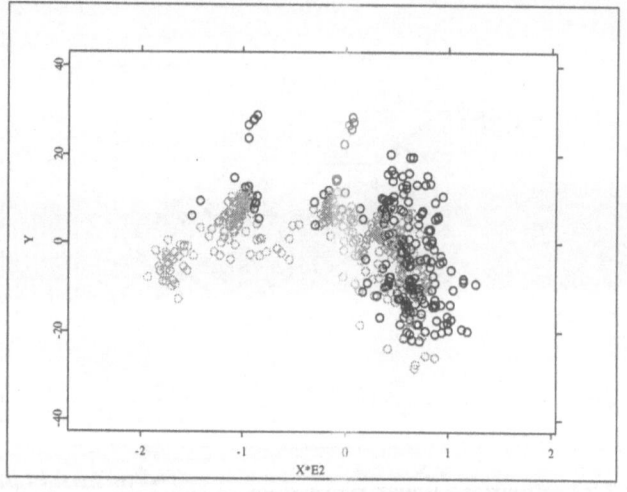

Figure 3.24. First two principal components based on the correlation
matrix of the Boston Housing data. ◌ graph47.xpl

When we load the library graphic, the library installs an object grc which
contains some often used graphical constants, e.g. grc.col.red for the color
red. grc.prep.none makes no transformation on the data and plots the first
two variables (here: per capita crime rate, CRIM, and proportion of residential
land zoned for lots over 25000 square feet, ZN).

grc.prep.standard
    standardizes the data,

`grc.prep.zeroone`
  transforms the data on the interval $[0, 1]$ before plotting,

`grc.prep.pcacov`
  takes the first two principal components based on the covariance instead of the two first two variables,

`grc.prep.pcacorr`
  takes the first two principal components based on the correlation instead of the two first two variables and

`grc.prep.sphere`
  spheres the data and the takes the first two components.

Since we assume a relation between all variables, we choose `grc.prep.pcacorr` and color the data points as described above. It seems to be an interesting (nonlinear) relationship for the house prices. A more complex example in Subsection 3.4.3 describes how the coloring works.

3.3.7   Star Diagrams

Star diagrams are used to visualize a multivariate data set. For each variable, we plot a star. Each axis of a star represents one variable. Obviously, we need to standardize the variables to a common interval which is internally done by $z_{i,j} = (x_{i,j} - \min_j)/(\max_j - \min_j)$. Depending on the method the user chooses, we select $\min_j$ and $\max_j$. The default method is **`grc.prep.zeroone`**.

```
library("plot")          ; loads library plot
data = read ("bostonh")  ; reads Boston Housing data
data = data[1:70,]       ; select first 70 observations
col  = grc.col.green-grc.col.blue
col  = grc.col.blue+col*(data[,14]<mean(data[,14]))
                         ; colors observations blue and green
plotstar (data, grc.prep.zeroone, col)
                         ; shows star diagram of the data
```

graph48.xpl

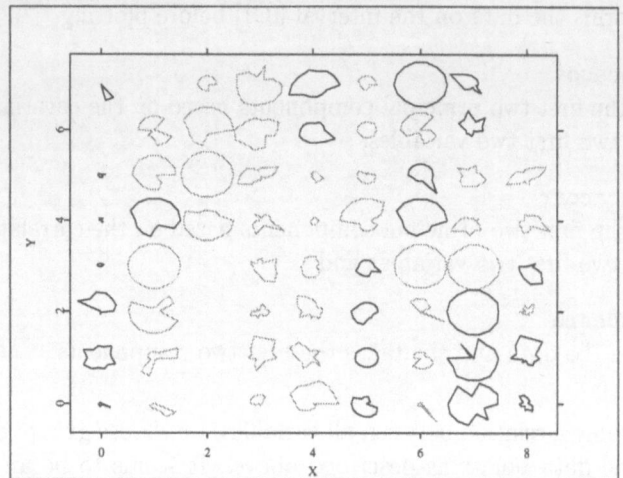

Figure 3.25. Star diagram of the Boston Housing data. Green stars represent observations below the average house price, blue stars represent house prices above the average house price. ⓠ graph48.xpl

We note immediately that we have several groups of similar looking data points. The Boston Housing data set is well-known for its outliers and subgroups in the data.

3.3.8   Scatter-Plot Matrices

Another possibility of analyzing multivariate data is the scatter-plot matrix. Here, we plot a set of scatter plots such that we see every possible variable combination. Obviously we should not throw too much variables in a scatter-plot matrix, since our screen size is limited. In fact, plotscml limits the number of variables to eight.

```
library("plot")                      ; loads library plot
data = read ("bostonh")              ; reads Boston Housing data
x = data[,5|6|13|14]                 ; selects columns 5, 6, 13 14
names="NOXSQ"~"RM"~"LSTAT"~"MEDV"    ; names of the variables
plotscml (x, names)                  ; shows scatter-plot matrix
```

🔍 graph49.xpl

We see a clear nonlinear relationship between RM and LSTAT as well as between LSTAT and MEDV.

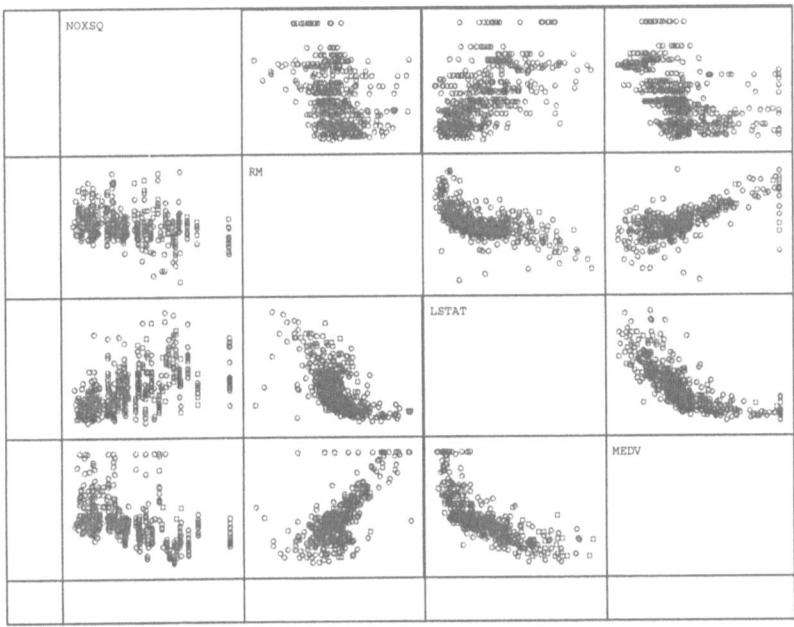

Figure 3.26. Scatter-plot matrix of the 5th (NOXSQ), 6th (RM), 13th (LSTAT) and 14th variables (MEDV) of the Boston Housing data. 🔍 graph49.xpl

3.3.9   Andrews Curves

The idea of Andrews curves is based on replacing each point by a curve such that some properties of the data points are transferred to properties of the curves. For example, it holds that

$$\int_{-\pi}^{\pi} (f_i(t) - f_j(t))^2 dt = \pi \|x_i - x_j\|$$

with x_i, x_j as data points and f_i, f_j representing curves generated from the data points. We see that distant points in p-space will generate quite different curves. The curve generation is defined by

$$f_i(t) = \frac{x_{i,1}}{\sqrt{2}} + x_{i,2} \sin(t) + x_{i,3} \cos(t) + x_{i,4} \sin(2t) + x_{i,5} \cos(2t) + \dots .$$

Again, we need the variables to be on comparable levels. In our example, we choose `grc.prep.pcacorr`. We see that the curves are quite different, but they cross near $t = 2$. For a fixed t, we can interpret the data points as a projection onto a very specific projection vector $(1/\sqrt{2}, \sin(2), \cos(2), ...)^T x_i$. Thus we conclude that one principal component of the correlation matrix is zero. In fact the fourth variable (CHAS), the Charles river index, has a rather small correlation with all other variables. Since the variable is not continuous, we should not include it in our picture.

```
library("plot")                ; loads library plot
data = read ("bostonh")        ; reads Boston Housing data
data = data[21:40]             ; observations 21 to 40
plotandrews (data, grc.prep.pcacorr)
                               ; shows Andrews curves based on
                               ;    principal components of
                               ; correlation matrix
```
graph4A.xpl

Note also that the order of the variables plays an important role. The last variable is represented with a rather high frequency. The human brain will not easily consider two high frequent curves as really distinct.

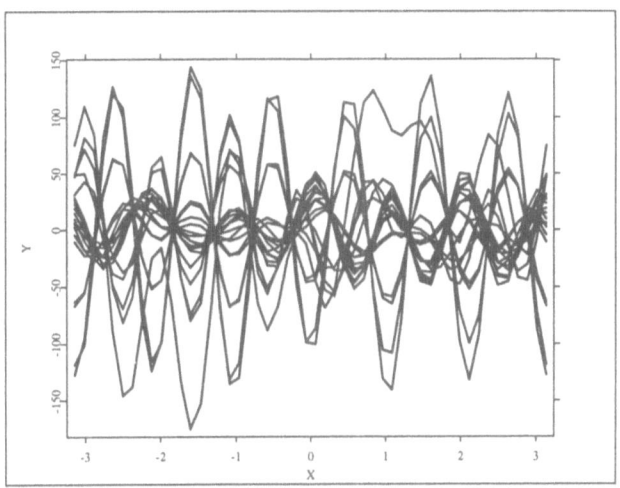

Figure 3.27. Andrews curves based on the principal components of 20 observations of the Boston Housing data. ⊙graph4A.xpl

3.3.10   Parallel Coordinate Plots

A completely different approach are parallel coordinate plots. Instead of insisting on orthogonality of the projections (e.g. in the scatter-plot matrix), we give it up. We plot on the jth parallel axis all data points x_{ij}. Then we connect the intersections such that each curve represents one data point. Some properties between the variables create specific patterns in the graphics. For example, a correlation of 1 between two variables is represented by parallel lines between the axes. A correlation of –1 results in a crossing of all lines in one point in the middle between two variables.

```
library("plot")            ; loads library plot
data = read ("bostonh")    ; reads Boston Housing data
data = data[21:40]         ; observations 21 to 40
x = data[,6|13|14]         ; selects columns 6, 13 and 14
plotpcp (x, grc.prep.standard) ; shows parallel coordinate plot
```

⊙ graph4B.xpl

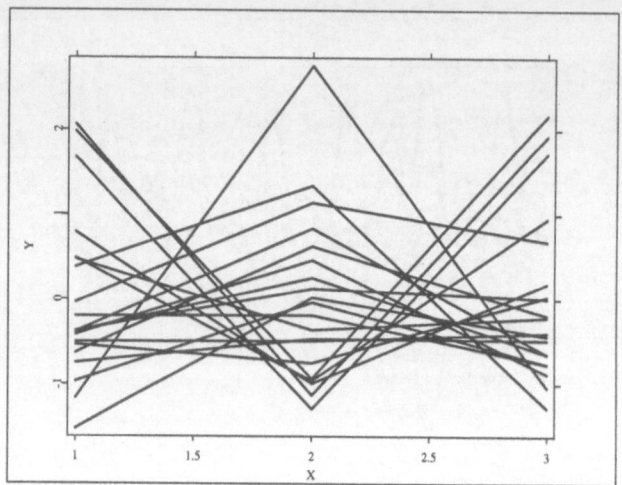

Figure 3.28. Parallel coordinate plot of 20 observations of the standardized 6th (RM), 13th (LSTAT) and 14th variables (MEDV) of the Boston Housing data. ◻graph4B.xpl

The Boston Housing data show some kind of negative correlation between RM and LSTAT and LSTAT and MEDV.

3.4   Advanced Graphics

```
grmove (grin, shf)
     moves a graphical object by shf

grrot (grin, rot)
     rotates a graphical object by rot times 90 degree rotations

grxline (x, v {, col})
     generates a vertical line at v

gryline (y, v {, col})
     generates a horizontal line at v

grcircle (radius {, col})
     generates a circle or ellipse.  The circle is centered at (0,0) and
     has the given radius.

hls2rgb (hls)
     generates RGB colors from the HLS color model

rgb2hls (rgb)
     generates HLS colors from the RGB color model

createcolor (rgb)
     sets a palette of colors
```

3.4.1   Moving and Rotating

The quantlets grmove and grrot allow the user to move graphic primitives without losing information including lines or line styles. grrot knows 4 counterclockwise rotations: 0 no rotation, 1 rotation by 90 degree, 2 rotation by 180 degree and 3 rotation by 270 degree. Since the rotation center is always (0,0), you should rotate first and then move the graphic primitive.

```
library("plot")              ; loads library plot
data = read ("bostonh")      ; reads Boston Housing data
gro1 = grbox (data[,11])     ; boxplot of 11th variable
gro2 = grbox (data[,13])     ; boxplot of 13th variable
gro3 = grbox (data[,14])     ; boxplot of 14th variable
```

```
gro1 = grrot (gro1, 1)        ; rotates first boxplot
gro2 = grrot (gro2, 1)        ; rotates second boxplot
gro2 = grmove (gro2, #(1.5,0)) ; moves second boxplot 1.5 right
gro3 = grrot (gro3, 1)        ; rotates the third boxplot
gro3 = grmove (gro3, #(3,0))  ; moves third boxplot 3.0 right
plot(gro1, gro2, gro3)        ; shows the boxplots
```

graph51.xpl

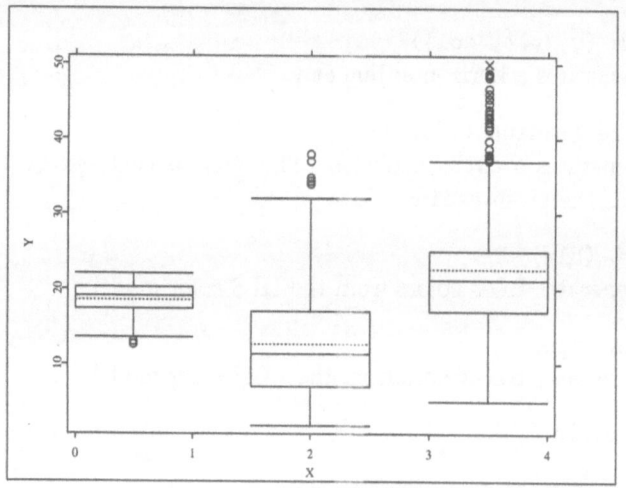

Figure 3.29.   Parallel boxplots of different variables (PTRATIO, LSTAT, MEDV) of the Boston Housing data. graph51.xpl

We see that all three variables, pupil–teacher ratio (PTRATIO), percentage lower status people (LSTAT) and median house prices (MEDV), are skewed, especially the last one (median house prices).

3.4.2   Simple Predefined Graphic Primitives

For the convenience of the user, some simple graphic primitives are predefined: grxline which draws a line from $(\min x_i, y)$ to $(\max x_i, y)$, gryline which draws a line from $(x, \min y_i)$ to $(x, \max y_i)$, and grcircle which draws a circle.

Let's draw a two-dimensional standard normal distributed data set. Furthermore, we include the coordinate system axes as well as the circle which contains 95% of the data of a bivariate standard normal distribution.

```
library("graphic")        ; loads library graphic
randomize(0)              ; initializes random generator
x  = normal(200,2)        ; generates 200 bivariate data
xl = grxline(0, x[,1])    ; computes horizontal axis
yl = gryline(0, x[,2])    ; computes vertical axis
cl = grcircle(2.44775)    ; computes circle with radius 2.44775
d  = createdisplay(1,1)   ; creates display
show(d,1,1,x,xl,yl,cl)    ; draws everything in one plot
```

Q graph52.xpl

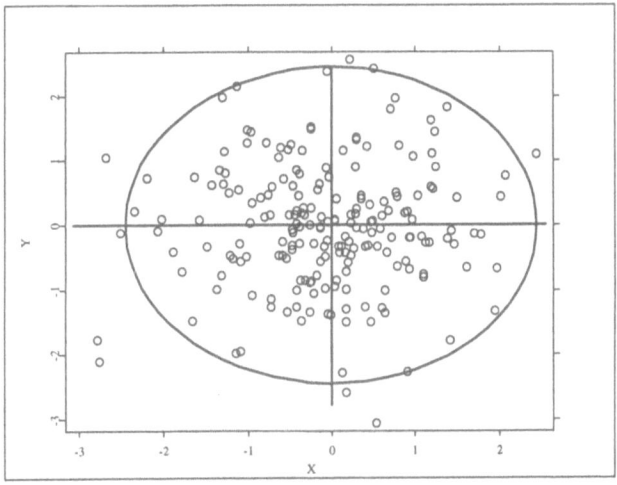

Figure 3.30. Coordinate axes, 95% mass circle and random sample of a bivariate standard normal distribution. Q graph52.xpl

If we count the number of observations outside the circle we will find nine. Since we have generated 200 points we would expect around ten (200×0.05) observations outside the circle.

3.4.3   Color Models

XploRe has eight standard colors (0 – black, 1 – blue, 2 – green, 3 – cyan, 4 – red, 5 – magenta, 6 – yellow and 7 – white) which sit in the corners of the RGB cube. We support RGB colors with Red, Green and Blue ranging from 0 to 255. The number of available colors depends on your monitor and windows system.

However, the RGB system is not well suited to construct a path through the color space, e.g. for contour lines indicating maxima by red lines and minima by blue lines. The HLS color model is much better suited for that. Thus we provide the quantlets rgb2hls and hls2rgb to transform a specific color from a color model into another.

Note that if we want to use colors other than the standard ones then it's necessary to create the colors by createcolor, since your output device supports xx-bit colors, but is not able to display them all simultaneously. Thus, we support a color palette of 256 colors (minus the 8 standard colors).

```
library("plot")              ; loads the library plot
h   = grid (120,26.6666,4)   ; creates grid in HLS double cone
l   = 0.5.*matrix(rows(h))
s   = matrix(rows(h))
rgb = hls2rgb(h~l~s)         ; transfers HLS model to RGB model
x0 = #(-3, -3)
h  = #(0.2, 0.2)
n  = #(31, 31)
x  = grid(x0, h, n)          ; generates a bivariate grid
f  = exp(-(x[,1]^2+x[,2]^2)/1.5)/(1.5*pi)
                             ; computes density of bivariate
                             ;    normal with correlation 0.5
c  = 0.2*(1:4).*max(f)       ; contour lines as 10%,...,90%
                             ;    times the maximum density
createcolor(rgb)             ; generates the necessary colors
gr = grcontour2(x~f, c, rgb) ; generates surface
plot(gr)                     ; plots the surface
```

🔍 graph53.xpl

Note that the object rgb is really a 4×3 matrix. The first vector represents the value for red, the second the value for blue and the last the value for green.

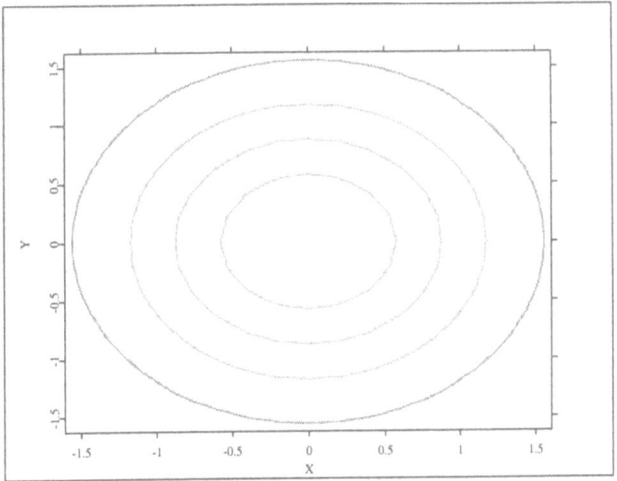

Figure 3.31. Contour plot of the density of the standard bivariate normal distribution colored after a HLS color scheme. ◓ graph53.xpl

3.5 Graphic Commands

setmaskp (data, color, layout, size)
    influences the layout, size and color of data points

setmaskl (data, lines, color, type, thickness)
    influences the layout, size, type and color between data points

setmaskt (data, labels, color, direction, size)
    influences the appearance of text at the data points

setgopt (d,row,col, optname,optval,..., optnameN,optvalN)
    influences several parameters of plots and displays

The displays created in the previous examples used either the default settings for displaying the data points, or were modified via setmask. However, setmask is a rather slow quantlet and it might be much quicker to use the built-in commands. We will show you how to change their color, graphical representation and size, and how to connect and label the data points. All these tasks are not performed by show itself, but rather by specific commands that have to preceed show. These commands are setmaskp, setmaskl and setmaskt. They are merged in the quantlet setmask.

3.5.1   Controlling Data Points

To control the color, the graphical representation (symbol) and the size of each data point, you must use setmaskp, which must preceed show. That is, first you call setmaskp to specify color, graphical representation and size of the data you want to plot. Then you call show to actually plot the data.

```
x   = 1:100                        ; creates variable x that takes
                                   ;    on the values 1, ..., 100
y   = sin(x/20)+uniform(100)/5 ; creates y-variable as sin(x/20)
                                   ;    and uniform error
data = x~y                         ; puts x, y together to form data
  ;
  ; now we call setmaskp to specify the layout for each point of
  ; the matrix, all data points are colored red (=4), shown as
  ; circles (=3) and have the default size (=8)
  ;
setmaskp(data, 4, 3, 8)       ; calls setmaskp
d = createdisplay(1, 1)       ; creates the display
show(d, 1, 1, data)           ; use show puts data into display
```

                                                      🔍graph61.xpl

You can replace setmaskp(data, 4, 3, 8) by

```
data = setmask(data, "red", "medium", "circle")
```

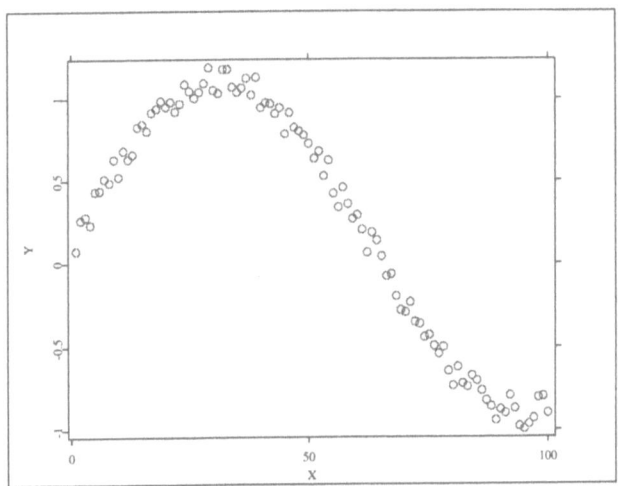

Figure 3.32. Coloring of data points by graphic commands or quantlets.
⊙ graph61.xpl

Note two remarks

- This example nicely illustrates that to use **setmaskp** you must give numerical values to specify color, style and size of the data points which speeds up the execution of the command but is inconvenient and not intuitive. **setmask**, on the other hand, uses strings rather than numbers to specify both the argument itself (like "color") and its value ("red"). This is intuitive, but slows down computations.

- Both **setmaskp** and **setmask** do not require you to specify all arguments if you are satisfied with the default settings. In the previous example, if you simply want to show red points but go along with the default settings of style and size, you can use **setmaskp(data, 4)** or **data = setmask(data, "red")**.

3.5.2   Color of Data Points

Each color is represented by an integer: 0 – black, 1 – blue, 2 – green, 3 – cyan, 4 – red, 5 – magenta, 6 – yellow and 7 – white. You may assign the same color

to each data point. In this case the second argument of `setmaskp` has to be a
single integer as in the previous example.

Alternatively, you can assign different colors to different points, in which case
the second argument of `setmaskp` has to be a column vector of integers with
the same number of rows as the data matrix. Here is an example with different
colors for different points:

```
x   = 1:100                    ; creates variable x that takes
                               ;     on the values 1, ..., 100
y   = sin(x/20)+uniform(100)/5 ; creates y-variable as sin(x/20)
                               ;     and uniform error
data = x~y                     ; puts x, y together to form data
;
; now we will create a vector of integers that assigns the
; color red (=4) to the first 50 data points and the color
; magenta (=5) to the next 50 data points
;
color = 4*matrix(50)|5*matrix(50)
;
; now we call setmaskp; as in the previous example, all points
; of data are shown as circles (=3) and have the default
; size (=8)
;
setmaskp(data, color, 3, 8)
d = createdisplay(1, 1)        ; creates the display
show(d, 1, 1, data)            ; use show puts data into display
```

🔍 graph62.xpl

To achieve the same result using `setmask`, you have to create the string vector
that is the equivalent to the numerical vector `color`.

```
library("plot")
mycolor = string("red", 1:50) | string("magenta", 51:100)
data    = setmask(data, mycolor)
d       = createdisplay(1,1)
show (d, 1, 1, data)
```

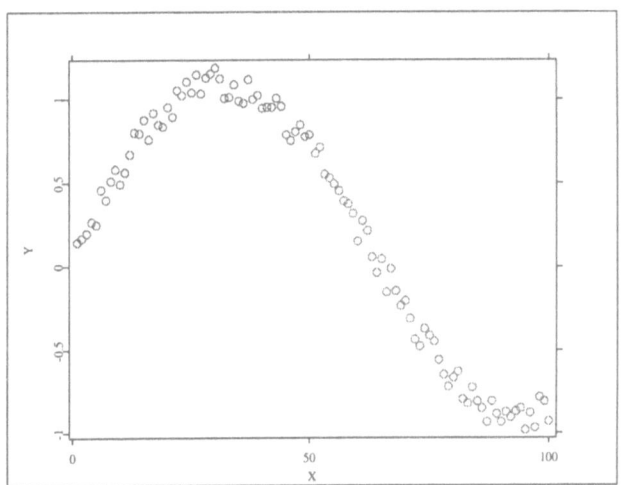

Figure 3.33. Different coloring of data points by graphic commands or
quantlets. ⊙ graph62.xpl

3.5.3 Symbol of Data Points

Similarly, you may represent each data point by the same graphical symbol (in
which case the third argument of setmaskp has to be a single integer), or you
may represent different data points by different graphical symbols (in which
case the third argument of setmaskp has to be a column vector of integers
with the same number of rows as the data matrix).

Here is the assignment of integers to graphical symbols:

       0 – empty, 1 – point, 2 – rectangle, 3 – circle, 4 – triangle, 5 – x-symbol,
       6 – rhombus, 7 – filled rectangle, 8 – filled circle, 9 – filled rhombus,
       10 – filled triangle, 11 – cross, 12 – star, 13 – grid, 14 – different cross.

Here is an example that assigns (through the vector layout) a circle to the first
25 data points, a triangle to the next 25 data points, a cross to the following
25 data points, and a rhombus to the last 25 data points.

```
x       = 1:100
y       = sin(x/20) +uniform(100, 1) /10
```

```
data    = x~y
color   = 4*matrix(50)|5*matrix(50)
layout = 3*matrix(25)|4*matrix(25)|5*matrix(25)|6*matrix(25)
setmaskp(data, color, layout, 8)
d       = createdisplay(1, 1)
show(d, 1, 1, data)
```

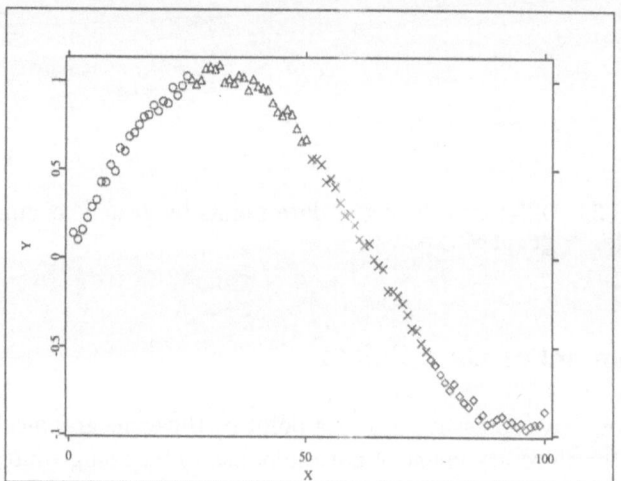

Q graph63.xpl

Figure 3.34. Different appearance of data points by graphic commands or quantlets. Q graph63.xpl

Again, we give the equivalent code using setmask:

```
library("plot")
x       = 1:100
y       = sin(x/20) +uniform(100, 1) /10
data    = x~y
mycolor = string("red", 1:50)|string("magenta", 51:100)
mystyle = string("circle", 1:25)|string("triangle", 26:50)
mystyle = mystyle|string("xsymbol",51:75)
mystyle = mystyle|string("rhomb",76:100)
data    = setmask(data, mycolor, mystyle)
```

```
d        = createdisplay(1, 1)
show(d, 1, 1, data)
```

3.5.4   Size of Data Points

Similar to the way setmaskp handles color and graphical representation, you can either give the same size to all data points (fourth argument of setmaskp is a scalar integer) or assign different sizes to different data points (fourth argument of setmaskp is a column vector of integers that has the same number of rows as the data matrix). The default size is assigned to the integer 8. You can choose any integer among 1, ..., 15 with 1 representing the smallest and 15 representing the largest possible size.

Here is an example that assigns (through the vector size) a rather small size (4) to the first 50 data points and the maximum size (15) to the last 50 data points:

```
x        = 1:100
y        = sin(x/20) + uniform(100, 1)/10
data     = x~y
color    = 4*matrix(50)|5*matrix(50)
layout = 3*matrix(25)|4*matrix(25)|5*matrix(25)|6*matrix(25)
size     = 4*matrix(50)|15*matrix(50)
setmaskp(data, color, layout, size)
d        = createdisplay(1, 1)
show(d, 1, 1, data)
```

🔍 graph64.xpl

Equivalently, you can replace the last five lines by

```
library("plot")
x        = 1:100
y        = sin(x/20) + uniform(100, 1)/10
data     = x~y
mycolor = string("red", 1:50)|string("magenta", 51:100)
mystyle = string("circle", 1:25)|string("triangle", 26:50)
mystyle = mystyle|string("xsymbol" ,51:75)
```

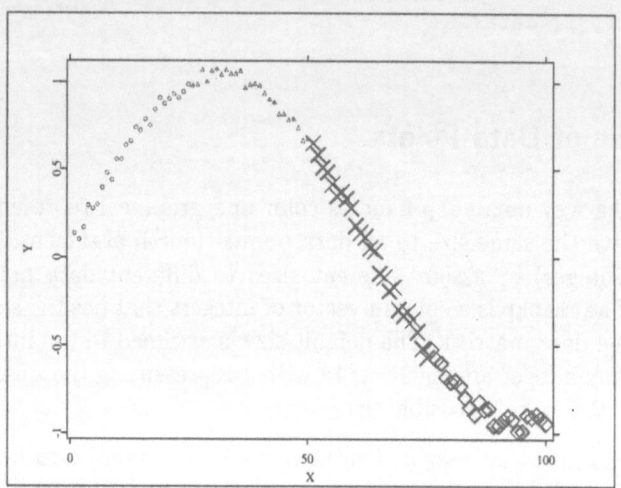

Figure 3.35. Different appearance of data points by graphic commands or quantlets. ⊙ graph64.xpl

```
mystyle = mystyle|string("rhomb", 76:100)
mysize  = string("small", 1:50)|string("huge", 51:100)
data    = setmask(data, mycolor, mystyle, mysize)
d       = createdisplay(1, 1)
show(d, 1, 1, data)
```

3.5.5   Connection of Data Points

Suppose you want to plot a data matrix and connect some or all of the points by lines. Then you have to use **setmaskl** to specify

- which points to connect,

- the color of the lines that connect the points,

- the appearance (solid or dashed) of the lines that connect the points and

- the thickness of the lines that connect the points.

Use `setmaskl` first to specify these options, then use **show** to actually generate the plot.

`setmaskl` is very flexible and allows you to specify in detail which points to connect and which kind of line to use. Here is a simple example that connects all points of a data matrix with a solid blue line:

```
randomize(666)            ; sets seed for random number generator
n = 6                     ; sample size
x = 4:(3+n)               ; generates x-variable as 4,...,9
y = 2*x+normal(n)         ; generates y-variable
z = x~y                   ; composes data matrix
d = createdisplay(1, 1)   ; creates display
;
; now we will create a row vector pm that tells setmaskl to
; connect all 6 points of the data matrix in the same order
; as they appear
;
pm    = (1:n)'            ; generates row vector 1,...,6
color = 1                 ; blue line
art   = 1                 ; solid line
thick = 5                 ; thick line
setmaskl(z, pm, color ,art ,thick) ; call setmaskl
show(d, 1, 1, z)          ; call show to plot the data
```

🔍 graph65.xpl

Again, `setmask` provides a convenient, more intuitive alternative:

```
library("plot")
randomize(666)
n = 6
x = 4:(3+n)
y = 2 *x +normal(n)
z = x~y
z = setmask(z, "line", "blue", "thick", "solid")
show(d, 1, 1, z)
```

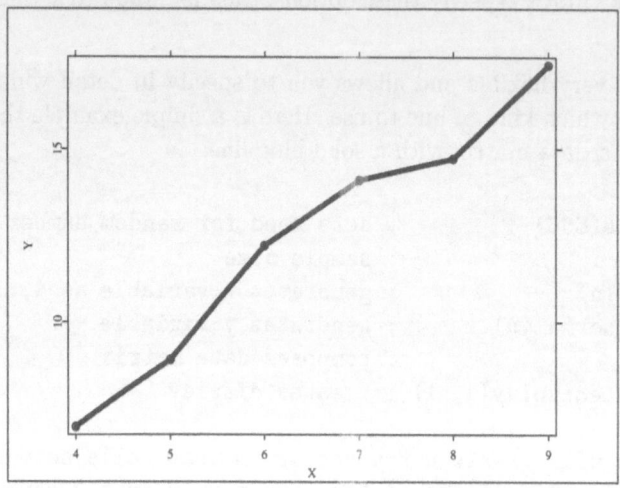

Figure 3.36.   Generating lines by graphic commands or quantlets.
◻ `graph65.xpl`

The crucial step in the previous example (and in the use of **setmask1** in general)
is the specification of the vector that tells **setmask1** which points to connect (it
is called **pm** in the example). We will give more examples for possible specifications of **pm** shortly. However, before we do that, we want to give the assignments
of integers to color, appearance and thickness of a line:

color:

> 0 – black, 1 – blue, 2 – green, 3 – cyan, 4 – red, 5 – magenta, 6 – yellow,
> 7 – white

appearance:

> 0 - invisible, 1 - solid, 2 - finely dashed, 3 - less finely dashed, 4 - even
> less finely dashed , and so on

thickness:

> 0 will produce the thinnest line and increasing integers up to 15 produce
> lines of increasing thickness

Now back to specifying pm. Suppose that in the previous example you want to connect only the 1st, 3rd and 6th points with a line. Then you have to generate pm as follows:

```
pm = 1~3~6
```

The points to be connected do not have to be in increasing order. For instance,

```
pm = 1~3~6~2
```

will connect the 1st point with the 3rd, the 3rd with the 6th and finally the 6th with the 2nd.

Suppose you want to draw a line, interrupt it and continue it again. Interrupting is done by including a 0 (zero):

```
pm = 1~3~0~5~6
```

will connect the 1st and the 3rd points, then interrupt the line and continue it by connecting the 5th and the 6th points.

Does **setmask1** allow to draw several lines? Yes, but in this case pm has to be specified as a matrix where each row corresponds to a line. For instance

```
pm = (1~3~5)|(2~4~6)
```

will connect the 1st, 3rd and 5th points with a line and the 2nd, 4th and 6th points with a different line. Note that if you want to draw two lines, you may also specify color, appearance and thickness for both lines — as shown in the following complete example:

```
randomize(666)        ; sets seed for random number generator
n = 6                 ; sample size
x = 4:(3+n)           ; generates x-variable as 4,...,9
y = 2*x+normal(n)     ; generates y-variable
z = x~y               ; composes data matrix
pm   = (1~3~5)|(2~4~6)
color = 1|4           ; sets color of first line to blue (=1)
                      ;     and of second line to red (=4)
art   = 1|2           ; first line will be solid (=1),
```

```
                          ;      second line will be dashed (=2)
setmaskl(z, pm, color, art)
                          ; calls setmaskl
d = createdisplay(1, 1) ; creates a display
show(d, 1, 1, z)         ; calls show to plot data
```

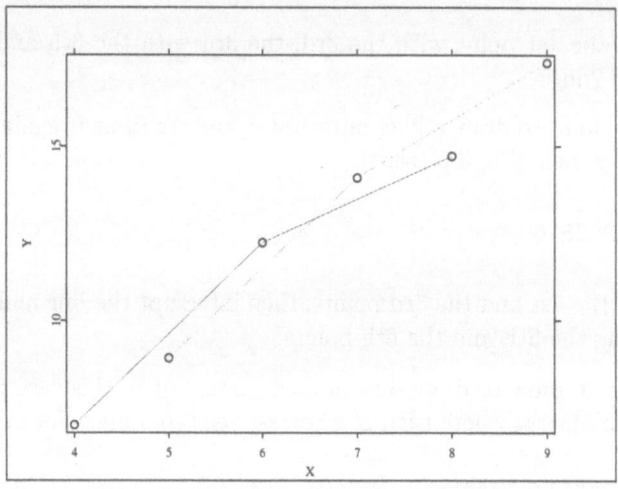graph66.xpl

Figure 3.37. Mixed lines and data points by graphic commands or
quantlets. graph66.xpl

The last five lines can be replaced by the following code which uses setmask
instead of setmaskl:

```
library("plot")
randomize(666)
n  = 6
x  = 4:(3+n)
y  = 2*x+normal(n)
z  = x~y
pm = (1~3~5)|(2~4~6)
c  = "blue"|"red"
```

```
t = "thin"|"thick"
a = "solid"|"dotted"
z = setmask(z, "line", pm, c, t, a)
d = createdisplay(1, 1)
show(d, 1, 1, z)
```

Of course, by adding more rows to the pm matrix, you can draw more than two lines.

3.5.6   Label of Data Points

To label the data points, you have to use setmaskt. setmaskt allows you to control the color, the position (relative to the associated data point) and the font size of the label. The labels themselves have to be collected in a text vector and given to setmaskt as an input. Color, position and size are controlled by the following integers:

color:
  0 – black, 1 – blue, 2 – green, 3 – cyan, 4 – red, 5 – magenta, 6 – yellow, 7 – white

position:
  −1 – no label at all, 0 – a centered label, 3 – a label to the right of the point, 6 – a label below the point, 9 – a label to the left of the point, 12 – a label above the point. Note that currently the positions are only considered for PostScript output.

font size:
  font sizes 12 to 16 are supported.

Here is a very simple example where each label gets the same color, position and size:

```
x      = 1:6
x      = x~x
text   = "Point1"|"Point2"|"Point3"|"Point4"|"Point5"|"Point6"
color = 1
position = 3
size  = 16
```

```
setmaskt(x, text, color, position, size)
d       = createdisplay(1, 1)
show(d, 1, 1, x)
```

Q graph67.xpl

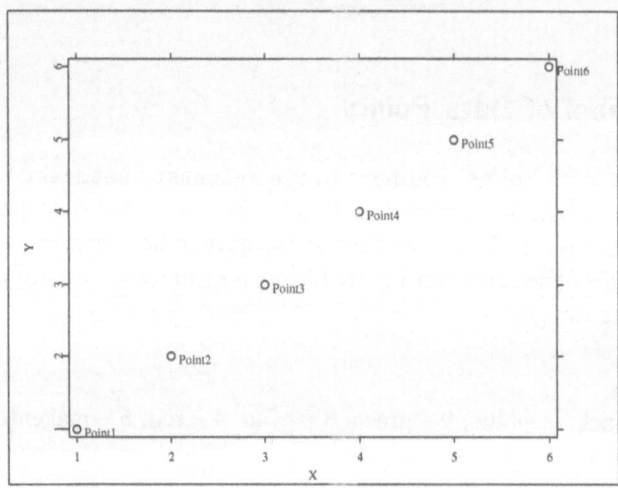

Figure 3.38. Labeling of data points by graphic commands or quantlets.
Q graph67.xpl

The equivalent code using **setmask** is given by

```
library("plot")
x = 1:6
x = x~x
mytext = "Point1"|"Point2"|"Point3"|"Point4"|"Point5"|"Point6"
x = setmask(x, "points","text",mytext,"blue","right","medium")
d = createdisplay(1, 1)
show(d, 1, 1, x)
```

If you want to assign different colors, positions, or sizes to different labels,
you must specify the respective arguments of **setmaskt** as column vectors of
integers. Here is an illustrative example where the text of the labels has been
chosen to match their positions as specified by the vector position:

```
x       = 1:6
x       = x~x
text    = "Right"|"Under"|"Left"|"Over"|"Center"|"No"
color   = 1|2|3|4|5|6
position = 3|6|9|12|0|(-1)
size    = 12|13|14|15|16|16
setmaskt(x, text, color,position, size)
d       = createdisplay(1, 1)
show(d, 1, 1, x)
```

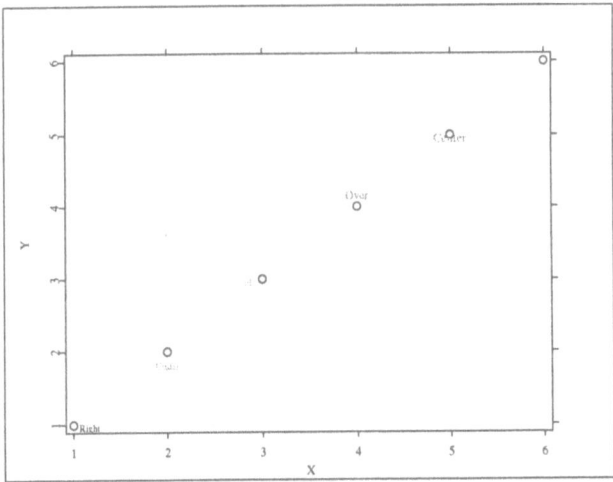

Q graph68.xpl

Figure 3.39. Labeling of data points at different positions by graphic commands or quantlets. Q graph68.xpl

Using `setmask` instead of `setmaskt`, we can write:

```
x = 1:6
x = x~x
mytext  = "Right"|"Under"|"Left"|"Over"|"Center"|"No"
mycolor = "blue"|"green"|"cyan"|"red"|"magenta"|"yellow"
mypos   = "right"|"below"|"left"|"above"|"center"|"request"
mysize  = "small"|"medium"|"medium"|"large"|"large"|"huge"
x = setmask(x, "points", "text",mytext,mycolor,mysize,mypos)
d = createdisplay(1, 1)
show(d, 1, 1, x)
```

3.5.7   Title and Axes Labels

The layout of the display is controlled by `setgopt`. In order to work, `setgopt` has to be called after you have called `show`.

Here is an example that uses three of `setgopt`'s seventeen options:

```
x    = 1:100
y    = sqrt(x)
data = x~y
d    = createdisplay(1, 1)
show(d, 1, 1, data)
title  = "Plot of Sqrt(x)"
ylabel = "y = sqrt(x)"
setgopt(d, 1, 1,"title",title, "xlabel","x", "ylabel",ylabel)
```
                                                            graph69.xpl

First we have to tell `setgopt` to which display (`di`) and to which window of that display (1 ,1) it ought to apply the requested layout options. Then we specify the name of the option we want to modify and — immediately following the name of the option — the value we want that option to take on. For instance, we used the option name `title` and the option value `"Plot of Sqrt(x)"` to produce the headline of our display. Similarly, the option name `"ylabel"` followed by the option value `"y = sqrt(x)"` produced the label of the vertical axis.

Table 3.1 lists all the available options available for `setgopt`.

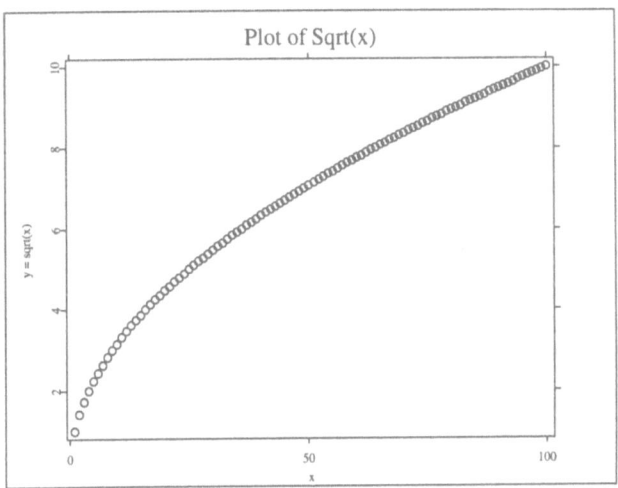

Figure 3.40. Example for manipulating a title and axes in a display.
🄰 graph69.xpl

3.5.8 Axes Layout

In the following example we illustrate how the xlim/ylim, the xmajor/ymajor
and the xoffset/yoffset options control the layout of the axes. Specifically,
we create a four window display and use show to put the same data matrix
into each window. Then we call to change the layout in each display. In the
upper left window we use the default settings for the axes options and we use
setgopt to create the headline "default". All other windows differ from this
default window by altering a single option of setgopt. This way you can see
the "isolated" effect of that option relative to the default settings:

```
x  = 1:100
y  = sin(x /20) +uniform(100, 1) /10
d = createdisplay(2, 2)
show(d, 1, 1, x~y)
show(d, 1, 2, x~y)
show(d, 2, 1, x~y)
show(d, 2, 2, x~y)
;
```

```
; now we use setgopt to give the upper left window the
; headline "default"
;
setgopt(d, 1, 1, "title", "default")
;
; note that the title tells you exactly what we have
; done in each of the other windows
;
title12 = "ylim = (-4) | 4, xlim = 0 | 50"
setgopt(d,1,2,"title",title12,"ylim",(-4)|4,"xlim",0|50)
title21 = "ymajor = 0.3, xmajor = 15"
setgopt(d,2,1,"title",title21,"ymajor",0.3, "xmajor",15)
title22 = "yoffset = 13 | 13, xoffset = 20 | 20"
setgopt(d,2,2,"title",title22,"yoffset",13|13,"xoffset",20|20)
```

Q graph6A.xpl

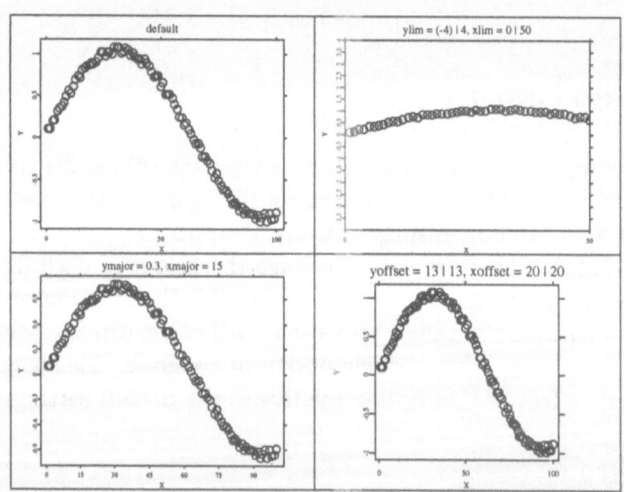

Figure 3.41.   Example for manipulating plots in a display.
Q graph6A.xpl

Option	Description	Option value
"title"	change headline	a string
"xlim"	change limits of x-axis	vector that contains two values, e.g. 0|10
"ylim"	change limits of y-axis	vector that contains two values, e.g. 0|10
"xoffset"	change the width of axis border	vector that contains two values to change the right and left widths of axis border (in %)
"yoffset"	change the height of axis border	vector that contains two values to change the upper and lower height of axis border (in %)
"xvalue"	change the values m and k of the transformation m+k*x	vector that contains one value for m and one value for k , e.g. 0|1 for 0+1*x
"yvalue"	change the values m and k of m+k*y	vector that contains one value for m and one value for k , e.g. 0|1 for 0+1*y
"xorigin"	change origin for tickmark of x-axis	scalar
"yorigin"	change origin for tickmark of y-axis	scalar
"xmajor"	change major for tickmark of x-axis	scalar
"ymajor"	change major for tickmark of y-axis	scalar
"xlabel"	change label of x-axis	a string
"ylabel"	change label of y-axis	a string
"rotpoint"	change rotation point	a rotation point vector
"rotcos"	change rotation matrix	a symmetric orthogonal matrix
"scal"	change scale matrix	a diagonal scale matrix
"transl"	change translation vector	a translation vector

Table 3.1. Parameter of **setgopt**.

4 Regression Methods

Jörg Aßmus

Simply speaking, a regression problem is a way to determine a function $\widehat{m}(\bullet)$ describing a functional relation $m(\bullet)$ between a p-dimensional variable $X = (X_1, \ldots, X_p)$ and an output variable Y

$$Y = m(X) + \varepsilon,$$

where ε is a random error term.

There are two different approaches of fitting the function m. In the first approach, we define m by a finite-dimensional parameter $\beta = (\beta_0, \ldots, \beta_p)$:

$$Y = m_\beta(X) + \varepsilon$$

where it is sufficient to estimate the parameter β. In this case, we are able to approximate the function at each point $x = (x_1, \ldots, x_p)$ using only the parameter estimate $\widehat{\beta}$ of β by

$$\widehat{Y}(x) = m_{\widehat{\beta}}(x).$$

This is called a **parametric** model. A very useful example is the linear regression

$$\begin{aligned} Y &= \beta_0 + \beta_1 X_1 + \ldots + \beta_p X_p + \varepsilon, \\ \widehat{Y}(x) &= \widehat{\beta}_0 + \widehat{\beta}_1 x_1 + \ldots + \widehat{\beta}_p x_p. \end{aligned}$$

On the other hand a model is called **nonparametric** if the dimension p of the parameter β is infinite. This means that we do not know anything about the function m. In this case we have to use the data set for the calculation of the estimated function $\widehat{m}$ at each point x. We investigate this question in Chapter 6.

Both models can be estimated with XploRe. A simple way of choosing the appropriate method is given here:

- If we know the functional form of m, we use a parametric model:

  - m is linear, X is one-dimensional,

  $$Y = \beta_0 + \beta_1 X + \varepsilon.$$

  We use **simple linear regression**. (linreg, gls)
  - m is linear, X is p-dimensional,

  $$Y = \beta_0 + \beta_1 X_1 + \beta_2 X_2 + \ldots + \beta_p X_p + \varepsilon.$$

  We use **multiple linear regression**.
  (linreg, gls, linregbs, linregfs, linregstep)
  - m is nonlinear, e.g.

  $$Y = \beta_0 + \beta_1 X^2 + \varepsilon \quad \text{or} \quad Y = \beta_1 X^{\beta_2}.$$

  We use **nonlinear regression**.

- If we don't know the functional form of m, then we use a nonparametric model. Nonparametric methods are introduced in Chapter 6.

The nonlinear methods are more general than the linear methods. This means that we can use the nonlinear regression for estimating linear models, but this cannot be recommended in general. In the same way, the nonparametric models are more general than the parametric ones.

In the following sections, the libraries stats containing the quantlets for the regression and graphic containing the plot quantlets will be used. We should load them before we continue:

```
library("graphic")              ; reads the library graphic
library("stats")                ; reads the library stats
```

4.1   Simple Linear Regression

> {b, bse, bstan, bpval} = linreg (x, y {,opt {,om}})
>   estimates the coefficients b for a linear regression of y on x and
>   calculates the ANOVA table
>
> gl = grlinreg (x, {,col})
>   creates a plot object of the regression line estimated
>
> b = gls (x, y {,om})
>   calculates the generalized least squares estimator from the data
>   x and y

In this section we consider the linear model

$$Y = \beta_0 + \beta_1 X + \varepsilon.$$

As an example, we use the Westwood data stored in `westwood.dat`. All XploRe
codes for this example can be found in ◻`regr1.xpl`. First we read the data
and take a look at them.

```
z=read("westwood.dat")    ; reads the data
z                         ; shows the data
x=z[,2]                   ; puts the x-data into x
y=z[,3]                   ; puts the y-data into y
```

gives as output

```
Contents of z
[ 1,]       1       30       73
[ 2,]       2       20       50
[ 3,]       3       60      128
[ 4,]       4       80      170
[ 5,]       5       40       87
[ 6,]       6       50      108
[ 7,]       7       60      135
[ 8,]       8       30       69
[ 9,]       9       70      148
[10,]      10       60      132
```

We use the quantlet linreg for simple linear regression. Since this quantlet has four values as output, we should put them into four variables. We store them in {beta, bse, bstan, bpval}. Their meaning will be explained below.

```
{beta,bse,bstan,bpval}=linreg(x,y)
          ; computes the linear regression and returns the
               variables of beta, bse, bstan and bpval
beta      ; shows the value of beta
```

gives the output

```
A   N   O   V   A              SS     df      MSS     F-test   P-value
---------------------------------------------------------------------
Regression              13600.000   1  13600.000   1813.333   0.0000
Residuals                  60.000   8      7.500
Total Variation         13660.000   9   1517.778


Multiple R         = 0.99780
R^2                = 0.99561
Adjusted R^2       = 0.99506
Standard Error     = 2.73861



PARAMETERS          Beta        SE     StandB     t-test   P-value
---------------------------------------------------------------------
b[ 0,]=          10.0000    2.5029     0.0000      3.995   0.0040
b[ 1,]=           2.0000    0.0470     0.9978     42.583   0.0000
```

and

```
Contents of beta
[1,]        10
[2,]         2
```

As a result, we have the ANOVA (ANalysis Of VAriance) table and the parameters. The estimates $(\widehat{\beta}_0, \widehat{\beta}_1)$ of the parameters (β_0, β_1) are stored in beta[1] and beta[2] and in this example we obtain

$$\widehat{Y}(x) = 10 + 2x.$$

It is not necessary to display the values of beta, bse, bstan, bpval separately because they are already written as Beta, SE, StandB and P-value in the parameter table created by linreg. Before considering the graphics we give a short overview of the values returned by the ANOVA and parameter tables.

- SS: Sums of Squares

  - Regression:

  $$SS_{reg} = \sum_{i=1}^{n} (\widehat{Y}_i - \bar{Y})^2 ,$$

  - Residuals:

  $$SS_{res} = \sum_{i=1}^{n} (Y_i - \widehat{Y}_i)^2 ,$$

  - Total Variation:

  $$SS_{tv} = \sum_{i=1}^{n} (Y_i - \bar{Y})^2 .$$

- df: Degrees of Freedom.

- MSS: Mean Squared Errors:

$$MSS = \frac{SS}{df} .$$

- Multiple R, R^2, Adjusted R^2: several correlation coefficients.

- Standard Error:

$$SE = \sqrt{MSS_{res}} .$$

- b[0,],b[1,]: Parameters of the regression

  - beta:

  $$\widehat{Y} = beta_{b[0,]} + beta_{b[1,]} X ,$$

  - SE: Standard Error

    ▷ b[0,]:

    $$SE_{b[0,]} = \sqrt{\frac{MSS_{res} \cdot \sum_{i=1}^{n} X_i^2}{n \sum_{i=1}^{n} (X_i - \bar{X})^2}} ,$$

▷ b[1,]:

$$SE_{b[1,]} = \sqrt{\frac{MSS_{res}}{\sum_{i=1}^{n}(X_i - \bar{X})^2}} \,.$$

The meaning of the t- and p-values is more apparent in the multiple case. That's why they are explained in the next section.

Let us now describe how to visualize these results. In the left window we show the regression result computed by linreg. In the right window we use the quantlet grlinreg to get the graphical object directly from the data set.

```
yq=(beta[1]+beta[2]*x[1:10]) ; creates a vector with the
                             ;    estimated values of y
data=sort(x~y)               ; creates object with the data set
setmaskp(data,1,11,4)        ; creates a graphical object for
                             ;    the data points
rdata=sort(x~yq)             ; creates an object with yq
rdata=setmask(rdata,"reset","line","red","thin")
                             ; sets the options for the
                             ;    regression function by linreg
regrdata=grlinreg(data,4)    ; creates the same graphical
                             ;    object directly from the data
regrdata=setmask(regrdata,"reset","line","red","thin")
                             ; sets options for the regression
                             ;    function by grlinreg
linregplot=createdisplay(1,2); creates display with 2 windows
show(linregplot,1,1,data,rdata)
                             ; shows rdata in the 1st window
show(linregplot,1,2,data,regrdata)
                             ; shows regrdata in the 2nd window
setgopt(linregplot,1,1,"title","linreg")
                             ; sets the title of the 1st window
setgopt(linregplot,1,2,"title","grlinreg")
                             ; sets the title of the 2nd window
```

This will produce the results visible in Figure 4.1. We create a plot of the regression function by grlinreg if we are only interested in a graphical exploration of the regression line.

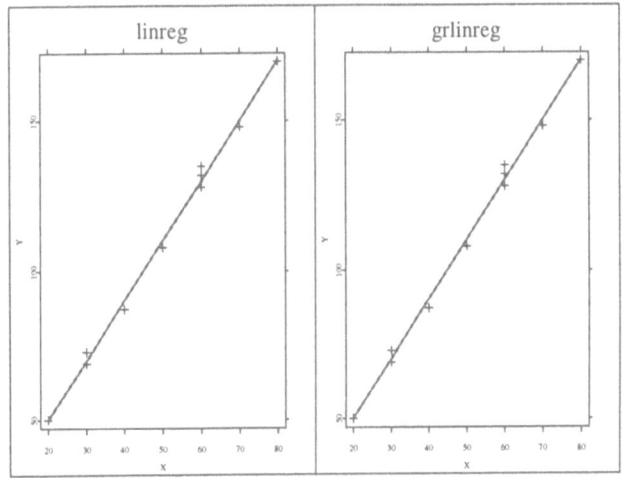

Figure 4.1. Linear regression of the Westwood data: Plot using the regression function computed by linreg (left) and plot using **grlinreg** (right).

Q regr1.xpl

A second tool for our simple linear regression problem is the **generalized least squares** (GLS) method given by the quantlet gls. Here we only consider a model

$$Y = bX + \varepsilon .$$

We take the Westwood data again and assume that it has already been stored in x and y using the unit matrix as weight matrix. This example is stored in
Q regr2.xpl.

```
b=gls(x,y)       ; computes the GLS fit and stores the
                 ;    coefficients in the variable b
b                ; shows b
```

shows

```
Contents of b
[1,]    2.1761
```

As a result, we get the parameter b. In our case we find that

$$\widehat{Y}(x) = 2.1761\,x\,.$$

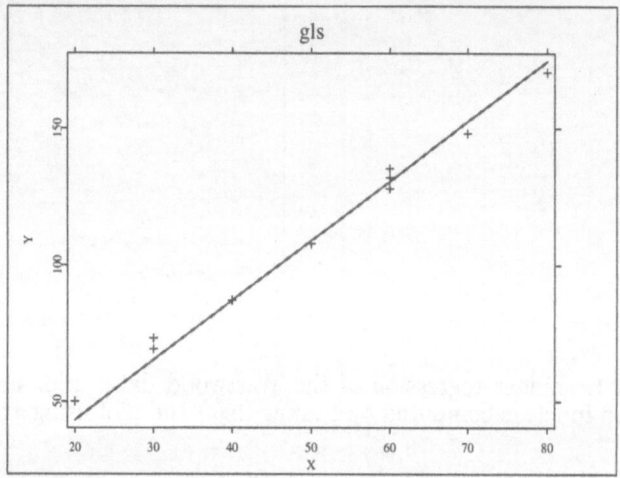

Figure 4.2. Linear regression of the Westwood data: Regression line using **gls**.
⊡ **regr2.xpl**

Note that we have got different results depending on the choice of the method. This is not surprising, as **gls** ignores the absolute value β_0. Now we also want to visualize this result.

```
yq=b*x[1:10]               ; creates a vector with the
                           ;    estimated values
data=sort(x~y)             ; creates object with the data set
setmaskp(data,1,11,8)      ; creates graphical object
                           ;    for the data
rdata=sort(x~yq)           ; creates object with yq
rdata=setmask(rdata,"reset","line","red","medium")
                           ; creates graphical object for yq
glsplot=createdisplay(1,1) ; creates display
show(glsplot,1,1,data,rdata) ; shows the graphical objects
setgopt(glsplot,1,1,"title","gls")
                           ; sets the window title
```

4.2 Multiple Linear Regression

{b,bse,bstan,bpval} = linreg ({x, y {,opt,om}})
    estimates the coefficients $\beta_0, \ldots, \beta_p$ for a linear problem from
    data x and y and calculates the ANOVA table

{xs,bs,pvalue} = linregfs (x, y {,alpha})
    computes the forward selection and estimates the coefficients
    $\beta_0, \ldots, \beta_p$ for a linear problem from data x and y and calculates
    the ANOVA table

{b,bse,bstan,bpval} = linregfs2 (x, y, colname {,opt})
    computes the forward selection and estimates the coefficients
    $\beta_0, \ldots, \beta_p$ for a linear problem from data x and y

{b,bse,bstan,bpval} = linregbs (x, y, colname {,opt})
    computes the backward elimination and estimates the coefficients
    $\beta_0, \ldots, \beta_p$ for a linear problem from data x and y and calculates
    the ANOVA table

{b,bse,bstan,bpval} = linregstep (x, y, colname {,opt})
    computes the stepwise selection and estimates the coefficients
    $\beta_0, \ldots, \beta_p$ for a linear problem from data x and y and calculates
    the ANOVA table

In this section, we consider the linear model

$$Y = \beta_0 + \beta_1 X_1 + \ldots + \beta_p X_p .$$

Looking at this model we are faced with two problems:

- Estimating the parameter vector $(\beta_0, \ldots, \beta_p)$

- Testing the significance of the components X_i

From the mathematical point of view, the second problem is about reducing the dimension of the model. Seen with the eyes of the user, it gives us information about building a parsimonious model. To this end, we have to find a way to handle the selection or removal of a variable.

We want to use a simulated data set to demonstrate the solution to these two problems. This example is stored in ⊠ regr3.xpl where we generate five uniform $[0, 1]$ distributed variables $X_1, \ldots, X_5$. Only three of them influence Y:

$$Y = 2 + 2 X_1 - 10 X_3 + 0.5 X_4 + \varepsilon .$$

Here, ε is a normally distributed error term.

```
randomize(1)            ; sets a seed for the random generator
eps=normal(10)          ; generates 10 standard normal errors
x1=uniform(10)          ; generates 10 uniformly distributed values
x2=uniform(10)
x3=uniform(10)
x4=uniform(10)
x5=uniform(10)
x=x1~x2~x3~x4~x5        ; creates the x data matrix
y=2+2*x1-10*x3+0.5*x4+eps/10 ; creates y
z=x~y                   ; creates the data matrix z
z                       ; returns z
```

This shows

```
Contents of z
[ 1,]   0.98028   0.35235   0.29969   0.85909   0.62176    1.3936
[ 2,]   0.83795   0.82747   0.13025   0.79595   0.59754    2.7269
[ 3,]   0.15873   0.93534   0.91259   0.72789   0.43156   -6.5193
[ 4,]   0.67269   0.67909   0.28156   0.20918   0.19878    0.69022
[ 5,]   0.50166   0.97112   0.39945   0.57865   0.19337   -0.66278
[ 6,]   0.94527   0.36003   0.77747   0.029797  0.40124   -3.9237
[ 7,]   0.18426   0.29004   0.24534   0.44418   0.35116    0.11605
[ 8,]   0.36232   0.35453   0.53022   0.4497    0.8062    -2.3026
[ 9,]   0.50832   0.00516   0.90669   0.16523   0.75683   -5.9188
[10,]   0.76022   0.17825   0.37929   0.093234  0.17747   -0.20187
```

Let us start with the first problem and use the the quantlet linreg to estimate the parameters of the model

$$Y = \beta_0 + \beta_1 X_1 + \ldots + \beta_5 X_5 + \varepsilon$$

```
{beta,bse,bstan,bpval}=linreg(x,y)  ; computes the linear
                                    ;      regression
```

produces

```
A N O V A              SS    df    MSS    F-test   P-value
-----------------------------------------------------------
Regression          87.241   5   17.448  4700.763  0.0000
Residuals            0.015   4    0.004
Total Variation     87.255   9    9.695

Multiple R       = 0.99991
R^2              = 0.99983
Adjusted R^2     = 0.99962
Standard Error   = 0.06092
```

```
PARAMETERS        Beta      SE    StandB    t-test   P-value
-----------------------------------------------------------
b[ 0,]=         2.0745   0.0941   0.0000   22.056    0.0000
b[ 1,]=         1.9672   0.0742   0.1875   26.517    0.0000
b[ 2,]=         0.0043   0.0995   0.0005    0.043    0.9677
b[ 3,]=       -10.0887   0.0936  -0.9201 -107.759    0.0000
b[ 4,]=         0.3991   0.1203   0.0387    3.318    0.0294
b[ 5,]=         0.0708   0.1355   0.0053    0.523    0.6289
```

We obtain the ANOVA and parameter tables which return the same values as found in the previous section. Substituting the estimated parameters $\widehat{\beta}_0, \ldots, \widehat{\beta}_5$, we get with our generated data set

$$\widehat{Y}(x) = 2.0745 + 1.9672x_1 + 0.0043x_2 - 10.0887x_3 + 0.3991x_4 + 0.0708x_5 .$$

We know that X_2 and X_5 do not have any influence on Y. This is reflected by the fact that the estimates of β_2 and β_5 are close to zero. Now we reach the point where we are faced with our second problem, how to eliminate these variables. We can get a first impression by considering the parameter estimates and their t-values in the parameter table. A t-value is small if there is no influence of the corresponding variable. This is reflected in the p-value, which is the significance level for testing the hypothesis that a parameter equals zero. From the above table, we can see that only the p-values for the constant, X_1, X_3 and X_4 are smaller than 0.05, the typical significance level for hypothesis testing. The p-values of X_2 and X_5 are much larger than 0.05 which means that they are not significantly different from zero.

The above way of choosing variables is convenient, but we want to know if the elimination or selection of a variable improves the result or not. This leads immediately to the stepwise model selection methods.

Let us first consider **forward selection**. The idea is to start from one "good" variable X_j and calculate the simple linear regression for

$$Y = \beta_0 + \beta_j X_j + \varepsilon.$$

Then we decide stepwise for each of the remaining variables if its inclusion to the model improves the fit of the model. The algorithm is:

FS1 Choose the variable X_j with the highest t- or F-value and calculate the simple linear regression.

FS2 Of the remaining variables, add the variable X_k which fulfills one of the three (equivalent) criteria below:

- X_k has the highest sample partial correlation.
- The model with X_k increases the R^2-value the most.
- X_k has the highest t- or F-value.

FS3 Repeat **FS2** until one of the stopping rules applies:

- The order p of the model has reached a predetermined p^*.
- The F-value is smaller then a predetermined value F_{in}.
- X_k does not significantly improve the model fit.

A similar idea leads to **backward elimination**. We start with the linear regression for the full model and eliminate stepwise variables without influence.

BE1 Calculate the linear regression for the full model.

BE2 Eliminate the variable X_k with one of the following (equivalent) properties:

- X_k has the smallest sample partial correlation among all remaining variables.
- The removing of X_k causes the smallest change of R^2.
- Of the remaining variables, X_k has the smallest t- or F-values.

BE3 repeat **BE2** until one of the following stopping rules is valid:

- The order p of the model has reached a predetermined p^*.
- The F-value is larger then a predetermined value F_{out}.
- Removing X_k does not significantly change the model fit.

A kind of compromise between forward selection and backward elimination is given by the **stepwise selection** method. Beginning with one variable just like in forward selection we have to choose one of the four alternatives:

1. Add a variable.

2. Remove a variable.

3. Exchange two variables.

4. Stop the selection.

This can be done with the following rules:

ST1 Add the variable X_k if one of the forward selection criteria **FS2** is satisfied.

ST2 Remove the variable X_k with the smallest F-value if there are (possibly more than one) variables with an F-value smaller than F_{out}.

ST3 Remove the variable X_k with the smallest F-value if this removal results in a larger R^2-value than it was obtained with the same number of variables before.

ST4 Exchange the variables X_k in the model and X_ℓ not in the model if this will increase the R^2-value.

ST5 Stop the selection if neither **ST1**, **ST2**, **ST3** nor **ST4** is satisfied.

Remarks:

- The rules **ST1**, **ST2** and **ST3** only make sense if there are two or more variables in the model. That is why they are only admitted in this case.
- Considering **ST3**, we see the possibility that the same variable can be added and removed in several steps of the procedure.

In XploRe we find the quantlets `linregfs` and `linregfs2` for forward selection, `linregbs` for backward elimination, and `linregstep` for stepwise selection. Whereas `linregfs` only returns the selected regressors X_i, the regression coefficients β_i and the p-values, the other three quantlets report each step, the ANOVA and the parameter tables. Because both the syntax and the output formats of these three quantlets are the same, we will only illustrate one of them with an example.

We use the data set generated above of the model

$$Y = 2 + 2X_1 - 10X_3 + 0.5X_4 + \varepsilon$$

to demonstrate the usage of stepwise elimination. Before computing the regression, we need to store the names of the variables in a column vector:

```
colname=string("X%.f",1:cols(x))
                    ; sets the column names to X1,...,X5
{beta,se,betastan,p} = linregstep(x,y,colname)
                    ; computes the stepwise selection
```

`linregstep` returns the same values as `linreg`. It shows the following output:

```
Contents of EnterandOut

Stepwise Regression
--------------------
F-to-enter 5.19
probability of F-to-enter 0.96
F-to-remove 3.26
probability of F-to-remove 0.90


Variables entered and dropped in the following Steps:

Step  Multiple R      R^2        F         SigF      Variable(s)
 1      0.9843       0.9688    248.658     0.000   In : X3
 2      0.9992       0.9984   2121.111     0.000   In : X1
 3      0.9999       0.9998  10572.426     0.000   In : X4


A N O V A       SS      df     MSS      F-test    P-value
----------------------------------------------------------------
Regression    87.239     3    29.080   10572.426  0.0000
Residuals      0.017     6     0.003
```

```
Total Variation        87    9    9.695
```

```
Multiple R       = 0.99991
R^2              = 0.99981
Adjusted R^2     = 0.99972
Standard Error   = 0.05245
```

```
Contents of Summary
```

```
Variables in the Equation for Y:
```

```
PARAMETERS   Beta    SE     StandB    t-test P-value Variable
------------------------------------------------------------
b[ 0,]=    2.0796  0.0742   0.0000   28.0417 0.0000  Constant
b[ 1,]=    1.9752  0.0630   0.1883   31.3494 0.0000  X1
b[ 2,]=  -10.0622  0.0690  -0.9177 -145.7845 0.0000  X3
b[ 3,]=    0.4257  0.0626   0.0413    6.8014 0.0005  X4
```

First, the quantlet `linregstep` returns the F_{in} values as `F-to-enter` and F_{out} as `F-to-remove`. Then each step is reported and we obtain again the ANOVA and parameter tables described in the previous section.

As expected, `linregstep` selects the variables X_1, X_3 and X_4 and estimates the model as

$$\widehat{Y}(x) = 2.0796 + 1.9752x_1 - 10.0622x_3 + 0.4257x_4 \,.$$

Recall the results of the previous ordinary regression. We can see that the accuracy of the estimated parameters has been improved by the selection method (especially for $\widehat{\beta}_4$). In addition, we obtained the information as to which variables can be ignored because the model does not depend on them.

4.3 Nonlinear Regression

A more difficult type of parametric models is the nonlinear regression model. Here we cannot find just one nice and easy algorithm to solve all problems, as in the linear case. We consider the general parametric model

$$Y = m_\beta(X) + \varepsilon,$$

where m_β is a nonlinear function depending on the parameter vector $\beta = (\beta_0, \dots, \beta_p)$ and ε a random error term. Examples can be

$$
\begin{aligned}
Y &= \beta X^2 + \varepsilon \text{ or} \\
Y &= \beta_1 X_1^2 + \beta_2 X_2^4 \text{ or} \\
Y &= \beta_1 X^{\beta_2}.
\end{aligned}
$$

Considering these three examples we can distinguish two types of nonlinear regression. The first two examples can be referred to as a curvilinear regression model leading to a simple linear regression model or multiple linear regression model by a transformation of the variables. The third can be called a "real nonlinear model". This will not be considered here.

We will face the general curvilinear model

$$
Y = \beta_0 + \sum_{i=1}^{p} \beta_i f_i(X_i) + \varepsilon
$$

where at least one of the functions f_i is assumed to be nonlinear. Using transformed variables $\widetilde{X}_i = f_i(X_i)$ we simply get the linear model

$$
Y = \beta_0 + \sum_{i=1}^{p} \beta_i \widetilde{X}_i + \varepsilon.
$$

That means that we only have to transform the data set in the same way and estimate the parameter β by the linear regression. As an example, we consider a model leading to the a simple linear regression. It is stored in ◘ regr4.xpl. The model is given as

$$
Y = 2X^2 + \varepsilon,
$$

where ε is normal distributed. As in the previous sections, we generate a data set accordingly:

```
randomize(0)          ; sets a seed for the random generator
x=2*uniform(20)-1     ; generates 20 on [-1,1] uniform
                      ;    distributed values
eps=normal(20)        ; generates 20 standard normal distributed
                      ;    values
y=2*x^2+eps/2         ; creates y
```

Next we transform the data $\widetilde{X} = X^2$ and compute the simple linear regression using the quantlet linreg.

```
xtrans=x^2                  ; transforms x into the variable xtrans
{beta,bse,bstan,bpval}=linreg(xtrans,y)
                            ; computes the linear regression
```

As in the linear case, we get the ANOVA and parameter tables.

A N O V A	SS	df	MSS	F-test	P-value
Regression	7.781	1	7.781	30.880	0.0000
Residuals	4.535	18	0.252		
Total Variation	12.316	19	0.648		

```
Multiple R      = 0.79483
R^2             = 0.63175
Adjusted R^2    = 0.61130
Standard Error  = 0.50196
```

PARAMETERS	Beta	SE	StandB	t-test	P-value
b[0,]=	-0.1143	0.1785	-0.0000	-0.640	0.7350
b[1,]=	1.7433	0.3137	0.7948	5.557	0.0000

Both tables are interpreted as in the previous sections. The parameter table shows the estimates for β_0 and β_1. From its high p-value, we see that the true value of the constant β_0 is estimated to be zero. The estimated regression function is

$$\widehat{Y}(x) = 1.7433\widetilde{x} = 1.7433x^2 .$$

Now we want to visualize this result.

```
data=sort(x~y)              ; creates the graphical object for
                            ;   the data
setmaskp(data,1,11,8)       ; sets the options for the graphical
                            ;   object data
rx=#(-10:10)/10             ; creates the vector (-1,-0.9,...,1)
yq=(beta[2]*rx^2)           ; creates vector with the regressed
                            ;   values
rdata=sort(rx~yq)           ; creates the graphical object for
                            ;   the regressed values
rdata=setmask(rdata,"reset","line","red","medium")
```

```
                                  ; sets the options for the graphical
                                  ;    object rdata
rf=sort(rx~2*rx^2)                ; creates the graphical object for
                                  ;    the original function
rf=setmask(rf,"reset","line","green","medium")
                                  ; sets the options for the graphical
                                  ;    object rf
nlplot=createdisplay(1,1)  ; creates the display nlplot
show(nlplot,1,1,data,rdata,rf) ;shows the display nlplot
setgopt(nlplot,1,1,"title","Nonlinear Regression")
                                  ; sets the title of the display
```

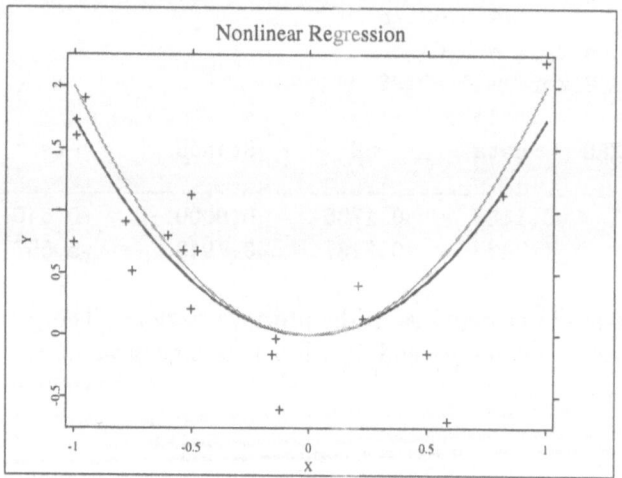

Figure 4.3. Nonlinear regression of $Y = 2X^2 + \varepsilon$ using the variable transformation.
🔵 regr4.xpl

Looking at this plot, we see that the transformation of the variables leads to a good reproduction of the shape of the true function, even when the errors are in the same range as the function itself. But we should remember that we have to know the functions f_i.

We don not want to consider the case leading to a multiple linear regression because the principle is exactly the same as with our example. We just have to use the methods of Section 4.2.

5 Teachware Quantlets

Nathaniel Derby

Teachware quantlets comprise a basic set of interactive, illustrative examples in introductory statistics. For the student, they provide an opportunity to understand some important basic concepts in statistics through trial and error. For the teacher, they can aid instruction by allowing the students to work independently on exploratory examples at their own pace. Additionally, with a modicum of understanding of the XploRe programming language, the teacher can modify these examples to fit his/her own preferences.

Overall, these quantlets can play a key role in bringing current technology into classroom instruction — something that is particularly crucial in beginning-level statistics classes. Statistics is often difficult for students since it requires the coordination of quantitative and graphical insights with mathematical ability. These quantlets provide a way in which this coordination can be made easier by allowing the student to develop his/her quantitative and graphical insight without getting bogged down in the mathematics (which can be learned later). That is, the underlying mathematics can be learned after the student has developed an intuitive understanding of certain concepts, which makes the learning process much easier. Moreover, these quantlets are designed for **interactive learning**, whereby the student actively participates in the learning process (i.e. try out different ideas and see what happens), rather than passively reads or hears about something outside of his/her control. This example of **learning by doing** has proven to be most effective in the learning process.

Furthermore, as statistical problems in general use mathematical formulas and data sets that grow in size and complexity, computer knowledge is essential to those who work with statistics. Learning with quantlets can give a student valuable comfort and skill with computers early on.

The teachware quantlets are part of the `tware` library. In order to access this, the user must call this library:

```
library("tware")
```

After this library has been loaded, the user has a choice of the following examples:

`tw1d`

> illustrates a variety of visual display devices for one-dimensional data.

`twrandomsample`

> illustrates that gathering a random sample is different from "just choosing some" by allowing the user to pick his/her own sample, then rejecting or not rejecting the hypothesis that it is a random sample.

`twpvalue`

> illustrates the concept of a *p*-value for hypothesis testing after having the user choose the parameters and an observed value for a binomial distribution.

`twnormalize`

> illustrates how the normal distribution provides a good approximation to the binomial for large n by allowing the user to transform binomial data and compare its distribution to that of the normal.

`twclt`

> illustrates the concept of the central limit theorem by showing the empirical distribution function of the averages of a large number of samples from a simulation using a user-specified underlying probability distribution.

`twpearson`

> illustrates how dependence is reflected in the formula for the estimated Pearson correlation coefficient, and why it is essential to normalize the data for the formula.

`twlinreg`

> illustrates the concept of linear regression by setting up a scatter plot and a line, and allowing the user to move the line to try to minimize the residual sum of squares.

The following pages will describe each example in detail.

5.1 Visualizing Data

```
tw1d()
        illustrates visual display devices for one-dimensional data
```

This quantlet illustrates a variety of visual display techniques for one-dimensional data. To activate this, the user should type in the following:

```
tw1d()
```

After this, the user should see the following windows:

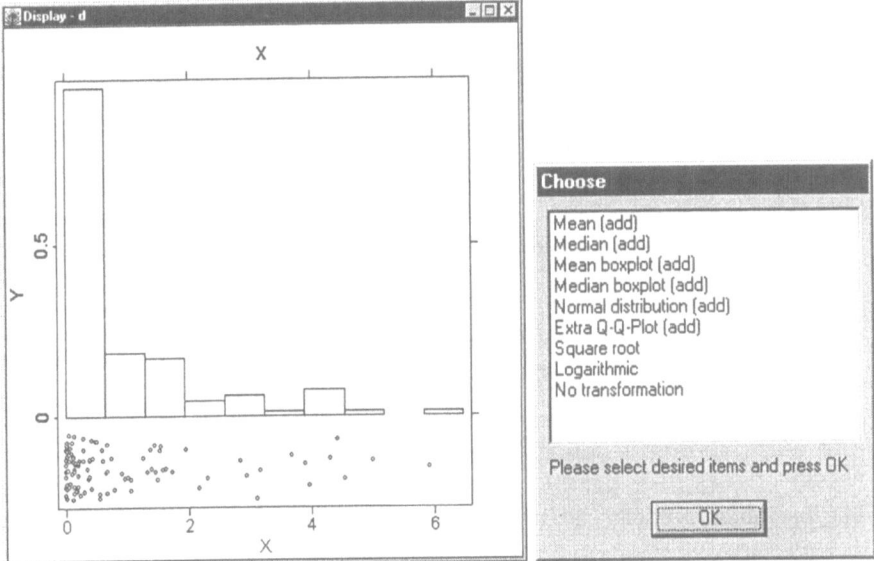

The Display window gives a visual representation of a sample data set already loaded into the computer — the individual data points are displayed at the bottom, and a histogram is displayed above it. The Choose window gives some choices of options available in data displays:

- Mean adds (or deletes) a bar in blue at the bottom of the Display window where the mean is located.

- Median adds (or deletes) a bar in red at the bottom of the Display window where the median is located.

- Mean boxplot adds (or deletes) a boxplot in blue at the bottom of the Display window, with the box centered on the mean, the box endpoints showing the mean plus and minus one standard deviation, and the whiskers (i.e. the lines extending to the left and right of the box) showing the mean plus and minus two standard deviations.

- Median boxplot adds (or deletes) a boxplot in red at the bottom of the Display window, with the middle line of the box showing the median, the box endpoints showing the quartiles (i.e. 25 and 75 percentiles), and the whiskers showing the 2.5 and 97.5 percentiles.

- Normal distribution adds (or deletes) a graph of the normal distribution in green over the histogram of the data.

- Extra Q-Q-Plot adds (or deletes) a quantile–quantile plot, which graphs the quantiles of a normal distribution function (with the sample mean and variance from the data) on the x-axis against the quantiles of the original data on the y-axis.

- Square root gives a graph of the square root transformation of the data (i.e. a histogram, etc., of the square root of each value will be shown, rather than of the actual value).

- Logarithmic gives a graph of the logarithmic transformation of the data (i.e. a histogram, etc., of the logarithm of each value will be shown, rather than of the actual value).

- No transformation gives a graph of the original data, to nullify the effects of the above two options.

The beginning student can use this quantlet to explore various data display techniques.

5.2   Random Sampling

```
twrandomsample()
     illustrates random sampling
```

This quantlet illustrates that "arbitrary human choice" is quite different from proper random sampling. To activate this, the user must type in the following:

```
twrandomsample()
```

After this, the user should see the following window:

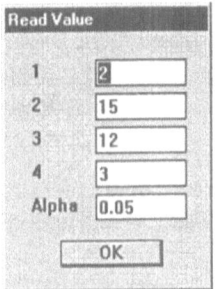

This corresponds to a classroom setting (real or virtual), where the students are asked to write down a "randomly chosen" number among 1, 2, 3, and 4. The numbers above are default values, chosen as such because most people choose 3, and most of the rest choose 2. The α level (denoted by **alpha**) is used to determine the level of significance for the hypothesis test that the numbers are randomly distributed. A default α level of 0.05 is indicated.

After entering the values, or using default values, clicking on the OK button will produce the following display (this one for the default values):

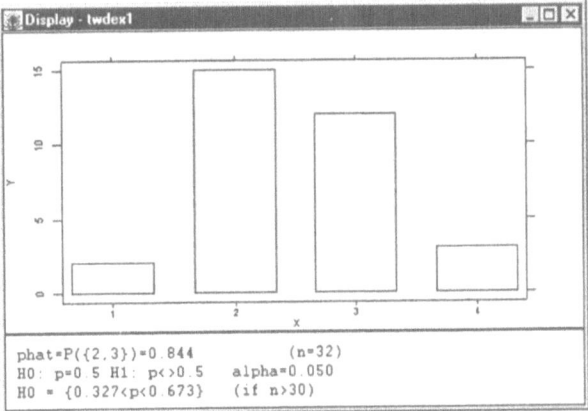

The top half of the display provides a bar graph of the data entered in the Read Value window. The bottom half gives information about the test of the hypothesis that the entered data are a random sample. Here, the meaning of "random sample" is that "all values are equally likely", or equivalently, that the data come from a uniform distribution. The test statistic used here, $\hat{p}$ (phat), is one of many possible test statistics for this hypothesis. This $\hat{p}$ is the empirical (i.e. observed) probability of getting a 2 or a 3, computed from the data entered by the user. If the data really are randomly distributed, we would expect this probability to be close to .5, since this is the probability of getting two choices out of four. Conversely, if this probability is "far" from 0.5, the data are most likely not randomly distributed. This is the idea behind this hypothesis test.

We want to test the hypothesis that this is a random (i.e. evenly distributed) sample. Thus, we have the following null and alternative hypotheses:

$$H_0 : p = 0.5, \qquad H_1 : p \neq 0.5.$$

On the computer screen, the alternative hypothesis is represented as p <> 0.5, but it means the same — that p could be less than or greater than 0.5. To test this hypothesis, we use the test statistic ($\hat{p}$) and see if it lies within our confidence interval. This interval is listed in the third line. For the example above, our confidence interval is the following:

$$0.327 < p < 0.673.$$

Since our $\hat{p}$ in the example is equal to 0.844, it does not lie within our confidence interval, so we can reject H_0. In other words, at the $\alpha = 0.05$ level of significance, the data are not randomly distributed.

Here, the user can see how the confidence interval and $\hat{p}$ changes for various values in the Read Value window.

The formula used for the computation of the confidence interval is the following:

$$0.5 \pm 1.96 \left(\frac{0.5}{\sqrt{n}} \right).$$

It uses the normal distribution as an approximation to the binomial distribution (valid if $n > 30$).

5.3   The p-Value in Hypothesis Testing

```
twpvalue()
       illustrates a p-value for hypothesis testing
```

This quantlet illustrates the concept of the p-value, which is used extensively for hypothesis testing. The main idea is that, if the probability of getting an observed value x is very small, then we most likely have the wrong assumptions in computing this probability.

More formally, suppose we take a sample value from the binomial probability distribution:

$$P(X = x) = \binom{n}{x} p^x (1 - p)^{(n-x)}, \quad P(X \geq x) = \sum_{i=x}^{n} \binom{n}{i} p^i (1 - p)^{(n-i)}.$$

Suppose we test the following null and alternative hypotheses:

$$H_0 : p = p_0, \qquad H_1 : p > p_0.$$

This is an example of a **one-tailed test**. That is, we rule out (usually because of peculiarities of the data) the possibility that $p < p_0$. To test this null hypothesis, we assume that H_0 is true (i.e. $p = p_0$) and using the above formulas, find the probability of getting the observed value. However, it turns out that for testing this hypothesis, $P(X \geq x)$ is more reliable to use than $P(X = x)$. This quantlet illustrates why this is so.

To activate this quantlet, the following should be typed in:

```
twpvalue()
```

After this, the user should see the following window:

Here, the user is asked to input the number of Bernoulli trials (n), the null hypothesis probability of success in the Bernoulli trials (p_0), and the observed binomial value (x). The values above are the default values. After choosing the desired values, clicking on the OK button results in the window below, asking which probability should be visually displayed.

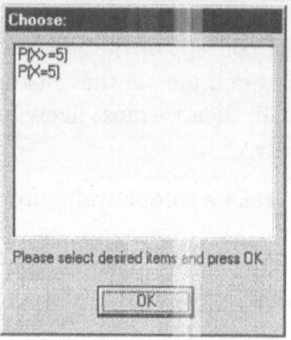

After choosing the desired choice, the next window is displayed (here, for choosing P(X=5)):

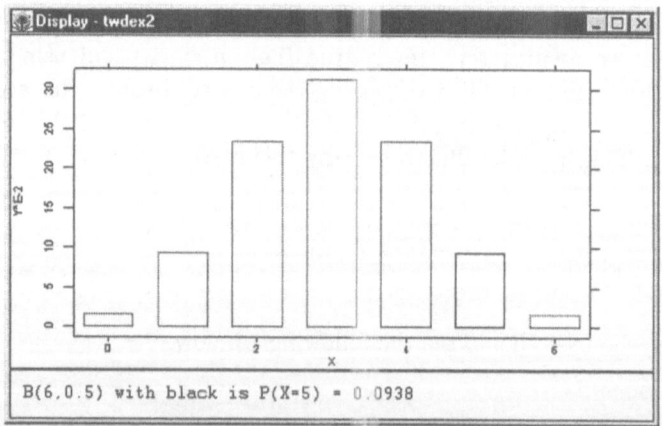

On the computer screen, the desired probability from the last window will be represented in black, with the rest of the distribution portrayed in red (the colors are not recognizable in the above illustration). In the above window, the second bar from the right, representing $P(X = 5)$, will be in black.

One intention of this quantlet is to demonstrate why the p-value is $P(X \geq x)$ and not $P(X = x)$. Indeed, through experimenting, the user can observe that $P(X = x)$ depends strongly on n, and gets small even when there is clearly no strong evidence against H_0. $P(X \geq x)$, on the other hand, is stable for increasing n, and stays large when there is no strong evidence against H_0.

Additionally, for testing the null hypothesis that $p = p_0 \ll n$ (i.e. for a given p_0 considerably smaller than n), this quantlet demonstrates that as the observed value x becomes larger, the p-value decreases (i.e. evidence against H_0 becomes stronger), and that when p becomes larger, the p-value becomes larger (i.e. evidence against H_0 becomes weaker).

5.4   Approximating the Binomial by the Normal Distribution

```
twnormalize()
        illustrates the approximation of the binomial by the normal dis-
        tribution
```

This quantlet illustrates that the normal distribution provides a good approximation to the binomial distribution for large n, as well as why it is important and natural to subtract the mean and divide by the standard deviation when doing such an approximation.

The mathematical formula used for the graphs is the probability function of the binomial distribution:

$$P(X = x) = f(x) = \binom{n}{x} p^x (1 - p)^{n-x} .$$

To activate this quantlet, the user should type in the following:

```
twnormalize()
```

After this, the following windows should be displayed:

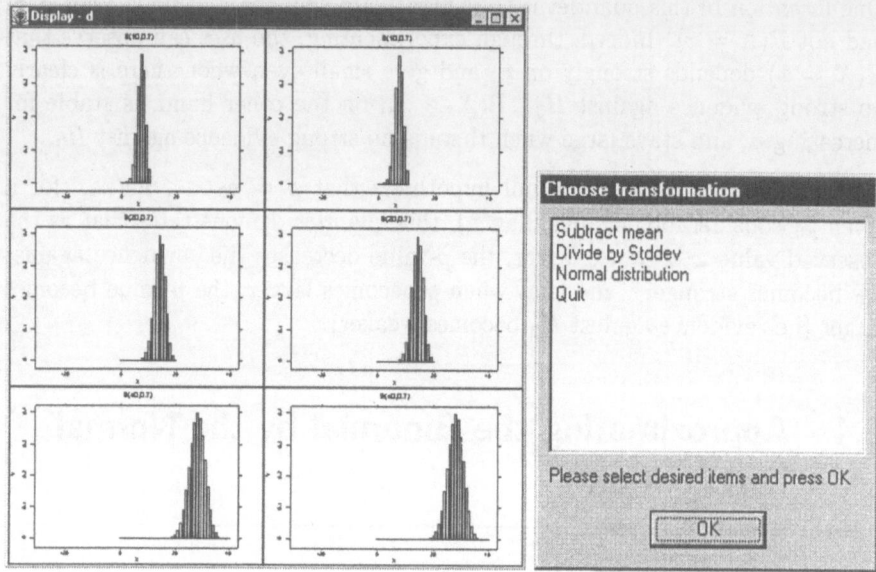

The Display window displays graphs of the distribution functions of three binomial distributions, all with $p = 0.7$, with $n = 10$, 20, and 40. The graphs on the left are the original graphs, those on the right (identical at first to the ones on the left) will become the transformed graphs.

In the Choose transformation window, the user is asked to select which transformation (subtract the mean or divide by the standard deviation) he/she wishes to apply to the data. Additionally, the Normal distribution choice will superimpose a graph of the standard normal distribution over the transformed data in red. Clicking on the OK button applies these transformations to the data, resulting in the transformed data on the right side of the Display window. For example, choosing the Subtract mean option will result in the right side of the Display window showing the data subtracted from the mean. By subtracting the mean from the data, dividing it by the standard deviation, and then superimposing the normal distribution over it, the user can see that the transformed data look very similar to the standard normal distribution. Furthermore, the user can see that as the sample size increases, the transformed data (i.e. subtracted from the mean and divided by the standard deviation) look more and more similar to the standard normal distribution. Thus, the standard normal distribution can be used to approximate the binomial distribution for large n.

5.5   The Central Limit Theorem

```
twclt()
      illustrates the central limit theorem for a user-specified underlying
      probability distribution
```

This quantlet illustrates the concept of the Central Limit Theorem by showing a
histogram of the averages of a large number of samples simulated from a user-
designated underlying probability distribution. That is, the user specifies a
probability distribution function, then a number of samples are simulated from
this distribution. For each sample, the average value is computed. Finally, a
histogram of these average values is displayed. The user can increase or decrease
the number of samples in the simulation to observe that as the number of
samples increases, the histogram approaches the density function of a standard
normal distribution.

To activate this quantlet, the user should type in the following:

```
twclt()
```

After this, the following window should be shown:

Here, the user can specify a probability distribution for a discrete random
variable of four values (i.e. $X = 1, 2, 3$, or 4). He/she should enter in four
values (making sure that they add up to 1), or use the default values of 0.25
for each one, then click on the OK button. The following display should follow
(here for the default values):

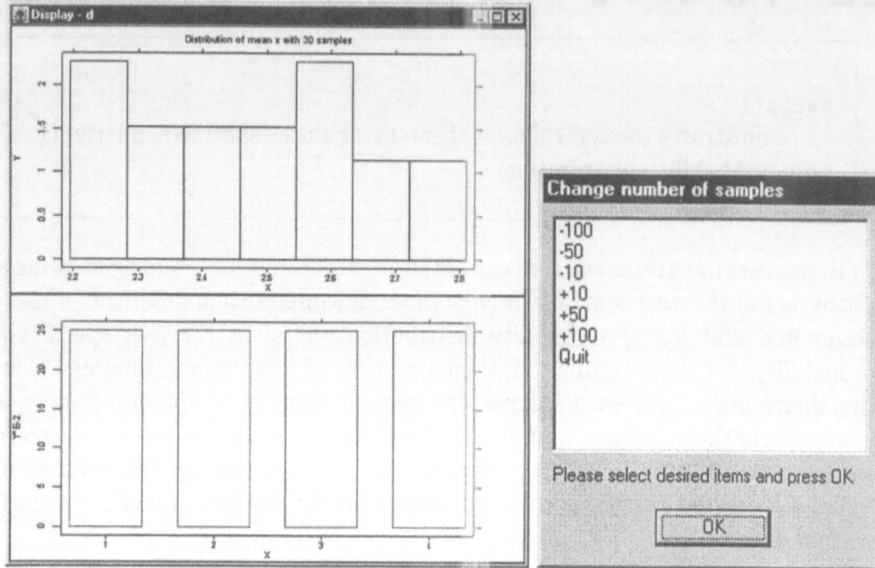

The lower portion of the Display window is the underlying probability distribution — a visual representation of what the user previously typed in for the last window. The upper part of this window is a histogram of averages of 30 simulated samples from the given distribution. That is, XploRe makes a simulation of thirty samples from the given probability distribution, computes the average value for each of them, and plots them in the histogram. Since this is a simulation of a random variable, the resulting histogram from two simulations of the same number of samples from the same underlying distribution will most likely not be the same.

The Change number of samples window above asks for increasing or decreasing the number of samples in the simulation (or quitting). As the user increases the number of samples, the resulting empirical distribution function should look more and more similar to a normal distribution. If the data were subtracted by the estimated mean and divided by the estimated standard deviation, the empirical distribution function would look more and more like a standard normal distribution.

For the beginning statistics student, this graphically shows that the distribution of the mean of a sample approaches the normal distribution as the number of samples increases.

5.6   The Pearson Correlation Coefficient

`twpearson()`
    illustrates the Pearson correlation coefficient

This quantlet illustrates how dependence is reflected in the formula for the estimated Pearson correlation coefficient, and why it is necessary to "normalize" the data (subtract means and divide by the standard deviations). More precisely, suppose we have a collection of points (x_i, y_i) from a bivariate standard normal distribution:

$$f(x, y) = \frac{1}{2\pi\sqrt{1 - \rho^2}} \exp\left\{-\frac{1}{2(1 - \rho^2)}(x^2 - 2\rho xy + y^2)\right\} .$$

The parameter ρ is the **correlation coefficient** and is usually unknown. When that is the case, we must use an estimator for ρ. A common estimator for this is the **Pearson correlation coefficient**, $\hat{\rho}$, with the following formula:

$$\hat{\rho} = \frac{\frac{1}{n-1}\sum_{i=1}^{n}(x_i - \bar{x})(y_i - \bar{y})}{\sqrt{S_{xx}S_{yy}}} , \quad \text{where } \ S_{xx} = \frac{\sum_{i=1}^{n}(x_i - \bar{x})^2}{n-1}, \ S_{yy} = \frac{\sum_{i=1}^{n}(y_i - \bar{y})^2}{n-1} .$$

In this quantlet, the user can see why the above formula is preferable to a couple of simpler formulas. To activate this quantlet, the user should type in the following:

`twpearson()`

After this, the user should see the following window:

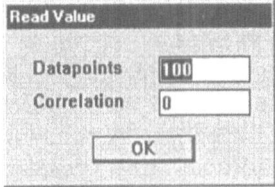

These are specifications for a simulation of the bivariate standard normal distribution, as described above. The **correlation** in the above window is the correlation coefficient ρ, and can sometimes be used as a measure of dependence — that is, two variables that are independent of each other will have a correlation of 0, while two variables that are totally dependent on each other will have a correlation of 1 or –1, depending on if an increase in one variable causes an increase or a decrease in the other, respectively. After the user enters his/her specifications and clicks on OK, the following window should appear (this is for the default values of 100 data points, 0 correlation):

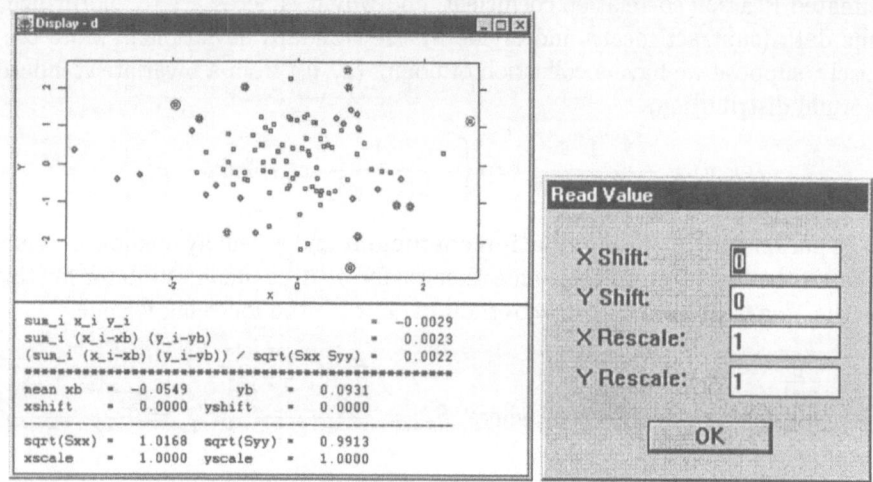

The upper half of the Display window is a scatter-plot diagram of the simulated data. On the computer screen, the black squares will be the original data points (x_i, y_i), and the colored circles will become the transformed data points (x_i', y_i') (the colors are not recognizable in the illustration above).

At the beginning, as in the above display, the transformed data are the same as the original data (since no transformation has been indicated yet), and only the outermost points are shown as colored circles around the black squares. Again on the computer screen (with colors not recognizable in the above illustrations) and imagining a grid formed around the central x' and y' values (i.e. $\bar{x}'$ and $\bar{y}'$), the circles will be red in the first and third quadrants, and blue in the second and fourth quadrants, with the outermost points larger than the others. In the lower half of this window, there are three formulas which are explained below, as well as the sample means and standard deviations for the x and y variables. The shifts and scales of the x and y variables begin with the default values of 0 and 1, respectively.

As mentioned above, the Pearson correlation coefficient is an estimate of the true correlation coefficient ρ, here given by the user — for real data (i.e. not a simulation), we wouldn't know the real correlation coefficient, and would have to estimate it.

The three formulas at the bottom of the Display window are three possible estimators of this correlation coefficient. Since these are estimates from a simulation, the numerical values of these estimator (i.e. the estimates) will most likely be different from the actual value entered in the first window, but they should not be too far off. In our example, the actual correlation is 0, but our estimates are -0.0029, 0.0023, and 0.0022, depending on which formula we use for an estimate.

Instinctively, if the two variables are independent of each other, then if we have the data centered around the origin (as we do in the diagram), we would expect the data points to be relatively evenly distributed among the four quadrants. If there is a measure of dependence, we would expect the data to be mainly in the first and third quadrants (for positive dependence), or in the second and fourth quadrants (for negative dependence). This can be expressed by adding the products of the two variables for each point:

$$\sum_{i=1}^{n} x_i y_i \, .$$

That is, if the data are mainly in the first and third quadrants, the above sum will be a large positive number (for positive correlation), and if the data are mainly in the second and fourth quadrants, the above sum will be a large negative number (for negative correlations). If the two variables are independent, there will be about the same number of data points evenly spaced within the first and third quadrants as within the second and fourth quadrants, and the products will cancel each other out for the most part, leaving the above sum relatively close to zero, as expected. The precise meanings of the words "large" and "close to zero" are dependent on various properties of the data, and cannot be defined exactly — for the moment, it is best to leave these definitions vague.

However, this is under the assumption that the data are centered at the origin. This is usually not the case, so we transform the data by subtracting the mean x-value from the x-variable of each data point, and do the same for the y-variable. Thus, the transformed data are now centered at the origin, and the above formula has been modified to the following:

$$\sum_{i=1}^{n}(x_i - \bar{x})(y_i - \bar{y}) \quad \text{with} \quad \bar{x} = \frac{1}{n}\sum_{1}^{n}x_i, \ \bar{y} = \frac{1}{n}\sum_{1}^{n}y_i$$

This formula is still not very reliable, however, since it depends on the scale of the variables. For example, if the x-variable is a length, we will get different results if we measure x by inches than if we measure x by feet. However, we can correct this by dividing the above formula by the square root of the variances of the two variables:

$$\frac{\sum_{i=1}^{n}(x_i - \bar{x})(y_i - \bar{y})}{\sqrt{S_{xx}S_{yy}}} \quad \text{with} \quad S_{xx} = \frac{\sum_{i=1}^{n}(x_i - \bar{x})^2}{n-1}, \ S_{yy} = \frac{\sum_{i=1}^{n}(y_i - \bar{y})^2}{n-1}.$$

The above formula is now "normalized" in the sense that we have subtracted the data by the means and divided by the standard deviation. Thus, we have a unitless measure of dependence, which lies between −1 and 1.

The user can prove to himself/herself that this last formula will give the same answer, even as the data are shifted and rescaled. In the Read Value window above, the user can indicate how he/she would like to shift and/or rescale the x or y variables. Through this, the user will see that the first two sums (in the bottom half of the Display window) change, while the third always remains the same. Thus, the third formula (the one used for the Pearson correlation coefficient) is the superior estimator.

5.7   Linear Regression

```
twlinreg()
        illustrates the concept of linear regression
```

This quantlet illustrates the concept of linear regression by setting up a scatter plot and a line, and allowing the user to change the slope or intercept of the line to try to minimize the residual sum of squares. To activate this, the following should be typed in:

```
twlinreg()
```

After this, the user should see the following windows:

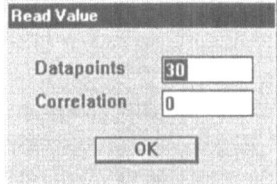

These are specifications for the scatter-plot diagram, the same as with the twpearson quantlet. After the specifications are entered and the OK button is clicked, the following window should appear (this is for the default values of 30 data points, 0 correlation):

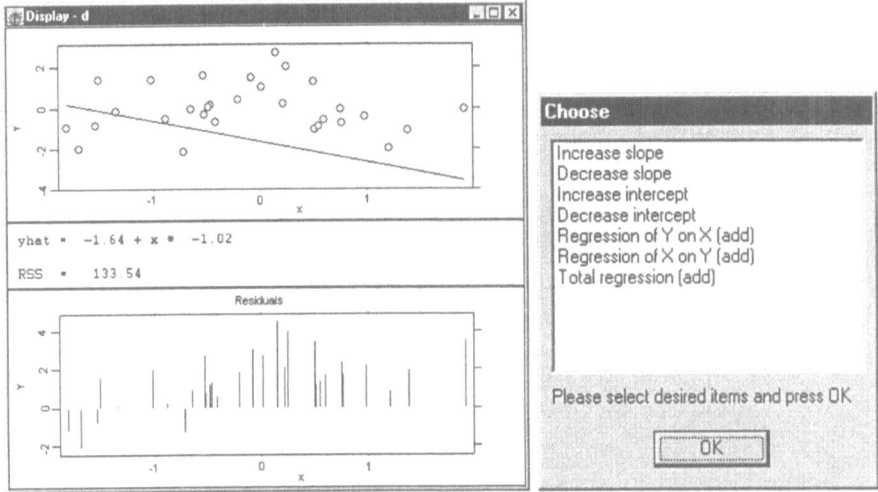

The upper frame of the Display window contains the scatter-plot diagram, as well as a line for which we would like to minimize the residual sum of squares. The middle frame contains the equation of this line ($\mathtt{yhat} = \hat{y}$), as well as the residual sum of squares (RSS). The lower frame contains a graph of the residuals. That is, each vertical line in the bottom graph represents the distance between one of the points and the line in the above graph. The RSS is the sum of the squares of these distances. The object is to find the line which gives the minimum of these sums, which can be done by changing the slope and/or intercept of the given line.

The first four selections in the **Choose** window are for changing the given line
— the user simply makes the appropriate choice(s), then clicks on **OK** to see
the result. Additionally, the user can request for the following to be shown on
the scatter plot:

- Regression of Y on X is the line that gives the minimum RSS in the sense
  described here. That is, the distances between the points and the line
  are measured along the direction of the y-axis.

- Regression of X on Y is the line that gives the minimum RSS if the distances between the points and the line are measured along the direction
  of the x-axis.

- Total Regression is the line that gives the minimum RSS if the distances
  between the points and the line are measured along the direction perpen-
  dicular to the line.

A user who is learning linear regression for the first time can change the slope
and/or intercept of the given line and see if this decreases the RSS. The lower
frame of the display gives a visual demonstration of the distances between the
points and the line. Showing the regression of x on y can verify if the user has
the minimum RSS, and show how far off he/she is. Showing the regression of
x on y, and the total regression, shows how different the result can be if the
distance is measured in two alternative ways.

The following formulas are used for the computations of $\widehat{y}$ and RSS, which are
taken from the Gauss–Markov theorem:

$$\widehat{y}_i = \widehat{\alpha} + \widehat{\beta} x_i,$$

$$\widehat{\beta} = \frac{\sum_{i=1}^{n}(x_i - \bar{x})(y_i - \bar{y})}{\sum_{i=1}^{n}(x_i - \bar{x})^2},$$

$$\widehat{\alpha} = \bar{y} - \widehat{\beta}\bar{x}, \quad \bar{y} = \frac{1}{n}\sum_{i=1}^{n} y_i, \ \bar{x} = \frac{1}{n}\sum_{i=1}^{n} x_i,$$

$$\text{RSS} = \frac{\sum_{i=1}^{n}(\widehat{y}_i - \bar{y})^2}{\sum_{i=1}^{n}(y_i - \bar{y})^2}.$$

Bibliography

Härdle, W., Klinke, S., and Marron, J. (1999). Connected Teaching of Statistics, Discussion Paper 24, SFB 373, Humboldt-Universität zu Berlin.

Müller, M. (1998). Computer-Assisted Statistics Teaching in Network Environment, *COMPSTAT 1998 Proceedings in Computational Statistics*, 77–88. (http://ise.wiwi.hu-berlin.de/~marlene/publications.html)

Müller, M. (1998). Teaching Statistics with XploRe, *Math&Stats Newsletter*, 21 – 24. (http://www.stats.gla.ac.uk/cti/activities/articles.html)

Redfern, E. J. and Bedford, S. E. (1994). Teaching and Learning through Technology: The Development of Software for Teaching Statistics to Non-Specialist Students, *COMPSTAT 1994 Proceedings in Computational Statistics*, 408–414.

Talbot, M. (1998). Statistics Training and the Internet, *COMPSTAT 1998 Proceedings in Computational Statistics*, 461–466. (http://www.bioss.sari.ac.uk/~mike)

West, R. W. (1997). Statistical Applications for the World Wide Web, *Bulletin of the International Statistical Institute, 51st Session Proceedings Book 2*, 7–10.

Part II: Statistical Libraries

6 Smoothing Methods

Marlene Müller

Nonparametric smoothing methods serve several needs in statistical data analysis: They provide a flexible analysis tool, often based on interactive graphical data representation. Also, they help in constructing a model from observations, for example by graphical comparison with already existing models.

This chapter focuses on function estimation by kernel smoothing methods. Of course, we cannot give a profound introduction into this topic. We refer therefore to textbooks that provide general introductions to smoothing techniques: Silverman (1986), Härdle (1990) and Scott (1992) for nonparametric density estimation, Härdle (1991), Wand and Jones (1995) for nonparametric regression.

Before proceeding to the next section, please type

```
library("smoother")
library("plot")
```

to load the necessary libraries. The `smoother` library automatically loads the `xplore` and the `kernel` library, which we will use as well. Additionally, we load the library `plot` which is used for graphing the resulting density and regression functions.

6.1 Kernel Density Estimation

The goal of density estimation is to approximate the probability density function $f(\bullet)$ of a random variable X. Assume we have n independent observations $x_1, \ldots, x_n$ from the random variable X. The **kernel density estimator** $\widehat{f}_h(x)$

for the estimation of the density value $f(x)$ at point x is defined as

$$\hat{f}_h(x) = \frac{1}{nh} \sum_{i=1}^{n} K\left(\frac{x_i - x}{h}\right),\tag{6.1}$$

$K(\bullet)$ denoting a so-called kernel function, and h denoting the bandwidth. A number of possible kernel functions is listed in Table 6.1.

Kernel	$K(u)$	XploRe function
Uniform	$\frac{1}{2} I(\|u\| \leq 1)$	uni
Triangle	$(1 - \|u\|) I(\|u\| \leq 1)$	trian
Epanechnikov	$\frac{3}{4}(1 - u^2) I(\|u\| \leq 1)$	epa
Quartic	$\frac{15}{16}(1 - u^2)^2 I(\|u\| \leq 1)$	qua
Triweight	$\frac{35}{32}(1 - u^2)^3 I(\|u\| \leq 1)$	tri
Gaussian	$\frac{1}{\sqrt{2\pi}} \exp(-\frac{1}{2}u^2)$	gau
Cosinus	$\frac{\pi}{4} \cos(\frac{\pi}{2}u) I(\|u\| \leq 1)$	cosi

Table 6.1. Kernel functions.

In order to illustrate how the kernel density estimator works, let us rewrite equation (6.1) for the Uniform kernel:

$$\begin{aligned}
\hat{f}_h(x) &= \frac{1}{2nh} \sum_{i=1}^{n} I\left(|x_i - x| \leq h\right) \\
&= \frac{1}{2nh} \{\text{number of observations } x_i \text{ that fall in } [x - h, x + h]\}.
\end{aligned}$$

It is easy to see that for estimating the density at point x, the relative frequency of all observations x_i falling in an interval around x is counted. The factor $1/(2nh)$ is needed to ensure that the resulting density estimate has integral $\int \hat{f}_h(x)\, dx = 1$.

The Uniform kernel implies that all observations x_i in the neighborhood of x are given the same weight, since observations close to x carry also information about $f(x)$. It is reasonable to attribute more weights to observations that are near to x and less weight to distant observations. All other kernels in Table 6.1 have this property. Figure 6.1 visualizes the construction of a kernel density estimate of 10 data points (red $+$) using the Gaussian kernel.

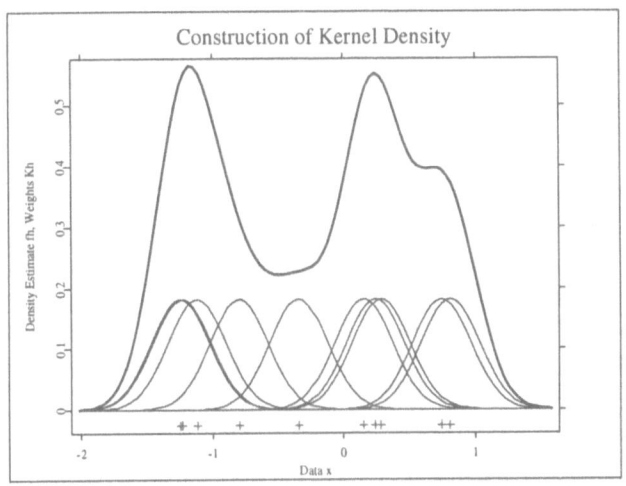

Figure 6.1. Kernel density estimate as a sum of bumps. ⌐ smoo01.xpl

6.1.1 Computational Aspects

To understand the structure of the **smoother** library, we need to explain shortly how kernel estimates can be computed in practice. Consider the equation (6.1) again. To obtain the kernel density estimates for all observations $x_1, \dots, x_n$ we must compute

$$\widehat{f_h}(x_j) = \frac{1}{nh} \sum_{i=1}^{n} K\left(\frac{x_i - x_j}{h}\right), \quad j = 1, \dots, n.$$

In the case of an exact computation of these expressions, the kernel function $K(\bullet)$ must be evaluated to $O(h \cdot n^2)$ times. The expression $O(h \cdot n^2)$ indicates a value proportional to n^2 (the squared sample size). This increases the computation time if the sample size n is large.

For exploratory purposes, i.e. for graphing the density estimate it is not necessary to evaluate the $\widehat{f_h}(\bullet)$ at all observations $x_1, \dots, x_n$. Instead, the estimate can be computed for example on an equidistant grid $v_1, \dots, v_m$:

$$v_k = x_{\min} + \frac{k}{m}\left(x_{\max} - x_{\min}\right), \quad k = 1, \dots, m \ll n.$$

The evaluation of $\widehat{f}_h(v_k)$, $k = 1, \ldots, m$, requires then only $O(h \cdot n \cdot m)$ steps.

The number of $O(h \cdot n \cdot m)$ evaluations of the kernel function $K(\bullet)$ is however time consuming if the sample size is large. An alternative and faster way is to approximate the kernel density estimate by the WARPing method (Härdle and Scott, 1992).

The basic idea of WARPing (Weighted Average of Rounded Points) is the "binning" of the data in bins of length d starting at the origin $x_0 = 0$. Each observation point is then replaced by the bincenter of the corresponding bin (that means in fact that each point is rounded to the precision given by d). A usual choice for d is to use $h/5$ or $(x_{\max} - x_{\min})/100$. In the latter case, the effective sample size r for the computation (the number of nonempty bins) can be at most 101.

For the WARPing method, the kernel function needs to be evaluated only at $\ell \cdot d/h$, $\ell = 1, \ldots, s$, where s is the number of bins which contains the support of the kernel function. The calculation of the estimated density reduces then to

$$\widetilde{f}_h(w_j) = \frac{n_j}{nh} \sum_{i=1}^{r} n_i K \left\{ \frac{(i-j) \cdot d}{h} \right\}, \quad j = 1, \ldots, r$$

computed on the grid $w_j = (j + 0.5) \cdot d$ (j integer) with n_i, n_j denoting the number of observations in the i-th and j-th bins, respectively. The WARPing approximation requires $O(h \cdot r/d)$ evaluations of the kernel function and $O(n) + O(h \cdot r/d)$ steps in total. This is considerably faster than the exact computation, when the sample size is large.

To have a choice between both the exact and the WARPing computation of kernel estimates and related functions, the smoother library offers most functionality in two functions:

Functionality	Exact	WARPing
density estimate	denxest	denest
density confidence intervals	denxci	denci
density confidence bands	denxcb	dencb
density bandwidth selection	–	denbwsel

6.1.2 Computing Kernel Density Estimates

fh = denest (x {,h {,K} {,d}})
   computes the kernel density estimate on a grid using the WARP-
   ing method

fh = denxest (x {,h {,K} {,v}})
   computes the kernel density estimate for all observations or on a
   grid v by exact computation

The easiest way to compute kernel density estimates is the function **denest**. This function is based on the WARPing method which makes the computation very fast. Let us apply the **denest** routine to the **nicfoo** data which contain observations on household netincome in the first column (see Appendix B.1). **denest** requires the data as a minimum input. Optionally, the bandwidth h, the kernel function K and the discretization parameter d which is used for the WARPing can be specified. For example,

```
netinc=read("nicfoo")
netinc=netinc[,1]
h=(max(netinc)-min(netinc))*0.15
fh=denest(netinc,h)
```

smoo02.xpl

calculates the density estimate using 15% of the range of the data as bandwidth h. If h is not given, the rule of thumb bandwidth (6.6) is used, see Subsection 6.1.4 for more information. Note that we did not specify the kernel function in the above code. Therefore the default Quartic kernel "qua" and the default binsize d are used.

To display the estimated function you may use these lines of code:

```
fh=setmask(fh,"line")
plot(fh)
tl="Density Estimate"
xl="netincome"
yl="density fh"
setgopt(plotdisplay,1,1,"title",tl,"xlabel",xl,"ylabel",yl)
```

The resulting density is plotted in Figure 6.2.

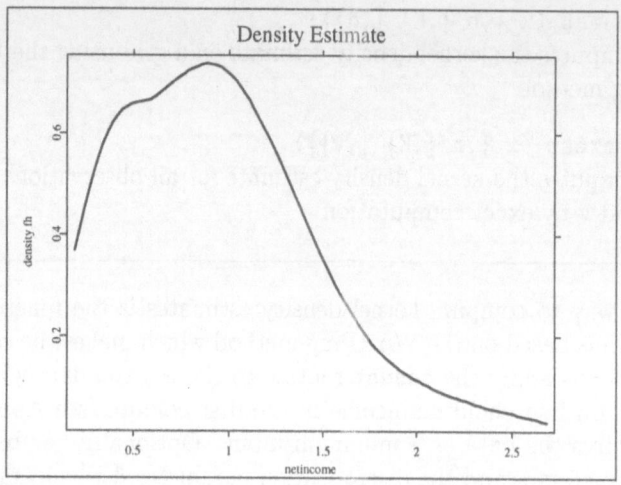

Figure 6.2. Kernel density estimate for the Netincome data.

As an alternative to **denest**, the function **denxest** can be used for obtaining a density estimate by exact computation. As with **denest** the data are the minimum input. Optionally, the bandwidth **h**, the kernel function **K** and a grid **v** for the computation can be specified. The following code computes the density estimate for all observations:

```
h=(max(netinc)-min(netinc))*0.15
fh=denxest(netinc,h)
```

To compute the density on a grid of 30 equidistant points within the range of **x**, the previous example can be modified as follows:

```
h=(max(netinc)-min(netinc))*0.15
v=grid( min(netinc), (max(netinc)-min(netinc))/29, 30)
fh=denxest(netinc,h,"qua",v)
```

Q smoo03.xpl

Finally, let us calculate the density estimate for a family of different bandwidths
in order to see the dependence on the bandwidth parameter. We modify our
first code in the following way:

```
hfac=(max(netinc)-min(netinc))
fh =setmask( denest(netinc,0.15*hfac) ,"line")
fh1=setmask( denest(netinc,0.05*hfac) ,"line","thin","magenta")
fh2=setmask( denest(netinc,0.10*hfac) ,"line","thin","red")
fh3=setmask( denest(netinc,0.20*hfac) ,"line","thin","blue")
fh4=setmask( denest(netinc,0.25*hfac) ,"line","thin","green")
plot(fh1,fh2,fh3,fh4,fh)
tl="Family of Density Estimates"
xl="netincome"
yl="density fh"
setgopt(plotdisplay,1,1,"title",tl,"xlabel",xl,"ylabel",yl)
```
                                                                    smoo04.xpl

As you can see from the resulting graph (Figure 6.3), smaller bandwidths reveal
more structure of the data, whereas larger bandwidths give smoother functions.

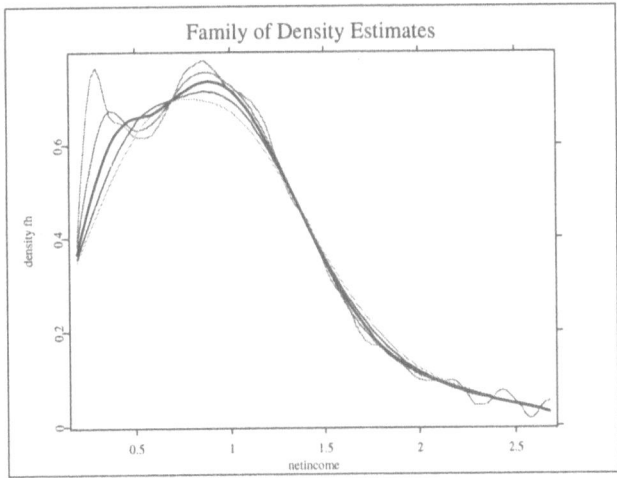

Figure 6.3. Family of kernel density estimates for the Netincome data.

6.1.3 Kernel Choice

```
h2 = canker (h1, K1, K2)
        does the canonical bandwidth transformation of a bandwidth
        value for kernel K1 into an equivalent bandwidth for kernel K2
```

Kernel density estimation requires two parameters: the kernel function K and the bandwidth parameter h. In practice, the choice of the kernel is not nearly as important as the choice of the kernel. The theoretical background of this observation is that kernel functions can be rescaled such that the difference between two kernel density estimates using two different kernels is almost negligible (Marron and Nolan, 1988).

Instead of rescaling kernel functions, we can also rescale the bandwidths. This rescaling works as follows: Suppose we have estimated an unknown density f using some kernel K_A and bandwidth h_A and we are considering estimating f with a different kernel, say K_B. Then the appropriate bandwidth h_B to use with K_B is

$$h_B = \frac{h^A}{\delta_0^A},\tag{6.2}$$

where the scale factors δ_0^A (the so-called canonical bandwidths) can be found in Table 6.2 for some selected kernel functions.

Kernel	XploRe function		δ_0
Uniform	uni	$\left(\frac{9}{2}\right)^{1/5}$	$\cong$ 1.3510
Epanechnikov	epa	$15^{1/5}$	$\cong$ 1.7188
Quartic	qua	$35^{1/5}$	$\cong$ 2.0362
Gaussian	gau	$\left(\frac{1}{4\pi}\right)^{1/10}$	$\cong$ 0.7764

Table 6.2. δ_0 for different kernels.

XploRe offers the function canker to do this recomputation of bandwidths:

```
hqua=canker(1.0,"gau","qua")
hqua
```

computes the bandwidth

```
Contents of hqua
[1,]    2.6226
```

that can be used with the Quartic kernel "qua" in order to find the kernel density estimate that is equivalent to using the Gaussian kernel "gau" and bandwidth 1.0.

6.1.4 Bandwidth Selection

hrot = denrot (x {,K {,opt}})
    computes a rule-of-thumb bandwidth for density estimation

{hcrit,crit} = denbwsel(x {,h {,K {,d}}})
    starts an interactive tool for kernel density bandwidth selection
    using the WARPing method

Now we can consider the problem of bandwidth selection. The optimal bandwidth for a kernel density estimate is typically calculated on the basis of an estimate for the **integrated squared error**

$$\text{ISE}(\widehat{f}_h) = \int \{\widehat{f}_h(x) - f(x)\}^2 \, dx$$

or the **mean integrated squared error**

$$\text{MISE}(\widehat{f}_h) = \int E\{\widehat{f}_h(x) - f(x)\}^2 \, dx.$$

Both criteria should be minimized to obtain a good approximation of the unknown density f. For ISE, it is easy to see that

$$\text{ISE} = \{\widehat{f}_h\} \int \widehat{f}_h^2(x) \, dx - 2 \int \{\widehat{f}_h f\}(x) \, dx + \int f^2(x) \, dx. \tag{6.3}$$

Since the third term does not depend on h, the classical **least squares cross-validation** criterion (LSCV) is an estimate for the first and second term of (6.3). Cross-validation variants that are based on MISE are the **biased cross-validation** criterion (BCV), the **smoothed cross-validation** criterion (SCV) and the **Jones, Marron and Park cross-validation** criterion (JMP).

Plug-in approaches attempt to estimate MISE. Under the asymptotic conditions $h \to 0$ and $nh \to \infty$, it is possible to approximate MISE by

$$\text{MISE}(\widehat{f}_h) \underset{\text{as.}}{\approx} \frac{1}{nh}\|K\|_2^2 + \frac{h^4}{4}\{\mu_2(K)\}^2 \|f''\|_2^2 . \tag{6.4}$$

The terms $\|K\|_2^2$ and $\{\mu_2(K)\}^2$ are constants depending on the kernel function K. Analogously, $\|f''\|_2^2$ denotes a constant depending on the second derivative of the unknown density f. Optimizing (6.4) with respect to h yields the following optimal bandwidth

$$h_{\text{opt}} = \left(\frac{\|K\|_2^2}{\|f''\|_2^2 \{\mu_2(K)\}^2 n} \right)^{1/5} . \tag{6.5}$$

$\|f''\|_2^2$ is the only unknown term in (6.5). The idea behind plug-in estimates is to replace f'' by an estimate. Silverman's rule of thumb computes f'' as if f where the density of the normal distribution $N(\mu, \sigma^2)$. This gives

$$\widehat{h}_0 = \left(\frac{4\widehat{\sigma}^5}{3n} \right)^{1/5} \cong 1.06\,\widehat{\sigma}\,n^{-1/5}$$

if the kernel function K is also assumed to be the Gaussian kernel.

If a different kernel function is used, the bandwidth $\widehat{h}_0$ can be rescaled by using equation (6.2). For example, the rule-of-thumb bandwidth for the Quartic kernel computes as

$$\widehat{h}_0 = 2.78\,\widehat{\sigma}\,n^{-1/5} \tag{6.6}$$

This rule of thumb is implemented in `denrot`.

Of course, $\widehat{h}_0$ is a very crude estimate, in particular if the true density f is far from looking like a normal density. Refined plug-in methods such as the **Park and Marron plug-in** and the **Sheater and Jones plug-in** use nonparametric estimates for f''.

The quantlet `denbwsel` needs the data vector as input:

```
netinc=read("nicfoo")
netinc=netinc[,1]
tmp=denbwsel(netinc)
```

Q smoo05.xpl

This opens a selection box which offers you the choice between the different bandwidth selectors:

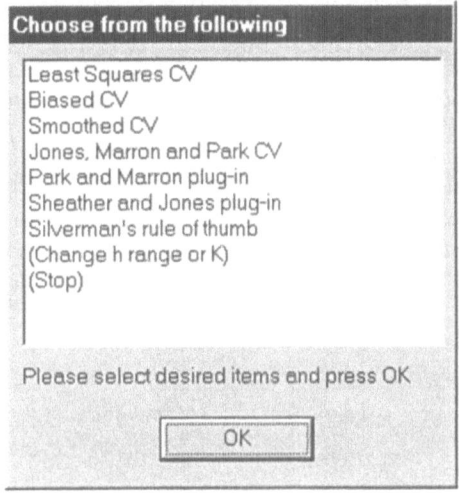

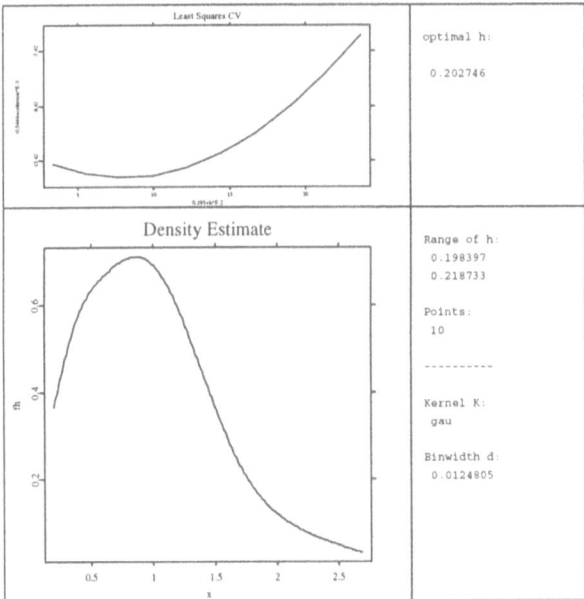

Figure 6.4. Least Squares CV bandwidth selector for Netincome data.

Choose for instance the **Least Squares CV** item to use LSCV. Here, a graphical display appears which shows the LSCV criterion in the upper left, the chosen optimal bandwidth in the upper right, the resulting kernel density estimate in the lower left, and some information about the search grid and the kernel in the lower right. This graphical display is shown in Figure 6.4.

You can play around now with the different bandwidth selectors. Try out the **Park and Marron plug-in** for example. Here, in contrast to the cross-validation methods, the aim is to find the root of the bandwidth selection criterion. The graphical display for the Park and Marron selector is shown in Figure 6.5.

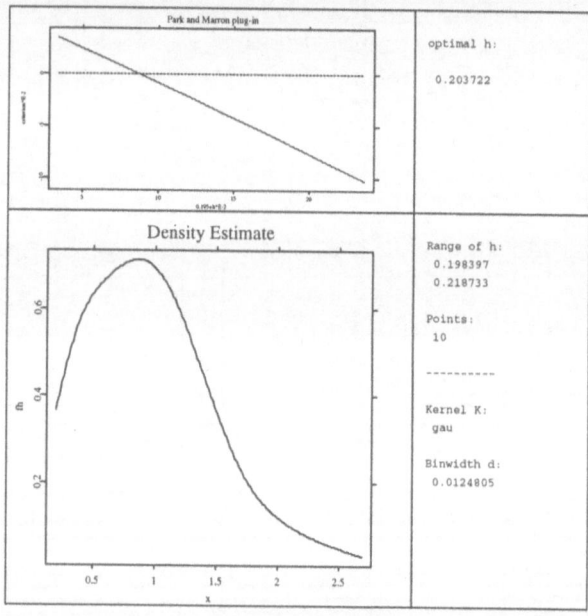

Figure 6.5. Park and Marron plug-in for Netincome data.

During the use of denbwsel, all currently obtained results are displayed in the output window. Note that for some selectors, it will be necessary to increase the search grid for h. This can be done by using the menu item (Change h range or K).

It is difficult to give a general decision of which bandwidth selector is the best among all. As mentioned above, Silverman's rule of thumb works well when the true density is approximately normal. It may oversmooth for bimodal or multimodal densities. A nice feature of the cross-validation methods is that the selected bandwidth automatically adapts to the smoothness of f. From a theoretical point of view, the *BCV* and *SCV* bandwidth selectors have the optimal asymptotic rate of convergence, but need either a very large sample size or pay with a large variance. As a consequence, we recommend determining bandwidths by different selection methods and comparing the resulting density estimates.

For detailed overviews on plug-in and cross-validation methods of bandwidth selection, see Härdle (1991), Park and Turlach (1992), and Turlach (1993). Practical experiences are available from the simulation studies in Marron (1989), Jones, Marron and Sheather (1992), Park and Turlach (1992), and Cao, Cuevas and González-Manteiga (1992).

6.1.5 Confidence Intervals and Bands

{fh, fhl, fhu} = denci (x {,h {,alpha {,K} {,d}}})
    computes the kernel density estimate and pointwise confidence intervals on a grid using the WARPing method

{fh, fhl, fhu} = dencb (x {,h {,alpha {,K} {,d}}})
    computes the kernel density estimate and uniform confidence bands on a grid using the WARPing method

{fh, fhl, fhu} = denxci (x {,h {,alpha {,K} {,v}}})
    computes the kernel density estimate and pointwise confidence intervals for all observations or on a grid v by exact computation

{fh, fhl, fhu} = denxcb (x {,h {,alpha {,K} {,v}}})
    computes the kernel density estimate and uniform confidence bands for all observations or on a grid v by exact computation

To derive confidence intervals or confidence bands for $\widehat{f}_h(x)$ one has to know its sampling distribution. The following result about the asymptotic distribution of $\widehat{f}_h(x)$ is known from the literature (Härdle, 1991, p. 62). Suppose that f''

exists and $h = cn^{-1/5}$. Then

$$n^{2/5} \left\{ \widehat{f}_h(x) - f(x) \right\} \underset{\text{as.}}{\sim} N \left(\frac{c^2}{2} f''(x) \mu_2(K), \frac{1}{c} f(x) \|K\|_2^2 \right), \quad (6.7)$$

which can be recalculated to

$$\left\{ \widehat{f}_h(x) - f(x) \right\} \underset{\text{as.}}{\sim} N \left(\frac{h^2}{2} f''(x) \mu_2(K), \frac{1}{nh} f(x) \|K\|_2^2 \right). \quad (6.8)$$

Assuming that the bias term $\frac{h^2}{2} f''(x) \mu_2(K)$ is negligible with respect to the standard deviation, the resulting **confidence interval** for $f(x)$ can be written as

$$\left[\widehat{f}_h(x) - z_{1-\frac{\alpha}{2}} \sqrt{\frac{\widehat{f}_h(x) \|K\|_2^2}{nh}} \, , \, \widehat{f}_h(x) + z_{1-\frac{\alpha}{2}} \sqrt{\frac{\widehat{f}_h(x) \|K\|_2^2}{nh}} \right]. \quad (6.9)$$

The value $z_{1-\frac{\alpha}{2}}$ is the $(1 - \frac{\alpha}{2})$-quantile of the standard normal distribution.

The functions `denci` and `denxci` compute pointwise confidence intervals for a grid of points (`denci` using WARPing) or all observations (`denxci` using exact computation). The following quantlet code computes the confidence intervals for bandwidth 0.2 (Gaussian kernel) and significance level $\alpha = 0.05$. For graphical exploration, the pointwise intervals are drawn as curves and displayed in a plot. The resulting graphics is shown in Figure 6.6.

```
{fh,fl,fu}=denci(netinc,0.2,0.05,"gau")
fh=setmask(fh,"line","blue")
fl=setmask(fl,"line","blue","thin","dashed")
fu=setmask(fu,"line","blue","thin","dashed")
plot(fl,fu,fh)
setgopt(plotdisplay,1,1,"title","Density Confidence Intervals")
```

         Q smoo06.xpl

Uniform **confidence bands** for f can only be derived under some rather restrictive assumptions, see Härdle (1991, p. 65). Suppose that f is a density on $[0, 1]$ and $h = n^{-\delta}$, $\delta \in (\frac{1}{5}, \frac{1}{2})$. Then the following formula holds under some

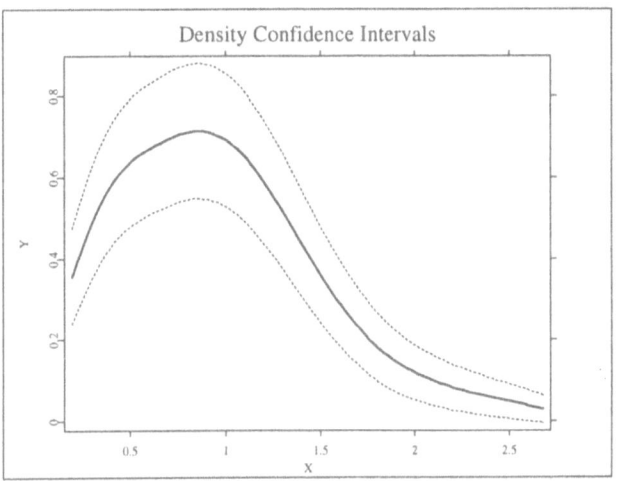

Figure 6.6. Pointwise confidence intervals.

regularity for all $x \in [0, 1]$ (Bickel and Rosenblatt, 1973):

$$P\left(\widehat{f}_h(x) - z_{n,\alpha}\sqrt{\frac{\widehat{f}_h(x)\|K\|_2^2}{nh}} \leq f(x) \leq \widehat{f}_h(x) + z_{n,\alpha}\sqrt{\frac{\widehat{f}_h(x)\|K\|_2^2}{nh}}\right) \underset{\text{as.}}{\sim} 1 - \alpha,$$

$$(6.10)$$

where

$$z_{n,\alpha} = \left\{\frac{-\log\{-\frac{1}{2}\log(1-\alpha)\}}{(2\delta\log n)^{1/2}} + d_n\right\}^{1/2}$$

and

$$d_n = (2\delta\log n)^{1/2} + (2\delta\log n)^{-1/2}\log\left(\frac{1}{2\pi}\frac{\|K'\|_2}{\|K\|_2}\right)^{1/2}.$$

In practice, the data $x_1, \ldots, x_n$ are transformed to the interval $[0, 1]$, then the confidence bands are computed and rescaled to the original scale of $x_1, \ldots, x_n$. The functions dencb and denxcb can hence be directly applied to the original data $x_1, \ldots, x_n$. The following quantlet code computes the confidence bands for the netinc data and displays them in a plot:

```
{fh,fl,fu}=dencb(netinc,0.2,0.05,"gau")
fh=setmask(fh,"line","blue")
fl=setmask(fl,"line","blue","thin","dashed")
fu=setmask(fu,"line","blue","thin","dashed")
plot(fl,fu,fh)
setgopt(plotdisplay,1,1,"title","Density Confidence Bands")
```

Q smoo07.xpl

The resulting graph is shown in Figure 6.7. It can be clearly seen that the confidence bands are wider than the confidence intervals (Figure 6.6). This is explained by the fact that here the whole function $f(\bullet)$ lies within the bands with probability 95%.

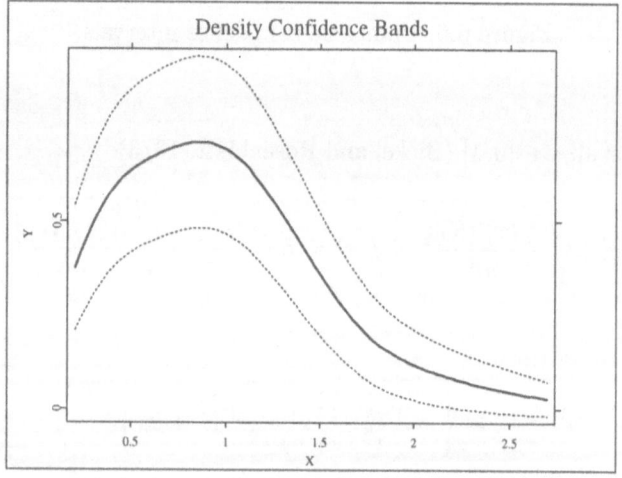

Figure 6.7. Uniform confidence bands.

6.2 Kernel Regression

Regression smoothing investigates the association between an explanatory variable X and a response variable Y. This section explains how to apply Nadaraya–Watson and local polynomial kernel regression.

Nonparametric regression aims to estimate the functional relation between Y and X, i.e. the conditional expectation

$$E(Y|X) = m(X).$$

The function $m(\bullet)$ is the regression function to estimate. Alternatively, we can write

$$Y = m(X) + \varepsilon, \quad E(\varepsilon) = 0.$$

It is not necessary that the variance of Y is a constant function. Typically one assumes

$$\mathrm{Var}(Y|X) = \sigma^2(X), \tag{6.11}$$

where $\sigma(\bullet)$ is a continuous and bounded function.

Suppose that we have independent observations $(x_1, y_1), \ldots, (x_n, y_n)$. The Nadaraya–Watson estimator is defined as

$$\widehat{m}_h(x) = \frac{\sum\limits_{i=1}^{n} K\left(\frac{x_i - x}{h}\right) y_i}{\sum\limits_{i=1}^{n} K\left(\frac{x_i - x}{h}\right)}.$$

It is easy to see that this kernel regression estimator is just a weighted sum of the observed responses y_i. The denominator ensures that the weights sum up to 1. The kernel function $K(\bullet)$ can be any kernel function from Table 6.1.

6.2.1 Computational Aspects

The computational effort for calculating a Nadaraya–Watson or local polynomial regression is in the same order as for kernel density estimation (see Section 6.1.1). As in density estimation, all routines are offered in an exact and in a WARPing version:

Functionality	Exact	WARPing
Nadaraya–Watson regression	`regxest`	`regest`
Nadaraya–Watson confidence intervals	`regxci`	`regci`
Nadaraya–Watson confidence bands	`regxcb`	`regcb`
Nadaraya–Watson bandwidth selection	`regxbwsel`	`regbwsel`
local polynomial regression	`lpregxest`	`lpregest`
local polynomial derivatives	`lpderxest`	`lpderest`

6.2.2 Computing Kernel Regression Estimates

```
mh = regest (x {,h {,K} {,d}})
    computes the kernel regression on a grid using the WARPing
    method

mh = regxest (x {,h {,K} {,v}})
    computes the kernel regression estimate for all observations or on
    a grid v by exact computation
```

The WARPing-based function **regest** offers the fastest way to compute the Nadaraya–Watson regression estimator for exploratory purposes. We apply this routine nicfoo data, which contain observations on household netincome in the first column and on food expenditures in the second column. The following quantlet computes and plots the regression curve together with the data:

```
nicfoo=read("nicfoo")
h=0.2*(max(nicfoo[,1])-min(nicfoo[,1]))
mh=regest(nicfoo,h)
mh=setmask(mh,"line","blue")
xy=setmask(nicfoo,"cross","small")
plot(xy,mh)
setgopt(plotdisplay,1,1,"title","regression estimate")
```
Q smoo08.xpl

The function **regest** needs to be replaced by **regxest**, if one is interested in the fitted values for all observations $x_1, \dots, x_n$. The following quantlet code computes the regression function on all observations and shows a residual plot together with the zero line:

```
mh=regxest(nicfoo,0.2)
res=nicfoo[,1] ~ (nicfoo[,2]-mh[,2])
res=setmask(res,"cross")
zline=(min(nicfoo[,1])|max(nicfoo[,1])) ~ (0|0)
zline=setmask(zline,"line","red")
plot(res,zline)
setgopt(plotdisplay,1,1,"title","regression residuals")
```

Q smoo09.xpl

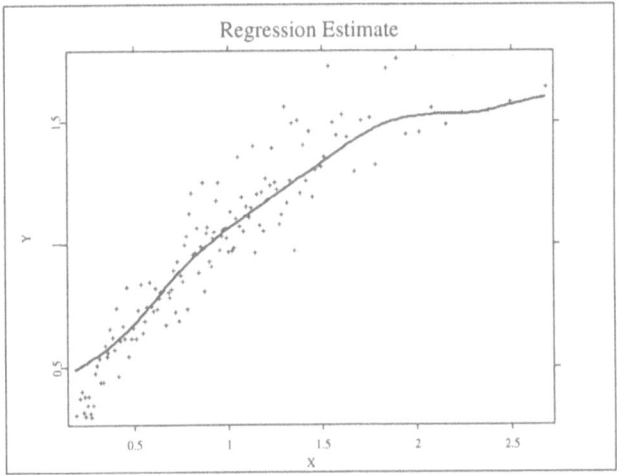

Figure 6.8.   Regression estimate for the netincome vs. food expenditures data.

The resulting regression function is shown in Figure 6.8. Figure 6.9 shows the resulting residual plot. We observe that most of the nonlinear structure of the data is captured by the nonparametric regression function. However, the residual graph shows that the data are heteroskedastic, in the way that the residual variance increases with increasing netincome.

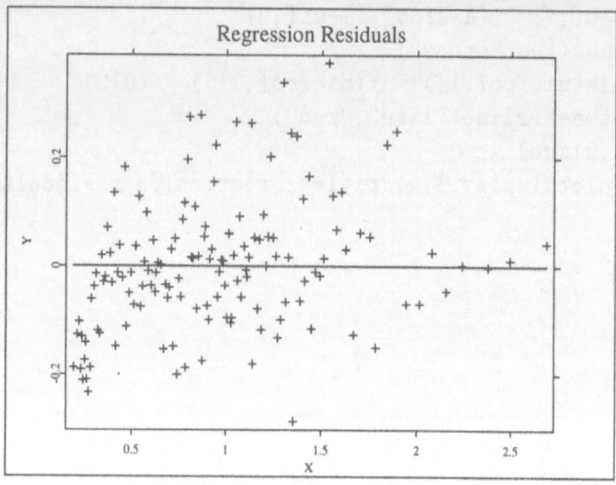

Figure 6.9. Residuals of kernel regression.

6.2.3   Bandwidth Selection

{hcrit,crit} = regbwsel ()
> starts an interactive tool for kernel regression bandwidth selection using the WARPing method

{hcrit,crit} = regxbwsel ()
> starts an interactive tool for kernel regression bandwidth selection using exact computation

As in kernel density estimation, kernel regression involves choosing the kernel function and the bandwidth parameter. One observes the same phenomenon as in kernel density estimation here: The difference between two kernel functions K is almost negligible when the bandwidths are appropriately rescaled. To make the bandwidths for two different kernels K comparable, the same technique as described in Subsection 6.1.3 can be used.

Consequently, we now concentrate on the problem of bandwidth selection. In the regression case, typically the **averaged squared error**

$$\text{ASE}(h) = \sum_{i=1}^{n} \{m(x_i) - \widehat{m}_h(x_i)\}^2 \qquad (6.12)$$

is used as the criterion for goodness of fit. A simple estimator for ASE would be to replace the unknown function values $m(x_i)$ by the observations y_i, which yields the residual sum of squares

$$\text{RSS}(h) = \sum_{i=1}^{n} \{y_i - \widehat{m}_h(x_i)\}^2 .$$

However, in the case of a nonparametric kernel regression, the values $\widehat{m}_h(x_i)$ approach y_i when the bandwidth h approaches zero. A more appropriate estimation of ASE is obtained by the **cross-validation** criterion

$$\text{CV}(h) = \sum_{i=1}^{n} \{y_i - \widehat{m}_{h,-i}(x_i)\}^2 , \qquad (6.13)$$

where $\widehat{m}_{h,-i}(\bullet)$ denotes the kernel regression estimate which is obtained without using the i-th observation (x_i, y_i). Some calculation shows (Härdle, 1990) that $\text{CV}(h)$ can be written as

$$\text{CV}(h) = \sum_{i=1}^{n} \{y_i - \widehat{m}_h(x_i)\}^2 \, \Xi \, (W_{hi}(x_i)) \qquad (6.14)$$

with $\Xi(u) = (1 - u)^{-2}$ and

$$W_{hi}(x_i) = \frac{K(0)}{\sum\limits_{j=1}^{n} K(\frac{x_i - x_j}{h})}.$$

Therefore, the cross-validation can also be interpreted as a residual sum of squares (RSS), where small values for h are penalized by means of including the function Ξ. The function $\Xi(\bullet)$ coincides with the penalty function that Craven and Wahba (1979) proposed for their generalized cross-validation criterion. Table 6.3 lists some more penalty functions.

All the mentioned penalty functions have the same asymptotic properties. In finite samples, however, the functions differ in the relative weight they give

Criterion	Penalty Function
(Generalized) Cross-validation	$\Xi_{GCV}(u) = (1-u)^{-2}$
Shibata's model selector	$\Xi_S(u) = 1 + 2u$
Akaike's Information Criterion	$\Xi_{AIC}(u) = \exp(2u)$
Akaike's Finite Prediction Error	$\Xi_{FPE}(u) = (1+u)/(1-u)$
Rice's T	$\Xi_T(u) = (1-2u)^{-1}$

Table 6.3. Penalizing functions.

to variance and bias of $\widehat{m}_h(x)$. Rice's T gives the most weight to variance reduction while Shibata's model selector stresses bias reduction the most.

In XploRe, all penalizing functions can be applied via the functions `regbwsel` and `regxbwsel`. As can be seen from (6.13), criteria like CV need to be evaluated at all observations. Thus, the function `regxbwsel` which uses exact computations is to be preferred here. `regbwsel` uses the WARPing approximation and may select bandwidths far from the optimal, if the discretization binwidth d large. Note that both `regbwsel` and `regxbwsel` may suffer from numerical problems if the studied bandwidths are too small.

An example for calling `regxbwsel` gives the following quantlet which uses the `nicfoo` data again:

```
nicfoo=read("nicfoo")
tmp=regxbwsel(nicfoo)
```

smoo10.xpl

At first, a selection box opens which offers the choice between the different penalty functions. Let us choose the first item **Cross Validation** which optimizes the CV criterion (6.13). A graphical display appears which shows the CV criterion in th upper left, the chosen optimal bandwidth in the upper right, the resulting kernel regression in the lower left, and information about the search grid and the kernel in the lower right.

Figure 6.10 shows this graphical display. The menu now allows the modification of the search grid and the kernel or the usage of other bandwidth selectors.

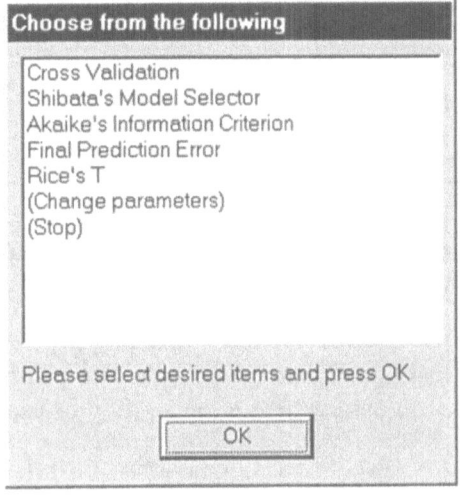

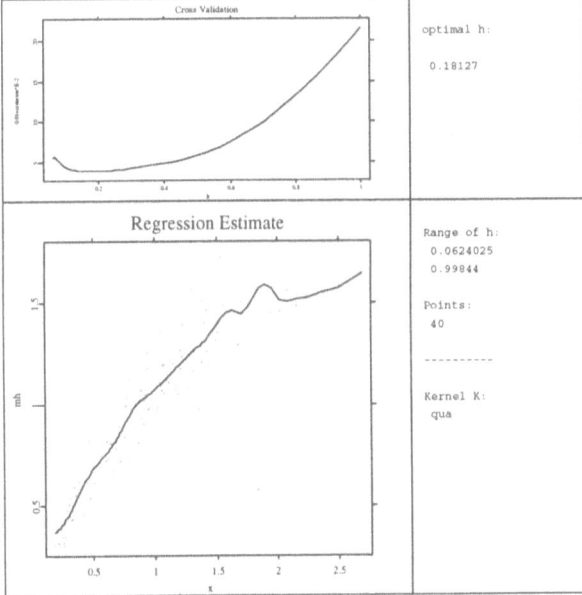

Figure 6.10. Cross-validation bandwidth selector for Netincome vs. Food data.

6.2.4 Confidence Intervals and Bands

{mh, mhl, mhu} = regci (x {,h {,alpha {,K} {,d}}})
computes the kernel density estimate and pointwise confidence
intervals on a grid using the WARPing method

{mh, mhl, mhu} = regcb (x {,h {,alpha {,K} {,d}}})
computes the kernel regression and uniform confidence bands on
a grid using the WARPing method

{mh, mhl, mhu} = regxci (x {,h {,alpha {,K} {,d}}})
computes the kernel density estimate and pointwise confidence
intervals for all observations or on a grid v by exact computation

{mh, mhl, mhu} = regxcb (x {,h {,alpha {,K} {,d}}})
computes the kernel regression and uniform confidence bands for
all observations or on a grid v by exact computation

As in the case of density estimation, it can be shown that the regression estimates $\widehat{m}_h(x)$ have an asymptotic normal distribution. Suppose that m and f
(the density of the explanatory variable X) are twice differentiable, and that
$h = cn^{-1/5}$. Then

$$n^{2/5}\left\{\widehat{m}_h(x) - m(x)\right\} \underset{\text{as.}}{\sim} N\left(\frac{c^2}{2}\, b(x)\, \mu_2(K),\, \frac{1}{c}\frac{\sigma^2(x)}{f(x)}\,\|K\|_2^2\right),\qquad (6.15)$$

with

$$b(x) = m''(x) + \frac{m'(x)f'(x)}{f'(x)}.$$

Thus, similar to the density case, an approximate confidence interval can be
obtained by

$$\left[\widehat{m}_h(x) - z_{1-\frac{\alpha}{2}}\sqrt{\frac{\widehat{\sigma}_h^2(x)\|K\|_2^2}{nh\widehat{f}_h(x)}},\ \widehat{m}_h(x) + z_{1-\frac{\alpha}{2}}\sqrt{\frac{\widehat{\sigma}_h^2(x)\|K\|_2^2}{nh\widehat{f}_h(x)}}\right],\qquad (6.16)$$

where $z_{1-\frac{\alpha}{2}}$ is the $(1-\frac{\alpha}{2})$-quantile of the standard normal distribution and the estimate of the variance $Var(Y|x) = \sigma^2(x)$ is given by

$$\widehat{\sigma}_h^2(x) = \frac{\frac{1}{n}\sum_{i=1}^{n} K\left(\frac{x_i-x}{h}\right) \{y_i - \widehat{m}_h(x)\}^2}{\sum_{i=1}^{n} K\left(\frac{x_i-x}{h}\right)}.$$

Also similar to the density case, uniform **confidence bands** for $m(\bullet)$ need rather restrictive assumptions (Härdle, 1990, p. 116). Suppose that f is a density on $[0,1]$ and $h = n^{-\delta}$, $\delta \in (\frac{1}{5}, \frac{1}{2})$. Then it holds under some regularity for all $x \in [0,1]$:

$$P\left(\widehat{m}_h(x) - z_{n,\alpha}\sqrt{\frac{\widehat{\sigma}_h^2(x)\|K\|_2^2}{nh\widehat{f}_h(x)}}\right.$$

$$\left. \leq m(x) \leq \widehat{m}_h(x) + z_{n,\alpha}\sqrt{\frac{\widehat{\sigma}_h^2(x)\|K\|_2^2}{nh\widehat{f}_h(x)}}\right) \underset{\text{as.}}{\approx} 1 - \alpha,$$

where

$$z_{n,\alpha} = \left\{\frac{-\log\{-\frac{1}{2}\log(1-\alpha)\}}{(2\delta\log n)^{1/2}} + d_n\right\}^{1/2},$$

$$d_n = (2\delta\log n)^{1/2} + (2\delta\log n)^{-1/2}\log\left(\frac{1}{2\pi}\frac{\|K'\|_2}{\|K\|_2}\right)^{1/2}.$$

Pointwise confidence intervals and uniform confidence bands using the WARPing approximation are provided by `regci` and `regcb`, respectively. The equivalents for exact computations are `regxci` and `regxcb`. The functions `regcb` and `regxcb` can be directly applied to the original data $x_1, \ldots, x_n$, the transformation to $[0,1]$ is performed internally. The following quantlet code extends the above regression function by confidence intervals and confidence bands:

```
{mh,mli,mui}=regci(nicfoo,0.18)  ; intervals
{mh,mlb,mub}=regcb(nicfoo,0.18)  ; bands
mh =setmask(mh,"line","blue","thick")
mli=setmask(mli,"line","blue","thin","dashed")
mui=setmask(mui,"line","blue","thin","dashed")
mlb=setmask(mlb,"line","blue","thin")
mub=setmask(mub,"line","blue","thin")
```

```
plot(mh,mli,mui,mlb,mub)
setgopt(plotdisplay,1,1,"title","Confidence Intervals & Bands")
```
🔍 smoo11.xpl

Both confidence intervals and bands were computed using bandwidth 0.2. The significance level and kernel are not specified, as the defaults 0.05 and "qua" are used. The resulting plot is shown in Figure 6.11.

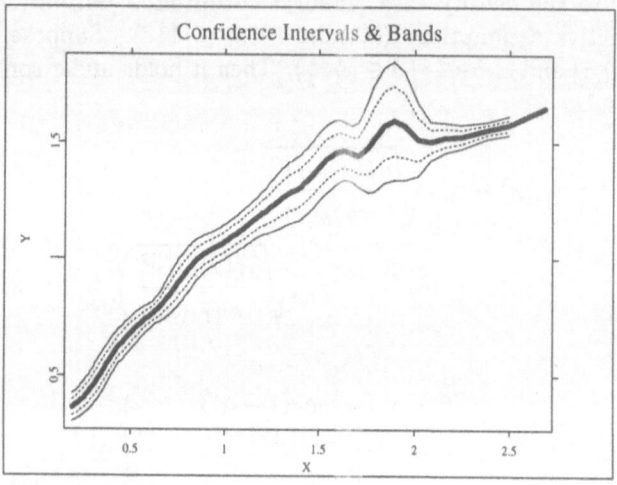

Figure 6.11. Pointwise confidence intervals (thin dashed) and uniform confidence (thin solid) bands in regression.

6.2.5   Local Polynomial Regression and Derivative Estimation

---

mh = **lpregest** (x {,h {,p {,K} {,d}}})
> computes the local polynomial kernel regression on a grid using the WARPing method

mh = **lpregxest** (x {,h {,p {,v}}})
> computes the local polynomial kernel regression estimate for all observations or on a grid v by exact computation

---

Note that the Nadaraya–Watson estimator is a local constant estimator, i.e. the solution of

$$\min_{\beta_0} \sum_{i=1}^{n} \{y_i - \beta_0\}^2 \, K\left(\frac{x_i - x}{h}\right).$$

Replacing β_0 by a p-th order polynomial in $x_i - x$ yields the **local polynomial kernel regression** estimator. For details, see Fan and Gijbels (1996). Let us illustrate this with the example of a local linear regression estimate ($p = 1$). The minimization problem is

$$\min_{\beta_0,\beta_1} \sum_{i=1}^{n} \{y_i - \beta_0 - \beta_1(x_i - x)\}^2 \, K\left(\frac{x_i - x}{h}\right).$$

The solution of the problem can be written as

$$\hat{\beta} = (\hat{\beta}_0, \hat{\beta}_1)^T = \left(\mathcal{X}^T \mathcal{W} \mathcal{X}\right)^{-1} \mathcal{X}^T \mathcal{W} \mathcal{Y} \tag{6.17}$$

using the notations

$$\mathcal{X} = \begin{pmatrix} 1 & (x_1 - x) \\ \vdots & \vdots \\ 1 & (x_n - x) \end{pmatrix}, \quad \mathcal{Y} = \begin{pmatrix} y_1 \\ \vdots \\ y_n \end{pmatrix},$$

and $\mathcal{W} = \mathrm{diag}\left\{K\left(\frac{x_1-x}{h}\right), \ldots, K\left(\frac{x_n-x}{h}\right)\right\}$. In (6.17), $\hat{m}_h(x) = \hat{\beta}_0$ estimates the regression function itself, whereas $\hat{m}_{1,h}(x) = \hat{\beta}_1$ estimates the first derivative of the regression function.

The functions `lpregest` and `lpregxest` for local polynomial regression have essentially the same input as their Nadaraya–Watson equivalents, except that an additional parameter p to specify the degree of the polynomial can be given. For local polynomial regression, an odd value of p is recommended since odd-order local polynomial regressions outperform even-order local polynomial regressions.

Derivatives of regression functions are computed with `lpderest` or `lpderxest`. For derivative estimation a polynomial order whose difference to the derivative order is odd should be used. Typically one uses the $p = 1$ (local linear) for the estimation of the regression function and $p = 2$ (local quadratic) for the estimation of its first derivative.

`lpdregxest` and `lpderxest` use automatically local linear and local quadratic estimation if no order is specified. The default kernel function is the Quartic kernel "qua". Appropriate bandwidths can be found by means of rule of thumbs that replace the unknown regression function by a higher-order polynomial (Fan and Gijbels, 1996). The following quantlet code estimates the regression function and its first derivative by the local polynomial method. Both functions and the data are plotted together in Figure 6.12.

```
motcyc=read("motcyc")
hh=lpregrot(motcyc)        ; rule-of-thumb bandwidth
hd=lpderrot(motcyc)        ; rule-of-thumb bandwidth
mh=lpregest(motcyc,hh)     ; local linear regression
md=lpderest(motcyc,hd)     ; local quadratic derivative
mh=setmask(mh,"line","black")
md=setmask(md,"line","blue","dashed")
xy=setmask(motcyc,"cross","small","red")
plot(xy,mh,md)
setgopt(plotdisplay,1,1,"title","Local Polynomial Estimation")
```
                                                              smoo12.xpl

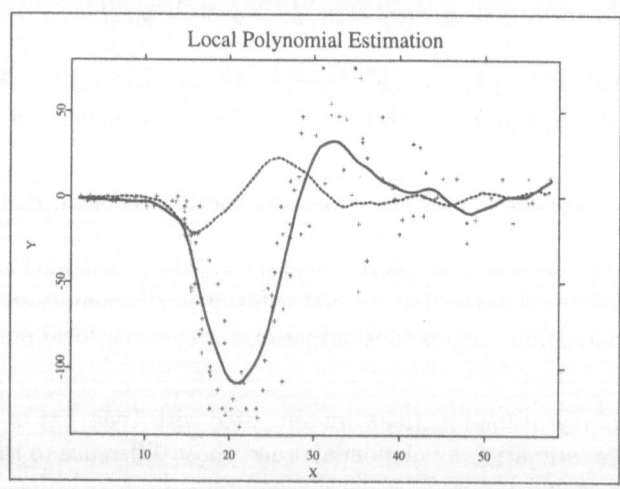

Figure 6.12. Local linear regression (solid), derivative (dashed) estimate and data.

6.3 Multivariate Density and Regression Functions

In this section we review kernel smoothing methods for density and regression function estimation in the case of multidimensional variables X.

6.3.1 Computational Aspects

As in the univariate case, density and regression functions can be estimated by exact computation or by WARPing approximation. However, the effect of WARPing is different in the multivariate case. WARPing is still relatively fast in the two-dimensional case. For three- and higher-dimensional estimates, exact estimation may be preferred. To have a choice between both the exact and the WARPing computation, all estimation routines are offered in two versions:

Functionality	Exact	WARPing
density estimation	`denxestp`	`denestp`
Nadaraya–Watson regression	`regxestp`	`regestp`
local linear regression	`lregxestp`	`lregestp`

6.3.2 Multivariate Density Estimation

```
hrot = denrotp (x {,K {,opt}})
       computes a rule-of-thumb bandwidth for multivariate density es-
       timation

fh = denestp (x {,h {,K} {,d}})
       computes the multivariate kernel density estimate on a grid using
       the WARPing method

fh = denxestp (x {,h {,K} {,v}})
       computes the multivariate kernel density estimate for all obser-
       vations or on a grid v by exact computation
```

The kernel density estimator can be generalized to the multivariate case in a straightforward way. Suppose we now have observations $x_1, \ldots, x_n$ where each of the observations is a d-dimensional vector $x_i = (x_{i1}, \ldots, x_{id})^T$. The multivariate kernel density estimator at a point $x = (x_1, \ldots, x_d)^T$ is defined as

$$\widehat{f_h}(x) = \frac{1}{n} \sum_{i=1}^{n} \frac{1}{h_1 \ldots h_d} \mathcal{K}\left(\frac{x_{i1} - x_1}{h_1}, \ldots, \frac{x_{id} - x_d}{h_d}\right), \qquad (6.18)$$

with $\mathcal{K}$ denoting a multivariate kernel function, i.e. a function working on d-dimensional arguments. Note that (6.18) assumes that the bandwidth h is a vector of bandwidths $h = (h_1, \ldots, h_d)^T$.

What form should the multidimensional kernel function $\mathcal{K}(u) = \mathcal{K}(u_1, \ldots, u_d)$ take on? The easiest solution is to use a multiplicative or product kernel

$$\mathcal{K}(u) = K(u_1) \cdot \ldots \cdot K(u_p)$$

with K denoting an univariate kernel function. This means if K is a univariate kernel with support $[-1, 1]$ (e.g. the Quartic kernel), observations in a cube around x are used to estimate the density at the point x. An alternative is to use a genuine multivariate kernel function $\mathcal{K}(u)$, e.g. the radial symmetric Quartic kernel

$$\mathcal{K}(u) \propto (1 - u^T u)^2 \ I(u^T u \leq 1).$$

Radial symmetric kernels can be obtained from univariate by defining $\mathcal{K}(u) \propto K(\|u\|)$, where $\|u\| = \sqrt{u^T u}$ denotes the Euclidean norm of the vector u. $\propto$ indicates that the appropriate constant has to be multiplied. Radial symmetric kernels use observations from a ball around x to estimate the density at x. Table 6.4 shows which product and which radial symmetric kernel functions are available in XploRe.

Kernel	Product	Radial symmetric
Uniform	uni	runi
Triangle	trian	rtrian
Epanechnikov	epa	repa
Quartic	qua	rqua
Triweight	tri	rtri
Gaussian	gau	gau

Table 6.4. Radial symmetric kernel functions.

The following quantlet computes a two-dimensional density estimate for the geyser data (see Appendix B.11). These are two-dimensional data featuring a bimodal density. The function **denxestp** can be called with only the data as input. In this case, the bandwidth vector is computed by Scott's rule (Scott, 1992). This rule of thumb is also separately implemented in **denrotp**. The default kernel function is the product Quartic kernel "qua". The resulting surface plot is shown in Figure 6.13.

```
geyser = read("geyser")
fh = denxestp(geyser)
fh = setmask(fh,"surface","blue")
axesoff()
cu = grcube(fh)              ; box
plot(cu.box,cu.x,cu.y, fh)   ; plot box and fh
setgopt(plotdisplay,1,1,"title","2D Density Estimate")
axeson()
```

Q smoo13.xpl

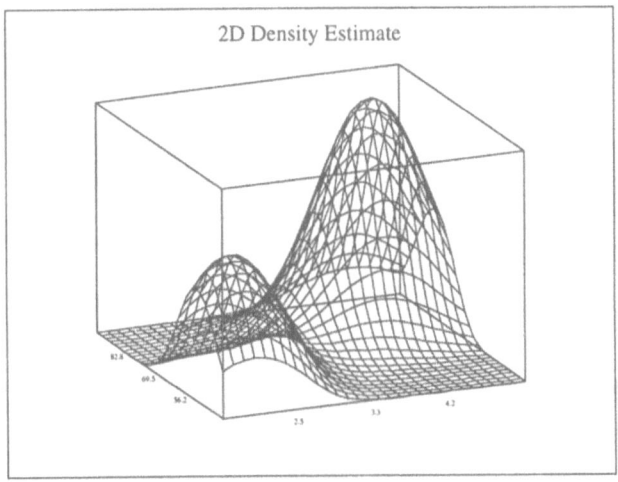

Figure 6.13. Two-dimensional density estimate.

The second example of this subsection shows a three-dimensional density estimate. This estimate can only be graphed in the form of a contour plot. See Chapter 3 for an introduction to contour plots. The estimated data are

columns 4 to 6 of the bank2 data (see Appendix B.7). This data set consists of
two clusters which can be easily detected from the contour plot in Figure 6.14.

```
bank    = read("bank2.dat")
bank456 = bank[,4:6]                    ; columns 4 to 6
fh = denxestp(bank456,1.5)
axesoff()
fhr = (max(fh[,4])-min(fh[,4]))         ; range of fh
cf1= grcontour3(fh,0.4*fhr,2)           ; contours
cf2= grcontour3(fh,0.6*fhr,4)           ; contours
cu = grcube(cf1|cf2)                    ; box
plot(cu.box, cf1,cf2)                   ; graph contours
setgopt(plotdisplay,1,1,"title","3D Density Estimate")
axeson()
```

Q smoo14.xpl

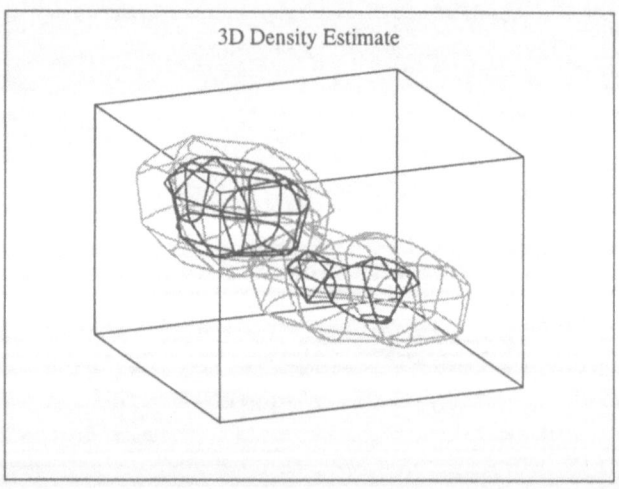

Figure 6.14. Contours of three-dimensional density estimate.

6.3.3   Multivariate Regression

```
mh = regestp (x {,h {,K} {,d}})
     computes the multivariate kernel regression on a grid using the
     WARPing method

mh = regxestp (x {,h {,K} {,v}})
     computes the multivariate kernel regression for all observations
     or on a grid v by exact computation

mh = lregestp (x {,h {,K} {,d}})
     computes the multivariate local linear kernel regression on a grid
     using the WARPing method

mh = lregxestp (x {,h {,K} {,v}})
     computes the multivariate local linear kernel regression for all
     observations or on a grid v by exact computation
```

Multivariate nonparametric regression aims to estimate the functional relation between a univariate response variable Y and a d-dimensional explanatory variable X, i.e. the conditional expectation

$$E(Y|X) = E(Y|X_1, \ldots, X_d) = m(X).$$

The multivariate Nadaraya–Watson estimator can then be written as a generalization of the univariate case. Suppose that we have independent observations $(x_1, y_1), \ldots, (x_n, y_n)$, then this estimator is defined as

$$\widehat{m}_h(x) = \frac{\sum\limits_{i=1}^{n} \mathcal{K}\left(\dfrac{x_{i1} - x_1}{h_1}, \ldots, \dfrac{x_{ip} - x_p}{h_p}\right) y_i}{\sum\limits_{i=1}^{n} \mathcal{K}\left(\dfrac{x_{i1} - x_1}{h_1}, \ldots, \dfrac{x_{ip} - x_p}{h_p}\right)}.$$

As in the univariate case, local polynomial approaches can be used. Due to the computational complexity one computes typically only local linear estimates.

The following quantlet compares the two-dimensional Nadaraya–Watson and the two-dimensional local linear estimate for a generated data set. For the bandwidth vector and the kernel function, we accept the defaults which are 20% of the range of the data and the product Quartic kernel "qua", respectively. Figure 6.15 shows the surface plots of both estimates.

```
randomize(0)
n=200
x=uniform(n,2)
m=sin(2*pi*x[,1])+x[,2]
y=m+normal(n)/4
mh= regestp(x~y)
ml=lregestp(x~y)
mh=setmask(mh,"surface","red")
ml=setmask(ml,"surface","blue")
c=grcube(mh)
d=createdisplay(1,2)
axesoff()
show(d,1,1,mh,c.box,c.x,c.y)
show(d,1,2,ml,c.box,c.x,c.y)
axeson()
setgopt(d,1,1,"title","Nadaraya-Watson")
setgopt(d,1,2,"title","Local Linear")
```

smoo15.xpl

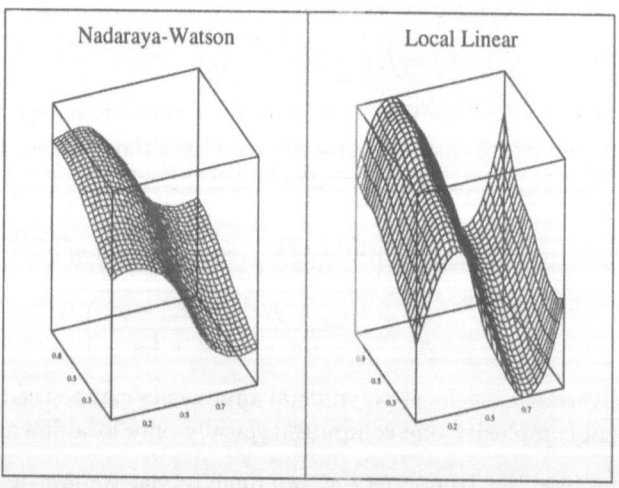

Figure 6.15. Bivariate Nadaraya–Watson and local linear estimate.

Bibliography

Bickel, P. and Rosenblatt, M. (1973). On some global measures of the deviations of density function estimators, *Annals of Statistics* **1**: 1071–1095.

Cao, R., Cuevas, A. and González-Manteiga, W. (1992). A comparative study of several smoothing methods in density estimation. manuscript.

Craven, P. and Wahba, G. (1979). Smoothing noisy data with spline functions, *Numerische Mathematik* **31**: 377–403.

Fan, J. and Gijbels, I. (1996). *Local Polynomial Modelling and Its Applications*, Vol. 66 of *Monographs on Statistics and Applied Probability*, Chapman and Hall, New York.

Härdle, W. (1990). *Applied Nonparametric Regression*, Econometric Society Monographs No. 19, Cambridge University Press.

Härdle, W. (1991). *Smoothing Techniques, With Implementations in S*, Springer, New York.

Härdle, W. and Scott, D. (1992). Smoothing in by weighted averaging using rounded points, *Computational Statistics* **7**: 97–128.

Jones, M. C., Marron, J. S. and Sheather, S. J. (1992). Progress in data–based bandwidth selection for kernel density estimation. manuscript.

Marron, J. (1989). Comments on a data based bandwidth selector, *Computational Statistics & Data Analysis* **8**: 155–170.

Marron, J. S. and Nolan, D. (1988). Canonical kernels for density estimation, *Statistics & Probability Letters* **7**(3): 195–199.

Park, B. U. and Turlach, B. A. (1992). Practical performance of several data driven bandwidth selectors, *Computational Statistics* **7**: 251–270.

Scott, D. W. (1992). *Multivariate Density Estimation: Theory, Practice, and Visualization*, John Wiley & Sons, New York, Chichester.

Silverman, B. W. (1986). *Density Estimation for Statistics and Data Analysis*, Vol. 26 of *Monographs on Statistics and Applied Probability*, Chapman and Hall, London.

Turlach, B. A. (1993). Bandwidth selection in kernel density estimation: A review, *Discussion Paper 9307*, Institut für Statistik und Ökonometrie, Humboldt-Universität zu Berlin.

Wand, M. P. and Jones, M. C. (1995). *Kernel Smoothing*, Vol. 60 of *Monographs on Statistics and Applied Probability*, Chapman and Hall, London.

7 Generalized Linear Models

Marlene Müller

McCullagh and Nelder (1989) summarized many approaches to relax the distributional assumptions of the classical linear model under the common term **Generalized Linear Models** (GLM). A generalized linear model (GLM) is a regression model of the form

$$EY = G(x^T \beta),$$

where EY denotes the expected value of the dependent variable Y, x is a vector of explanatory variables, β an unknown parameter vector and $G(\bullet)$ a known link function.

An essential feature of the GLM is that the expectation $\mu = EY$ is directly dependent on a function of the index $\eta = x^T \beta$. Additionally, one assumes that $\text{Var}(Y) = \sigma^2 V(\mu)$. The function G which relates μ and η is called the **link function**. (Note that McCullagh and Nelder (1989) actually denote G^{-1} as the link function.)

It is easy to see that GLM covers a range of widely used models, e.g.

- **Linear regression (OLS)**
  The model
  $$Y = x^T \beta + \varepsilon, \quad \varepsilon \sim N(0, \sigma^2)$$
  implies that
  $$EY = x^T \beta, \quad \text{Var}(Y) = \sigma^2.$$
  Hence the classical linear model falls into the GLM framework with an identity link function and a variance function $V(\mu) = 1$.

- **Binary response models (Logit, Probit)**
  The probability $P(Y = 1)$ for Bernoulli distributed Y is identical to the expectation EY. Hence Logit or Probit models
  $$P(Y = 1) = G(x^T \beta) = \mu,$$

where $G(\bullet)$ is the Logistic or Gaussian distribution function, can be estimated within the GLM framework as well. The variance function is $V(\mu) = \mu(1-\mu)$ in this case.

7.1   Estimating GLMs

It is known that the least squares estimator $\widehat{\beta}$ in the classical linear model coincides with the maximum-likelihood estimator under the imposed normal distribution. By using appropriate distributional assumptions for Y in GLM, one may stay in the framework of maximum-likelihood in this case.

7.1.1   Models

For maximum-likelihood estimation, one assumes that the distribution of Y belongs to an **exponential family**. Exponential families cover a broad range of distributions, for example discrete as the Binomial and Poisson distribution or continuous as the Gaussian (normal) and Gamma distribution.

A distribution is said to belong to an exponential family if its probability function (if Y discrete) or its density function (if Y continuous) has the structure

$$f(y, \theta, \phi) = \exp\left\{ \frac{y\theta - b(\theta)}{a(\phi)} + c(y, \phi) \right\}, \tag{7.1}$$

with some special functions $a(\bullet)$, $b(\bullet)$ and $c(\bullet)$. These functions vary for the distributions contained in this model class.

Generally speaking, we are interested in estimating $\theta = \theta(x^T\beta)$, the **canonical parameter**. ϕ is a **nuisance** parameter (as the variance σ^2 in linear regression for example). Apart from the distribution of Y, the link function is another essential part of the generalized linear model. Recall the notations

$$\eta = x^T\beta \quad \text{and} \quad \mu = G(\eta).$$

For each distribution, one special link function exists, namely if

$$x^T\beta = \eta = \theta.$$

If this holds, the link function is called the **canonical link** function. For models with a canonical link, some theoretical and practical problems are easier

Notation	Range of y	$b(\theta)$	$\mu(\theta)$	Canonical link $\theta(\mu)$	Variance $V(\mu)$	$a(\phi)$
Normal $N(\mu, \sigma^2)$	$(-\infty, \infty)$	$\theta^2/2$	θ	identity	1	σ^2
Poisson $P(\mu)$	$[0, \infty)$ integer	$\exp(\theta)$	$\exp(\theta)$	log	μ	1
Binomial $B(m, \pi)$	$[0, m]$ integer	$m \log(1 + e^\theta)$	$\dfrac{e^\theta}{1 + e^\theta}$	logit	$m\pi(1 - \pi)$	1
Gamma $G(\mu, \nu)$	$(0, \infty)$	$-\log(-\theta)$	$-1/\theta$	reciprocal	μ^2	$1/\nu$
Inverse Gaussian $IG(\mu, \sigma^2)$	$(0, \infty)$	$-(-2\theta)^{1/2}$	$\dfrac{-1}{\sqrt{(-2\theta)}}$	squared reciprocal	μ^3	σ^2
Negative Binomial $NB(\mu, k)$	$[0, \infty)$ integer	$\dfrac{-\log\left(1 - e^\theta\right)}{k}$	$\dfrac{e^\theta}{k(1 - e^\theta)}$	$\log\left(\frac{k\mu}{1+k\mu}\right)$	$\mu + k\mu^2$	1

Table 7.1. Distribution implemented in GLM.

to solve. Table 7.1 summarizes characteristics for some exponential functions together with canonical parameters θ and their canonical link functions. Note that the Negative Binomial distribution only fits into the framework described above if we assume that the parameter k is known.

7.1.2   Maximum-Likelihood Estimation

All models in the `glm` library are estimated by maximum-likelihood. The default numerical algorithm is the Newton–Raphson iteration (except for ordinary regression where no iteration is necessary). Optionally, a Fisher Scoring can be chosen, which uses the expectation of the Hessian matrix instead of the Hessian itself. In the case of a canonical link function, the Newton–Raphson algorithm and the Fisher scoring algorithm coincide.

7.2   Computing GLM Estimates

Currently six types of distributions are supported by the `glm` library: Binomial, Normal (Gaussian), Poisson, Gamma (includes Exponential), Inverse Gaussian and Negative Binomial (includes Geometric).

The functions in the `glm` library which are mainly responsible for GLM estimation are `doglm` (interactive, menu controlled) and `glmest` (noninteractive). We will explain `doglm` in Subsection 7.2.2 and `glmest` in Subsection 7.2.3.

7.2.1   Data Preparation

The estimation functions in the `glm` library expect at least two input parameters: A matrix x containing the observations of the explanatory variables and a vector y containing the observed responses.

The vector y should have n rows, each corresponding to one observation. The matrix x should have n rows and p columns, i.e. the rows correspond to the individual observations and the columns to the variables.

A nx1 vector of 1 can be concatenated to x to allow for a constant in the model:

```
x = matrix(rows(x))~x
```

This is not necessary for the interactive estimation by `doglm` (see Subsection 7.2.2).

Neither the matrix x nor the vector y should contain missing values (`NaN`) or infinitesimal values (`Inf,-Inf`). Those should be identified by `isNumber` and removed by `paf` (or replaced by something) before the GLM estimation.

7.2.2 Interactive Estimation

```
doglm (x, y {, opt})
        starts the interactive GLM estimation tool
```

To have a "real" example let us generate some pseudo random data. Type at the command line or in an editor window

```
randomize(0)
n=100
b=1|2
p=rows(b)
x=2.*uniform(n,p)
y=x*b+normal(n)./2
```

Q glm01.xpl

Do not forget to call the glm library:

```
library("glm")
```

The interactive estimation with doglm is simply invoked by

```
doglm(x,y)
```

A selection box appears which starts the interactive fitting procedure.

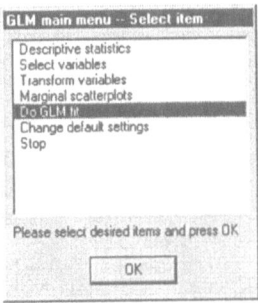

The first three items (Descriptive statistics, Select variables and Transform variables) offer some possibilities to select and manipulate the matrix x. We choose

directly to start the fit (select Do GLM fit). The next selection box presents
the available distributions for y.

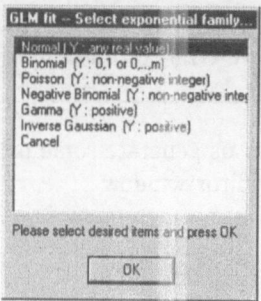

Simply select the appropriate distribution. Depending on the choice of the
distribution a second selection box will ask for the link function. The following
will be shown if Normal was selected for the distribution.

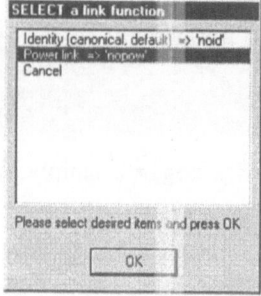

Here it is possible to press just OK to accept the canonical link function. Codes
such as noid and nopow in the above selection box point to the short codes for
the models, see Subsection 7.2.3.

Again, depending on the choice, a third section box may appear which offers
to change control parameters for the selected model. Suppose we have pressed
the button for Power link. Then we will be asked if we want to change the
power parameter:

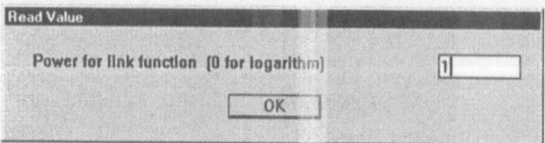

Note that the power corresponds to the inverse link function. Thus when power 0 is chosen, G is in fact the exponential function. Power 0.5 makes G the quadratic function, consequently. If we select power 1, we get the identity link.

We choose a power of 1 at this point. (We could have chosen this one selection box earlier by selecting the identity link already!) The underlying estimation XploRe function realizes that for the ordinary least squares regression no iteration needs to be performed and shows the estimation result quite immediately. A graphical output display appears.

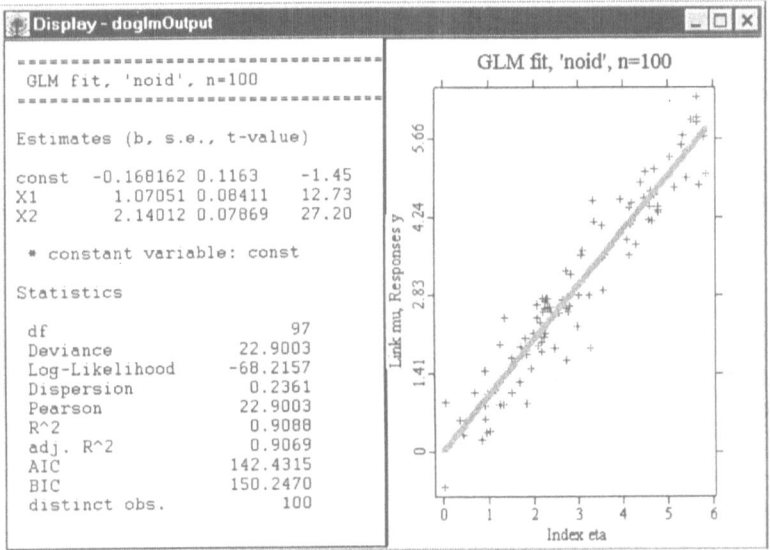

The output display shows in the left panel the estimation results. The chosen model (here **noid** which is **no**rmal with **id**entity link) is recalled in the headline together with the sample size. The upper table gives the estimated coefficient vector b together with the estimated standard errors and t-values. **doglm** includes automatically a constant in the model, hence the first component of b relates to this constant.

The lower table gives some statistics for this fit, such as the degrees of freedom (**df**), the deviance, Pearson's χ^2 statistic, the coefficient of determination R^2, an Akaike criterion, and the number of distinct rows in x. The right panel shows a plot of the index x*b vs. y (red +) and a plot of x*b vs. the predicted regression function (green line). The latter is just a straight line here since the identity link function was chosen.

Besides the output display `doglmOutput`, two additional output objects are produced (after stopping `doglm`):

`doglm`
> A list, which contains the list elements
>
> `doglm.b`
>> the parameter estimates,
>
> `doglm.bv`
>> the estimated covariance,
>
> `doglm.stat`
>> statistics (itself a list containing the different statistics as components, see Subsection 7.5.1).

`doglmtxt`
> A string vector, which contains the contents of the left panel of the output display.

The estimation in `doglm` can be controlled by a number of options and parameters. All of them can be set beforehand by defining a list of options `opt` with `glmopt`. For example, with

```
opt=glmopt("code","noid")
doglm(x,y,opt)
```

we would specify the model for `doglm` and are not asked interactively anymore. The `glmopt` and how to set up optional parameters is shown in Section 7.4.

Alternatively we can set all optional parameters interactively from the main menu of `doglm`:

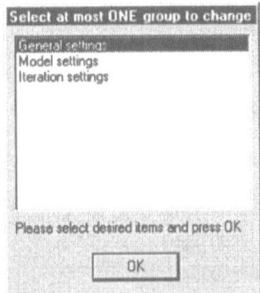

The default settings are grouped into three fields. With selecting General settings, one can change general properties (intercept, search for replications, name for output to store several estimated models etc.). Model settings allow us to specify the model for the estimation and possible additional model parameters (e.g. power for power link). Finally Iteration settings covers everything for tuning the iterative estimation process. Note: Some of the settings are not applicable for some models (e.g. a power can be specified, although a model without power link can be chosen). Those settings are simply ignored.

7.2.3 Noninteractive Estimation

```
g = glmest (code, x, y {, opt})
        estimates a GLM noninteractively
```

The function `glmest` provides a noninteractive way to estimate a GLM. This is useful for using GLM estimates as pilot estimates for other procedures. The standard call is quite simple, for example

```
g=glmest("noid",x,y)
```

estimates the normal model with identity link. For `glmest` the short code of the model (e.g. `"noid"`) needs always to be given, it is not an optional parameter as for `doglm`. The following models are supported:

Binomial

"bilo" Logistic link (Logit, canonical)

"bipro" Gaussian link (Probit)

"bicll" complementary log-log link

Poisson

"polog" Poisson with logarithm (inverse) link (canonical)

"popow" Poisson with power (inverse) link

Gamma

"gacl" reciprocal (inverse) link (canonical)

"gapow" power (inverse) link

Inverse Gaussian

"igcl" squared reciprocal (inverse) link (canonical)

"igpow" power (inverse) link

Negative Binomial

"nbcl" canonical link

"nbpow" power (inverse) link

The list of all available models can be consulted by typing glmmodels.all at the command line when the glm library is loaded. Let us go back to the example from the previous section and replace doglm by the noninteractive function glmest. The XploRe codes for this subsection can be found in the quantlet ◘ glm02.xpl:

```
g=glmest("noid",matrix(rows(x))~x,y)
```

Here, a vector of 1 is appended to x to include a constant in the model. The result of the estimation is assigned to the variable g. Thus g is a list containing the output:

g.b

the estimated parameter vector

g.bv
  the estimated covariance of **g.b**

g.stat
  contains the statistics (see Subsection 7.5.1).

In our running example **g.b** presents in the XploRe output window:

```
Contents of b
[1,] -0.16816
[2,]  1.0705
[3,]  2.1401
```

A graphical output can be created from the result of a noninteractive GLM estimation as well. This is done by

```
glmout("noid",matrix(rows(x))~x,y,g.b,g.bv,g.stat)
```

in the current example. For more features of **glmout**, see Subsection 7.5.2.

Optional parameters must be given to **glmest** in a list of optional parameters. A detailed description of what is possible can be found in Section 7.4. For an illustration, recall our first estimation attempt with **doglm**. There we had chosen interactively the power parameter for the power link. In noninteractive estimation, this can be done by

```
opt=glmopt("pow",1)
g=glmest("nopow",matrix(rows(x))~x,y,opt)
```

The first statement creates a list **opt** containing the parameter **opt.pow** with value 1. The next line passes this option list **opt** to the estimation routine **glmest**.

7.3 Weights & Constraints

The estimation functions **glmest** and **doglm** are able to handle special cases as prior weights, replications (automatic search for replications) and constraints on parameters (fix parameters). We will demonstrate this by means of a second running example.

7.3.1   Prior Weights

Consider the lines

```
x = read("lizard")
x = paf(x,(x[,6]!=0))
y = x[,5]
m = x[,6]
x = matrix(rows(x))~x[,1:3]~(x[,4]==1)~(x[,4]==2)
```

🔍 glm03.xpl

which read the lizard data (McCullagh and Nelder, 1989) containing the daytime habits of the two species of lizards (see Appendix B.7). Columns 1 to 4 are the explanatory variables x. Note that we create dummies from x[,4] and include a constant by adding matrix(rows(x)). The fifth column (the dependent variable y) is the number of grahami lizards who decided for a certain site. Finally, the sixth column of this dataset contains the total number of lizards who decided for the site. Hence, the sixth column is a prior weight vector for this data. Note that it is just the binomial index vector which can also be interpreted as the number of replications in x.

We proceed as follows in the estimation here:

```
opt=glmopt("wx",m)
g=glmest("bilo",x,y,opt)
```

🔍 glm03.xpl

The code "bilo" stands for the logit model (**binomial** with **lo**git linkfunction). The list opt of optional parameters now contains the weight vector as component opt.wx. The resulting vector of estimated coefficients is

```
Content of object g.b
[1,]   1.94469
[2,]   1.12999
[3,]  -0.76263
[4,]  -0.84727
[5,]   0.22711
[6,]  -0.73681
```

An output display can be obtained manually in this case, see Subsection 7.5.2.

7.3.2 Replications in Data

There are two types of replications (or ties) that can occur:

- replications in x with different y values,

- replications in x and y jointly.

The first case (replications in x with different y values) can be directly handled only in Binomial models. In this case, the matrix should contain the different realizations of the explanatory variables, the response vector y should contain the summed response values corresponding to each line of x. The number of replications needs to be given as prior weights opt.wx in the list of optional parameters opt.

For models other than the Binomial, the original matrix x (with multiple entries) and the original response vector y should be given to **glmest** or **doglm**. Both will search automatically for replications in data.

The second case (replications in x and y jointly) can simply be handled by giving the number of replications as component opt.wx in the list of optional parameters opt. opt.wx may be a single number if the number of replications in x and y is constant. For an example, see the previous Subsection 7.3.1.

7.3.3 Constrained Estimation

Returning to the running lizard example of Subsection 7.3.1, the resulting vector of estimated coefficients is

```
Content of object g.b
[1,]   1.94469
[2,]   1.12999
[3,]  -0.76263
[4,]  -0.84727
[5,]   0.22711
[6,]  -0.73681
```

Suppose now that we are interested in estimating the same model under the constraint that the last coefficient equals exactly –1. The new estimate could be used in hypothesis testing.

This type of constrained estimation can be done by introducing an offset into the model and omitting the last variable instead. Continuing with

```
offset=-x[,6]
opt=glmopt("off",offset,opt)
c=glmest("bilo",x[,1:5],y,opt)
```

Q glm04.xpl

will now yield the estimated coefficients under the constraint $b_6 = -1$:

```
Content of object c.b
[1,]   1.56509
[2,]   1.05173
[3,]  -0.72229
[4,]  -0.76795
[5,]   0.52239
```

7.4 Options

```
opt = glmopt (string1, value1, ...  {, opt})
        creates a list of options for GLM estimation or appends options
        to an existing list
```

Options for the algorithm and optional parameters should be collected in a list object. This allows us to set or to modify those options which are necessary. Almost all functions in the glm library allow options. It is possible to give the same list of options to different functions. For example,

```
opt=glmopt("miter",20,"name","MyDisplay")
```

will set the maximal number of iteration to 20 and the name of the output display to MyDisplay. Now, one can call first glmest and then glmout with the list opt:

```
l=glmest("bilo",x,y,opt)
glmout("bilo",x,y,l.b,l.bv,l.stat,opt)
```

Both `glmest` and `glmout` only consider those optional parameters which are intended for them. Hence `glmest` will only care about `miter` and `glmout` will present a display with the title `MyDisplay`.

7.4.1 Setting Options

Principally, it is possible to define the list of optional parameters with the XploRe command `list`. However, it is recommended to use the `glmopt` tool to set the options. The first call of `glmopt` will create a list of options. To append a further component to `opt`, we have to repeat the name `opt` as the last argument of `glmopt`. For example,

```
opt=glmopt("miter",20,"title","MyDataset")
opt=glmopt("name","MyDisplay",opt)
```

creates the list `opt` with a component `miter` containing the value 20, a component `title` containing the string `"MyDataset"` and a component `name` containing the string `"MyDisplay"`. The resulting list has three components as one can check with `names(opt)`.

The next sections will explain which options can be used in GLM estimation.

7.4.2 Weights and Offsets

Prior weights and offsets can always be given as an optional parameter. The corresponding components of the list of optional parameters are

`wx`
> weights, nx1 vector or scalar. Default is `wx=1`.

`off`
> offset, nx1 vector or scalar. Default is `off=1`.

Neither of both parameters should contain missing or infinite values.

7.4.3 Control Parameters

There is a number of control parameters which modify the used algorithm:

cnv

> convergence criterion. The iteration stops when the relative change of the coefficients vector b or the relative change in deviance is less than cnv. Default is cnv=0.0001. This parameter is ignored in noniterative estimation (model code "noid").

miter

> maximal number of iterations. The iteration stops when this maximal number of iterations is reached. Default is miter=10. This parameter is ignored in noniterative estimation (model code "noid").

fscor

> indicator for Fisher scoring (instead of Newton–Raphson optimization). fscor=1 means that the Fisher scoring is used. Default is fscor=0 for Newton–Raphson. This parameter is ignored for canonical link functions.

norepl

> norepl=1 forces not to search for replications in x. Default is norepl=0, i.e. to search for replications.

The following parameters switch on/off information during the computation.

shf

> shows how the iteration proceeds, if shf=1 set. Default is shf=0.

shm

> shows how the model selection proceeds, if shm=1 set. Default is shm=0. This parameter is only recognized for model selection.

The model selection functions glmselect, glmforward and glmbackward also provide two extra parameters:

crit

> single string, "aic" or "bic" for Akaike or Schwarz criterion to use in model selection.

fix

> indicates which columns in x are held fixed in model selection.

7.4.4 Output Modification

Functions which provide graphical output (`glmout`, `glmplot`, `doglm`) accept special options to change output.

`nopic`
> suppresses output display in `glmout` or `doglm`, if `nopic=1`. Default is `nopic=0`.

`xvars`
> string vector, containing variable names for the columns of x.

`name`
> single string, name for output and prefix for output displays.

`title`
> single string, title to be used in `glmout` or `doglm`.

7.5 Statistical Evaluation and Presentation

To assess the estimated model it is useful to check the significance of single parameter values or of linear combinations of parameters. To compare two nested models a likelihood ratio test can be performed. Last but not least, an optimal submodel can be selected by model selection via Akaike's AIC or Schwarz' BIC.

7.5.1 Statistical Characteristics

```
stat = glmstat (code, x, y, b, bv {, opt})
        computes statistical characteristics for an estimated GLM
```

The functions `doglm` and `glmest` provide a number of statistical characteristics of the estimated model in the output component `stat`. Alternatively, the function `glmstat` can be used to create the above-mentioned statistics by hand. Suppose we have input x, y and have estimated the vector of coefficients b with covariance bv by the model "nopow". Then the list of statistics can be found from

```
stat=glmstat("nopow",x,y,b,bv)
```

Of course, a list of options opt can be added. If options from opt have been used for the estimation, these should be included in **glmstat**.

The following characteristics are contained in the output **stat**. This itself is a list and covers the components

df
>   degrees of freedom (typically sample size minus number of estimated parameters).

deviance
>   the deviance of the estimated model.

pearson
>   the Pearson statistic.

loglik
>   the log-likelihood of the estimated model, using the estimated dispersion parameter.

dispersion
>   an estimate for the dispersion parameter (**deviance/df**).

aic, bic
>   Akaike's AIC and Schwarz' BIC criterion, respectively.

r2, adr2
>   the (pseudo) coefficient of determination and its adjusted version, respectively.

it
>   the number of iterations needed.

ret
>   the return code, which is 0 if everything went without problems, 1 if the maximal number of iterations was reached, and -1 if missing values have been encountered. In the latter case, the parameter estimates and its covariance come from the penultimate iteration step.

nr
   the number of replicated observation in **x**, if they were searched for.

Sometimes one or the other statistic may not be available when it is not applicable. This can always be checked by searching for the components in **stat**:

   names(stat)

The function **names** will report all components of the list **stat**.

7.5.2   Output Display

> glmout (code, x, y, b, bv, stat {, opt})
>   creates a nice output display for an estimated GLM

Recall the example ☐ glm03.xpl, which estimated the lizard data. The last line of this quantlet creates the output display shown in Figure 7.1:

   glmout("bilo",x,y,g.b,g.bv,g.stat,opt)

Note that the option list **opt** should also be given to **glmout** to adjust the resulting estimated curve by the weights. In the binomial case (as in our lizard example), the right panel shows the predicted probabilities. For all other distributions, the estimated regression function (the index $x^T\beta$ vs. $G(x^T\beta)$) will be shown.

7.5.3   Significance of Parameters

Let us continue with the lizard example from Subsection 7.5.2. As a result of the estimation, we have an output list **g** containing components **g.b** (the estimated parameter vector), **g.bv** (the estimated covariance of **g.b**), and **g.stat** (contains the statistics).

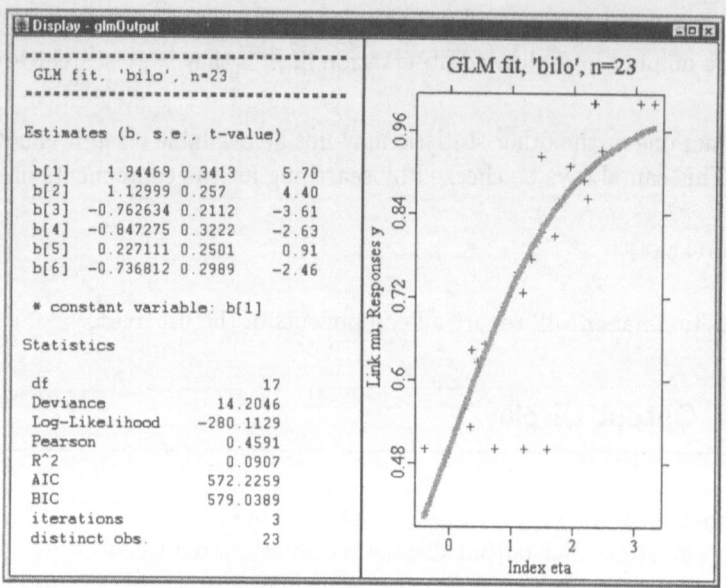

Figure 7.1. Output display for lizard example.

The significance of coefficients can be measured by a t-test. To obtain t-values and p-values, simply calculate:

```
tvalue=g.b/sqrt(xdiag(g.bv))
pvalue=2.*cdfn(-abs(tvalue))
```

🔍 glm05.xpl

The **xdiag** extracts the diagonal of a quadratic matrix, **sqrt** takes the square root. The functions **cdfn** (or **cdft** for small samples) provide the cumulative distribution functions of the Gaussian distribution (t-distribution). In our running example, the latter instruction yields

```
Content of object pvalue
[1,] 1.2155e-06
[2,] 1.0972e-05
[3,] 0.0003059
[4,] 0.0085538
[5,] 0.3639
[6,] 0.013712
```

which means that all except the 5th coefficient are significant (at the 5% level).

For linear hypotheses on the parameters, a Wald test can be used. Suppose we want to test if $2\beta_1 = \beta_2$. This can be written as $Ab = a$ with $A = (2, -1, 0, ..., 0)$. Hence, define

```
A=2 ~ (-1) ~ 0.*matrix(rows(g.b)-2)'
a=0
W=(A*g.b-a)'*(A*g.bv*A')*(A*g.b-a)
pvalue=1-cdfc(W,1)
```

🔍 glm05.xpl

W denotes the test statistic which has an asymptotic χ^2 distribution with rank(A) degrees of freedom. The p-value for this test problem is calculated by means of cdfc, the cumulative distribution function of the χ^2 distribution, the rank of the matrix A is 1 in this case. For our running example, this produces a p-value of

```
Content of object pvalue
[1,] 0.033318
```

i.e. the relation $2\beta_1 = \beta_2$ is significant at 3.3318%.

7.5.4 Likelihood Ratio Tests for Comparing Nested Models

{lr,alpha} = glmlrtest (loglik0,dim0,loglik1,dim1)
     computes a likelihood ratio test for two nested GLMs on the basis
     of the χ^2 distribution

Suppose now we have two (nested) models estimated and obtained two estimation results c (the smaller model) and g (the larger model). To compare both models, one needs to calculate the likelihood ratio test statistic.

In some cases, the distribution of this test statistic can be derived exactly. Otherwise, the (negative) doubled logarithm of the likelihood ratio can be computed which has an asymptotic χ^2 distribution. In this case, the test statistic lr and the p-value pvalue can be obtained from glmlrtest. Recall the lizard example, where we estimated the full model g and the constrained model c. Now we determine if the difference between both models is significant. Computing

```
lc=c.stat.loglik
lg=g.stat.loglik
pc=rows(c.b)
pg=rows(g.b)
{lr,pvalue}=glmlrtest(lc,pc,lg,pg)
```

🔍 glm05.xpl

gives

```
Contents of pvalue
[1,]   0.37944
```

i.e. there is no statistically significant difference between the two models.

7.5.5   Subset Selection

```
select = glmselect (code, x, y {,opt})
     performs a complete search model selection by choosing the best
     of all subset models with respect to the AIC or BIC criterion

select = glmforward (code, x, y {,opt})
     performs a forward search model selection by choosing the best
     of all subset models with respect to the AIC or BIC criterion

select = glmbackward (code, x, y {,opt})
     performs a backward search model selection by choosing the best
     of all subset models with respect to the AIC or BIC criterion
```

The model selection functions glmselect, glmforward and glmbackward have essentially the same syntax as glmest. Additionally, the optional parameters shm (show model selection going on), crit ("aic" or "bic" for Akaike or Schwarz criterion) and fix (columns of x to be held fixed) will be recognized.

In the following we generate a 200 × 4 matrix x and a response y which only depends on the first two columns of x:

```
randomize(0)
n=200
b=1|2|0|0
p=rows(b)
x=normal(n,p)
y=x*b+normal(n)
```

Q glm06.xpl

Now add a constant column to x and set options shm and fix. The optional parameter fix is such that the first two columns of x (constant and first explanatory variable) are always in the model. The last line of the following quantlet

```
x=matrix(n)~x
opt=glmopt("shm",1,"fix",1|2)
g=glmselect("noid",x,y,opt)
```

Q glm06.xpl

now starts the model selection. For many possible submodels, this can take a while. In our case we have 3 free variables (we fixed two out of five), hence the total number of models to estimate is 7. The output list **g** consists of 5 components:

best
> the 5 best models.

bestcrit
> a list containing bestcrit.aic and bestcrit.bic, the Akaike and the Schwarz criteria for the 5 best models.

bestord
> the best models of each order.

bestordcrit
> like bestcrit, but for the best model of each order.

bestfit
> containing bestfit.b, bestfit.bv and bestfit.stat, the estimation results for the best model.

Hence, **g.best** will display the five best models in our example. The contents of **g.best** reads columnwise:

```
Content of object g.best
[1,]    1   1  1   1   1
[2,]    2   2  2   2   2
[3,]    3   3  3   3   0
[4,]    0   0  4   4   4
[5,]    0   5  0   5   0
```

Those components which are not in a submodel are indicated by the value 0. Hence the model selection procedure found indeed that the last two columns of x do not explain y.

The functions glmforward and glmbackward have the same functionality as glmselect, except that they do a forward and backward search, respectively.

Bibliography

McCullagh, P. and Nelder, J. A. (1989). *Generalized Linear Models*, Vol. 37 of *Monographs on Statistics and Applied Probability*, 2 edn, Chapman and Hall, London.

8 Neural Networks

Wolfgang Härdle and Heiko Lehmann

A neural network consists of many simple processing units that are connected by communication channels. Much of the inspiration for the field of neural networks came from the desire to perform artificial systems capable of sophisticated, perhaps intelligent computations similar to those of the human brain.

Neural networks usually learn from examples and exhibit some capability for generalization beyond the data used for training. They are able to approximate highly nonlinear functional relationships in data sets.

The smallest part of a neural network is one single neuron as shown in Figure 8.1. It takes a set of individual **inputs** $x = (x_1, \ldots, x_I)$ and determines (through the learning algorithm) the optimal **connection weights** $w = (w_1, \ldots, w_I)$ that are appropriate to each input. Next, the neuron aggregates these weighted values to a single value

$$u = w_0 + \sum_{i=1}^{I} w_i x_i \, .$$

An **activation function** $F(\bullet)$ is then applied to the aggregated weighted value to produce an individual output

$$F(u)$$

for the specific neuron. A typical activation function is the logistic distribution function

$$F(u) = \frac{1}{1 + \exp(-u)} \, .$$

The aim of a neural network is to explain the **outputs** $y = (y_1, \ldots, y_Q)$ by the input variables $x = (x_1, \ldots, x_I)$. More exactly, we want to find functions $f_k(\bullet)$ such that $f_k(x)$ explains the output variable y_k.

Output of the Neuron

$$F_j\left(w_0 + \sum x_i \cdot w_i\right)$$

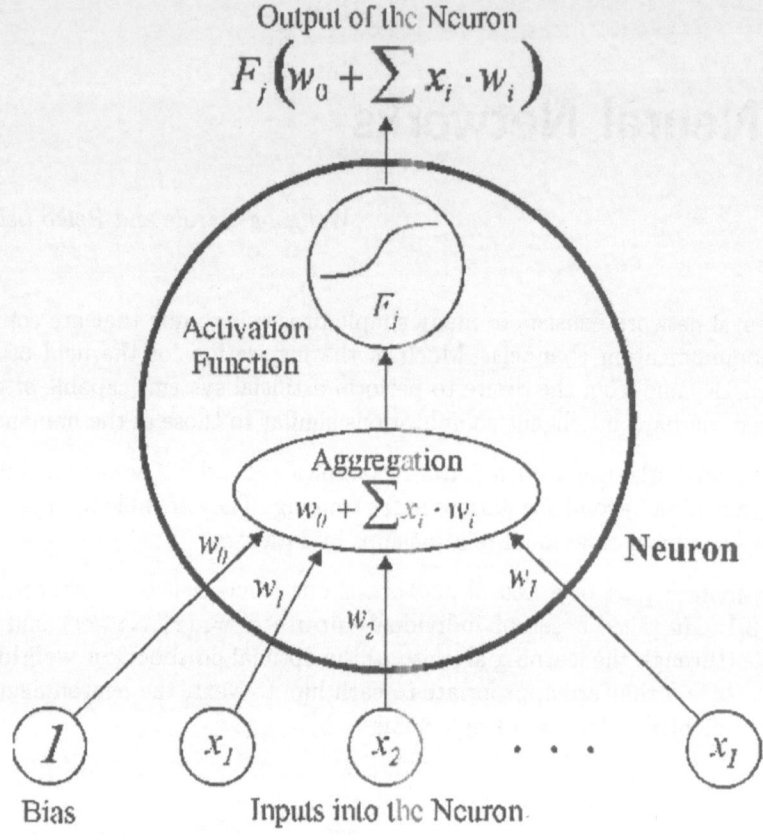

Figure 8.1. A neuron within a neural network.

A neural network with one **hidden layer** (single hidden layer) consists of neurons of three basic types:

- The **input neurons** collect the external information and send it to the layer of hidden units.

- The **hidden neurons** aggregate the information and send it to the output neuron(s).

- The **output neurons** contain the aggregated information passed through the activation function.

8.1   Feed-Forward Networks

Figure 8.2 shows a **feed-forward** network with one hidden layer. This network attempts to fit the model

$$f_k(x) = F\left\{ w_{0k}^{(2)} + \sum_{j=1}^{J} w_{jk}^{(2)} F\left(w_{0j}^{(1)} + \sum_{i=1}^{I} w_{ij}^{(1)} x_i \right) \right\}$$

for the output unit y_k. Feed-forward means that information can only flow forward from the input units to the first hidden layer, from the first hidden layer to the second hidden layer, and so on. Information cannot flow between the units of one layer.

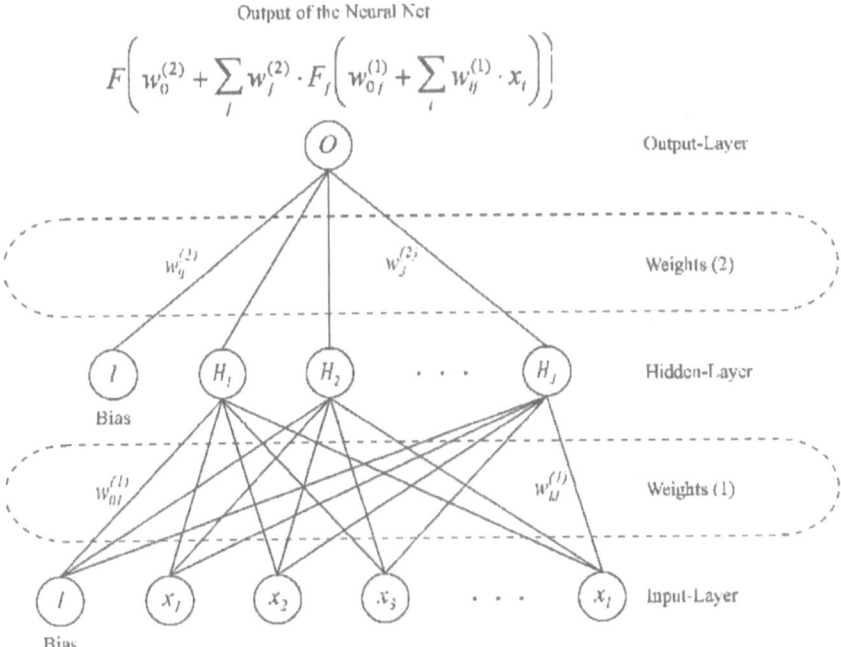

Figure 8.2. Feed-forward network with one hidden layer.

8.2   Computing a Neural Network

---

net = nnrnet (x, y, w, size, {, param, wts})
>   trains a single layer feed-forward network with input x, output
>   y, prior weights w, and number of hidden units size; optionally
>   the type of the network can be determined by param and initial
>   weights wts can be given

net = nnrpredict (x, net)
>   predicts the responses for given variables x and network net

net = nnrinfo (net)
>   shows information about network net

nnrsave(net, "nnfile")
>   saves network net to files nnfile.*

net = nnrload ("nnfile")
>   loads network net from files nnfile.*

The function nnrnet allows for constructing and training a single hidden layer
network with maximal 100 units. The call looks like

        net = nnrnet (x, y, w, size, param, wts)

where x and y are the input and output variables. Note that x as well as y can
consist of several variables (columns). We assume that x and y have dimensions
$n \times I$ and $n \times Q$, respectively.

With the w parameter, we can associate a prior weight to each observation.
This is useful, e.g. for ties in the data. Note that the prior weights w have
nothing in common with the weights calculated in the net.

The parameter size determines the number of units in the hidden layer. The
total number of units must not exceed 100, i.e.

        columns of x + columns of y + units in hidden layer ≤ 100.

The default network is a classification network: logistic output units, no softmax, no "skip-layer" connections, no weight decay and the training stops after

100 iterations. The default model for the output units y_k, $k = 1, \ldots, Q$, is
hence

$$f_k(x) = F \left\{ w_{0k}^{(2)} + \sum_{j=1}^{size} w_{jk}^{(2)} F \left(w_{0j}^{(1)} + \sum_{i=1}^{I} w_{ij}^{(1)} x_i \right) \right\}$$

with $F(\bullet)$ the logistic function. If a model different to the default is fitted,
the parameter **param** needs to be modified. We explain this in more detail in
Subsection 8.2.1.

The result of **nnrnet** is a composed object **net**. More information on the
components of **net** can be found in Subsection 8.2.2. The function **nnrinfo**
shows a short information about the fitted network. The result of

```
nnrinfo(net)
```

could for example print the following information in the output window:

```
[ 1,] "A 2 - 1 - 1 network:"
[ 2,] "# weights      : 5"
[ 3,] "linear output : no"
[ 4,] "error function: least squares"
[ 5,] "log prob model: no"
[ 6,] "skip links     : no"
[ 7,] "decay          : 0"
[ 8,] ""
[ 9,] " From    To Weights"
[10,] "   0     3  -0.751"
[11,] "   1     3   0.81"
[12,] "   2     3   0.575"
[13,] "   0     4  -4.95"
[14,] "   3     4   14.8"
```

The abbreviation $2 - 1 - 1$ means two input units, one hidden layer and one
output unit. Altogether five weights w_{st} have been calculated, the values of
these weights are given in the last lines. The other items show which parameters
have been specified for the network.

Typically, a neural network is applied to a subsample of the data which is used
as a training data set. The remaining observations are then used to validate
the network. To compute predicted values for the validation set, **nnrpredict**
is used:

```
ypred = nnrpredict (xval, net)
```

Since the result of a neural network fitting is a composed object, two convenient functions for saving and loading neural networks are provided. The network net can be stored into a set of files by

```
nnrsave (net, "mynet")
```

All created files start with the prefix mynet. The network can be reloaded by

```
net = nnrload ("mynet")
```

8.2.1   Controlling the Parameters of the Neural Network

The type of a network and the control parameters for the iteration are determined by the parameter param of nnrnet. If, for instance, a model different to the default is fitted, this parameter needs to be modified. param is a vector of eight elements:

param[1]
: determines if the activation function for the output is the logistic function (default value 0). Setting param[1] to the value 1 changes the activation function of the output unit to the identity function.

param[2]
: determines the error function (the optimization criterion). The default value 0 indicates the quadratic least squares error function

$$\sum_{k=1}^{size}\sum_{i=1}^{n}\{f_k(x_i) - y_{i,k}\}^2 \; .$$

Setting param[2] to the value 1 changes the error function to the entropy for the classification case

$$\sum_{k=1}^{size}\sum_{i=1}^{n}\left\{f_k(x_i)\log\left(\frac{f_k(x_i)}{y_{i,k}}\right) + \{1 - f_k(x_i)\}\log\left(\frac{1 - f_k(x_i)}{1 - y_{i,k}}\right)\right\} \; .$$

`param[3]`

If `param[3]` is set to the value 1, then the softmax activation function is used for the outputs. This means the output is

$$f_k(x_i) = \frac{\exp\{f_k^*(x_i)\}}{\sum_{\ell=1}^{Q} \exp\{f_\ell^*(x_i)\}}.$$

The default value is 0, which means no softmax.

`param[4]`

includes "skip-layer" connections. Setting `param[4]` to the value 1 generates "skip-layer" connections, i.e.

$$f_k(x) = w_{0k}^{(2)} + \sum_{i=1}^{p} w_{ij}^{(2)} x_i + \sum_{j=1}^{\text{size}} w_{kj}^{(2)} F\left(w_{0j}^{(1)} + \sum_{i=1}^{p} w_{ij}^{(1)} x_i \right).$$

The default value is 0, which means no "skip-layer" connections.

`param[5]`

sets the maximal value δ for the initial weights. If the optional input parameter **wts** is not given, uniform random numbers from $[-\delta, \delta]$ are used. The default value is $\delta = 0.7$.

`param[6]`

sets the weight decay, the default is 0.

`param[7]`

sets the maximal number of iterations, the default is 100.

`param[8]`

shows information about the iteration. Setting `param[8]` to the value 1 produces control output in the output window during the optimization. The default is 0, i.e. not to show control output.

8.2.2 The Resulting Neural Network

The result of **nnrnet** is a composed object, the list **net**, which contains the resulting fit and information about the network. The components are the following:

net.n
> three-dimensional vector that contains the number of input, hidden and output units, respectively

net.nunits, net.nconn, net.conn
> internal information about the network topology

net.decay
> scalar, the weight decay parameter (=param[6])

net.entropy
> scalar, the value of the entropy

net.softmax
> scalar, softmax indicator (=param[3])

net.value
> scalar, the value of the error function

net.wts
> vector of final weights

net.yh.result
> $n \times Q$ matrix, the estimated outputs

net.yh.hess
> the Hessian matrix

8.3   Running a Neural Network

In the following two sections we run simple neural nets on clustered data. Before proceeding to the examples, the following libraries need to be loaded:

```
library ("plot")
library ("nn")
```

The nn library contains the functions for running the networks. The plot library is used to produce scatter plots of the clusters.

8.3.1 Implementing a Simple Discriminant Analysis

In the following, we will use a single hidden layer network with one hidden unit to perform a discriminant analysis on an artificially generated data set with two clusters.

All XploRe codes for this subsection can be found in ✿ nn1.xpl. The first step is to generate the training data set:

```
randomize(0)
n  = 200
xt = normal(n,2)+#(-1,-1)' | normal(n,2)+#(+1,+1)'
```

Here, a mixture of two two-dimensional normal distributions is generated. Each cluster consists of n = 200 observations. The variances are identical (equal to 1 in both directions) whereas the means are shifted by (+1,+1) and (−1,−1), respectively. The following code lines can be used to display the data set graphically:

```
color  = string("red",1:n) | string("blue",1:n)
symbol = string("circle",1:n) | string("triangle",1:n)
xt     = setmask(xt, color, symbol)
plot(xt)
xl="x1"
yl="x2"
tl="Training Data Set"
setgopt(plotdisplay,1,1,"title",tl,"xlabel",xl,"ylabel",yl)
```

The generated two-dimensional data are shown in Figure 8.3. We have labeled the observations from the first cluster by red circles, whereas the observation from the second cluster are labeled as blue triangles.

To apply the neural network, we need to create now the output variable y and the prior weights w. For y, we use a value of 0 for the first and a value of 1 for the second cluster. The prior weights are all set to 1. The last statement of the following code computes the neural network using one hidden unit and assigns the result to **net**.

```
yt = (matrix(n)-1)|matrix(n)
w  = matrix(2*n)
```

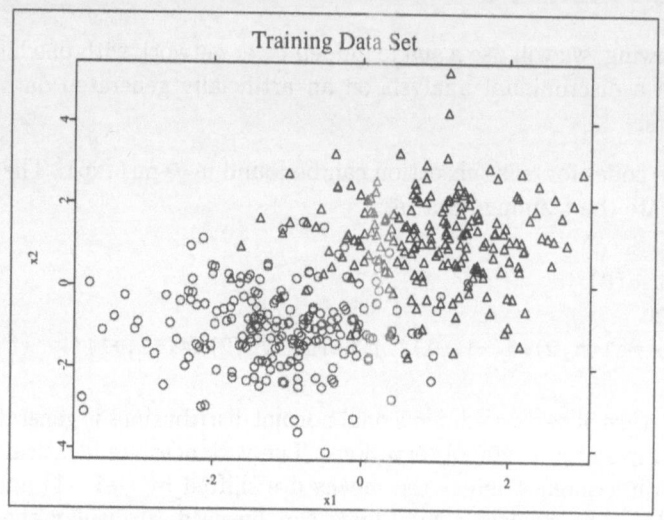

Figure 8.3. A generated training data set with two clusters.

```
param = 1
net = nnrnet(xt,yt,w,1)
```

We can obtain a summary of the fitted network from

```
nnrinfo(net)
```

which prints into the output window:

```
Contents of ts
[ 1,] "A 2 - 1 - 1 network:"
[ 2,] "# weights     : 5"
[ 3,] "linear output : no"
[ 4,] "error function: least squares"
[ 5,] "log prob model: no"
[ 6,] "skip links    : no"
[ 7,] "decay         : 0"
[ 8,] ""
[ 9,] " From    To Weights"
```

```
[10,] "    0    3    -1.18"
[11,] "    1    3    -0.285"
[12,] "    2    3    -0.198"
[13,] "    0    4     99.8"
[14,] "    3    4     44.8"
```

To validate the obtained network, we generate new random data from the same mixture of two-dimensional normal distributions. The classification of these data using the network **net** is done by **nnrpredict**.

```
x = normal(n,2)+#(-1,-1)' | normal(n,2)+#(+1,+1)'
pred  = nnrpredict(x, net)
prob  = pred.result
```

The macro **nnrpredict** calculates the predicted values and the Hessian matrix. **pred.result** extracts the predicted values.

Now we compute the misclassified observations and show them in comparison with the original data x.

```
y   = (matrix(n)-1)|matrix(n) ; true
yp = prob > 0.5                ; predicted
misc = paf(1:2*n,y!=yp)        ; misclassified
good = paf(1:2*n,y==yp)        ; correctly classified
nm = rows(misc)
sm = string("fill",1:nm)+symbol[misc]
xm = setmask(x[misc],color[misc],sm,"huge")
xg = setmask(x[good],color[good],symbol[good])

pm = 100*nm/(2*n)              ; percentage of misclassified
spm = string("%1.2f",pm)+"%"
Network = createdisplay(1,1)
show(Network,1,1,xg,xm)
tl="Network: misclassified = "+spm
setgopt(Network,1,1,"title",tl,"xlabel",xl,"ylabel",yl)
```

Figure 8.4 shows the two-dimensional data that we used for validation. As before, observations from the first cluster are labeled by red circles, whereas the observation from the second cluster are labeled as blue triangles. All misclassified data are labeled by large filled symbols.

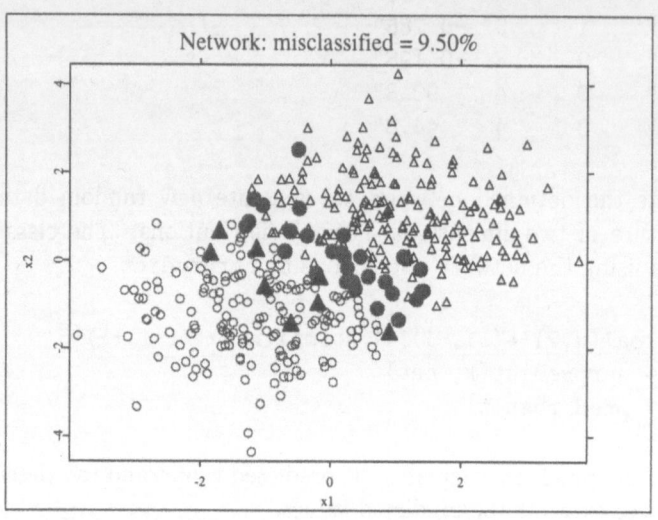

Figure 8.4. Neural network classification.

Let's compare the classification obtained from the neural network with that from a classical linear discriminant analysis. Apart from the discrimination rule that is used for the prediction here, the code is almost identical to the above.

```
mu0 = mean(xt[1:n])
mu1 = mean(xt[n+1:2*n])
mu  = (mu0+mu1)/2
lin = inv(cov(xt))*(mu0-mu1)'

y   = (matrix(n)-1)|matrix(n) ; true
yp  = (x-mu)*lin<=0           ; predicted
misc = paf(1:2*n,y!=yp)       ; misclassified
good = paf(1:2*n,y==yp)       ; correctly classified
nm  = rows(misc)
sm  = string("fill",1:nm)+symbol[misc]
xm  = setmask(x[misc],color[misc],sm,"huge")
xg  = setmask(x[good],color[good],symbol[good])
x   = setmask(x, color, symbol)
```

```
pm = 100*nm/(2*n)                ; percentage of misclassified
spm = string("%1.2f",pm)+"%"
Discrim = createdisplay(1,1)
show(Discrim,1,1,xg,xm)
tl="Linear misclassified = "+spm
setgopt(Discrim,1,1,"title",tl,"xlabel",xl,"ylabel",yl)
```

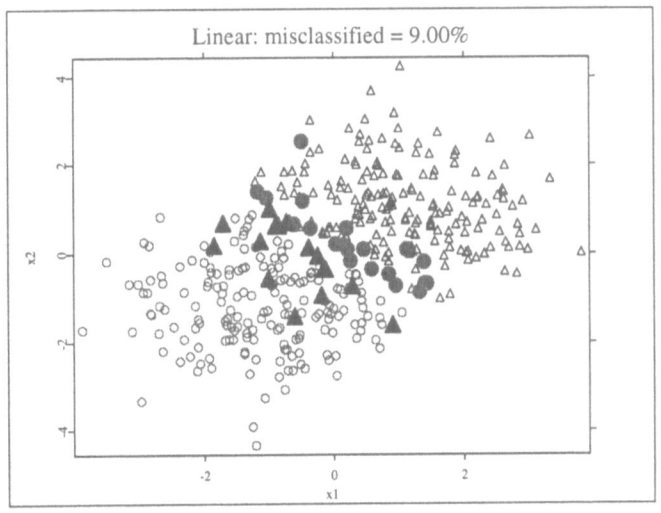

Figure 8.5. Linear discriminant analysis.

Figure 8.5 shows the resulting classification. Again, all misclassified data are labeled by large filled symbols. Comparing Figures 8.4 and 8.5 shows that the percentage of misclassification is nearly equal for both methods. The linear discriminant analysis performs slightly better. This is not astonishing, since the linear discriminant analysis is designed to handle the data that we generated.

8.3.2 Implementing a More Complex Discriminant Analysis

In contrast to the previous subsection, we will now consider a generated data set where the linear discriminant analysis performs worse than the neural network.

The XploRe codes are largely identical to the previous examples and can be found in ❏ nn2.xpl.

As before we generate a training data set, which features two clusters.

```
randomize(0)
n  = 100
xt = normal(n,2)+#(-1,-1)' | normal(n,2)+#(+1,-2)'
xt = xt | normal(n,2)+#(+4, 0)' | normal(n,2)+#(+1,+1)'

color  = string("red",1:3*n) | string("blue",1:n)
symbol = string("circle",1:3*n) | string("triangle",1:n)
xt     = setmask(xt, color, symbol)
plot(xt)
xl="x1"
yl="x2"
tl="Training Data Set"
setgopt(plotdisplay,1,1,"title",tl,"xlabel",xl,"ylabel",yl)
```

The generated two-dimensional data are shown in Figure 8.6. It is obvious that here the points from the second group (labeled by blue triangles) overlap the points from the first group (red circles) in a more complicated way.

We proceed in the same way as before, i.e. we create the output variable y and set all prior weights w to 1. Then the neural network is fitted. In contrast to the previous section, we now use 3 hidden layers to take the more complex structure of the data into account.

```
yt = (matrix(3*n)-1)|matrix(n)
w  = matrix(4*n)
param = 1
net = nnrnet(xt,yt,w,3)
nnrinfo(net)
```

The resulting fit is summarized as follows:

```
Contents of ts
[ 1,] "A 2 - 3 - 1 network:"
[ 2,] "# weights     : 13"
[ 3,] "linear output : no"
```

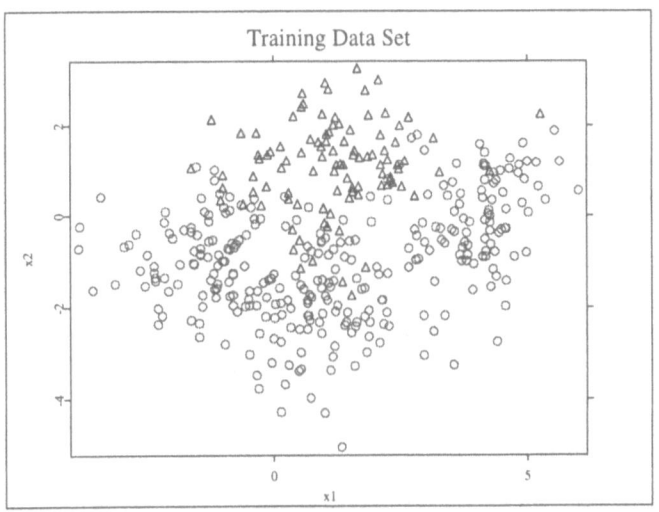

Figure 8.6. A generated training data set with two clusters.

```
[ 4,] "error function: least squares"
[ 5,] "log prob model: no"
[ 6,] "skip links    : no"
[ 7,] "decay         : 0"
[ 8,] ""
[ 9,] " From    To Weights"
[10,] "   0     3    1.26"
[11,] "   1     3   -0.106"
[12,] "   2     3   -5.15"
[13,] "   0     4    2.92"
[14,] "   1     4   -1.32"
[15,] "   2     4    0.37"
[16,] "   0     5   -7.61"
[17,] "   1     5   -56.1"
[18,] "   2     5   -28.3"
[19,] "   0     6   -2.76"
[20,] "   3     6   -4.38"
[21,] "   4     6    7.64"
[22,] "   5     6   -4.71"
```

Again, we assess the quality of the obtained network by counting the misclassified observations for a validation data set.

```
x = normal(n,2)+#(-1,-1)' | normal(n,2)+#(+1,-2)'
x = x | normal(n,2)+#(+4, 0)' | normal(n,2)+#(+1,+1)'
pred  = nnrpredict(x, net)
prob  = pred.result

y  = (matrix(3*n)-1)|matrix(n) ; true
yp = prob > 0.5                ; predicted
misc = paf(1:4*n,y!=yp)        ; misclassified
good = paf(1:4*n,y==yp)        ; correctly classified
nm = rows(misc)
sm = string("fill",1:nm)+symbol[misc]
xm = setmask(x[misc],color[misc],sm,"huge")
xg = setmask(x[good],color[good],symbol[good])

pm = 100*nm/(4*n)                      ; percentage of misclassified
spm = string("%1.2f",pm)+"%"
Network = createdisplay(1,1)
show(Network,1,1,xg,xm)
tl="Network: misclassified = "+spm
setgopt(Network,1,1,"title",tl,"xlabel",xl,"ylabel",yl)
```

Figure 8.7 shows the resulting plot of the two-dimensional data that we used for prediction, with misclassified data labeled by large filled symbols.

The comparison with the classical linear discriminant analysis is implemented in the following lines:

```
mu0 = mean(xt[1:3*n])
mu1 = mean(xt[3*n+1:4*n])
mu  = (mu0+mu1)/2
lin = inv(cov(xt))*(mu0-mu1)'

y  = (matrix(3*n)-1)|matrix(n) ; true
yp = (x-mu)*lin<=0             ; predicted
misc = paf(1:4*n,y!=yp)        ; misclassified
good = paf(1:4*n,y==yp)        ; correctly classified
nm = rows(misc)
```

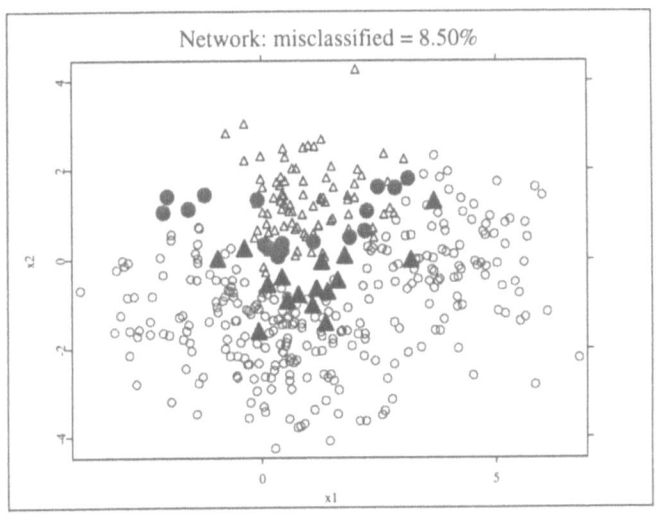

Figure 8.7. Neural network classification.

```
sm = string("fill",1:nm)+symbol[misc]
xm = setmask(x[misc],color[misc],sm,"huge")
xg = setmask(x[good],color[good],symbol[good])
x  = setmask(x, color, symbol)

pm = 100*nm/(4*n)                  ; percentage of misclassified
spm = string("%1.2f",pm)+"%"
Discrim = createdisplay(1,1)
show(Discrim,1,1,xg,xm)
tl="Linear misclassified = "+spm
setgopt(Discrim,1,1,"title",tl,"xlabel",xl,"ylabel",yl)
```

Figure 8.8 shows the resulting classification. The comparison of Figures 8.7
and 8.8 reveals now that the neural network separates the clusters more accu-
rately. This is due to the fact that the neural network with three hidden units
can better adapt to a nonlinear discrimination rule.

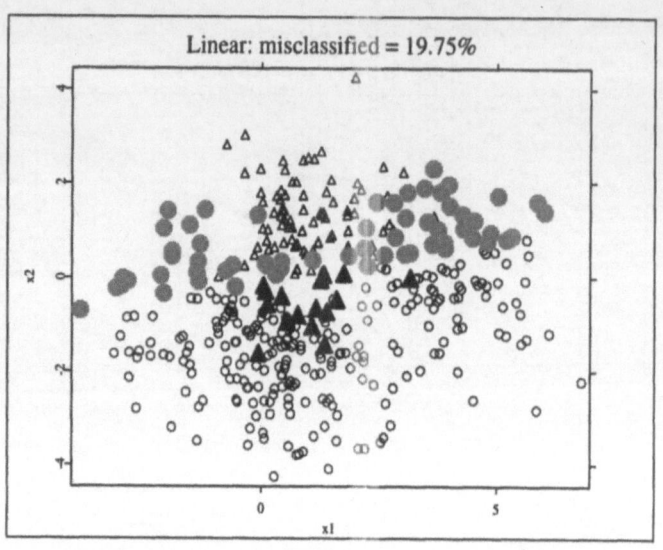

Figure 8.8. Linear discriminant analysis.

Bibliography

Bishop, C. (1995). *Neural Networks for Pattern Recognition*, Clarendon Press, Oxford.

Ripley, B. (1996). *Pattern Recognition and Neural Networks*, Cambridge University Press.

9 Time Series

Petr Franěk and Wolfgang Härdle

The purpose of this chapter is to show how XploRe may be used by practitioners for analyzing observed time series.

Some of the time series tools are standard in the literature. The more elaborated nonlinearity tests based on artificial neural networks are implemented for the nonadvanced use.

9.1 Time Domain and Frequency Domain Analysis

```
x = acf (y {,k})
        computes the autocorrelation function of a time series

acfplot(y {,k})
        plots the autocorrelation function of a time series

x = pacf (y,k)
        computes the partial autocorrelation function of a time series

pacfplot(y {,k})
        plots the partial autocorrelation function of a time series

x = pgram (x { opt})
        computes and plots the raw (log) periodogram of a time series

x = spec (y {,width {,opt}})
        estimates and plots the spectral density of a time series
```

```
timeplot(y {,len {,header}})
        plots a time series in multiple windows with user-specified maxi-
        mum length per window
```

A time series represents a path (realization) of a stochastic process $\{Y_t\}$. The subscript t $(t = 1, \ldots, T)$ is usually understood as a time index (e.g. days or years). In this section we will consider that the underlying stochastic process is **weakly stationary**, i.e. we will assume that it satisfies the conditions

$$
\begin{aligned}
E\,|Y_t|^2 &< \infty \quad \forall t, \\
E\,Y_t &= \mu \quad \forall t, \\
E\,(Y_t - \mu)(Y_s - \mu) &= E\,(Y_{t+\tau} - \mu)(Y_{s+\tau} - \mu) \quad \forall s, t, \tau.
\end{aligned}
$$

All functions for the time series analysis are part of the **times** library and become available after loading this library:

```
library("times")
```

To display a time series we use **timeplot**. The following example plots the first 250 observations of the time series **dmus58** (Deutschmark–Dollar FX rates in 1982):

```
y = read("dmus58")
y = y[1:250,]
timeplot(y)
```

🔍 <code>times01.xpl</code>

The resulting graph is shown in Figure 9.1.

9.1.1   Autocovariance and Autocorrelation Function

The sample **autocovariance** function at lag τ of a process Y_t is defined for $\tau \in \{0, \ldots, T-1\}$ as

$$
\widehat{\gamma}(\tau) = T^{-1} \sum_{t=\tau+1}^{T} (Y_t - \overline{Y})(Y_{t-\tau} - \overline{Y})
$$

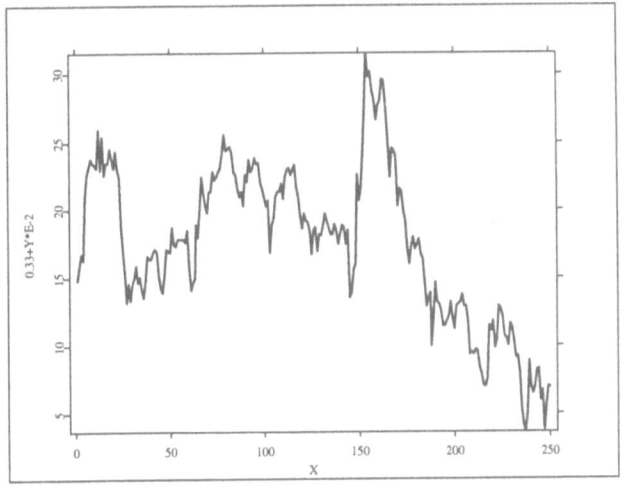

Figure 9.1. First 250 observations of dmus58.

where $\overline{Y} = T^{-1}\sum_{t=1}^{T} Y_t$ is the arithmetic mean of the time series Y_t. We define the sample **autocorrelations** $\widehat{\rho}(\tau)$ of the process by standardizing the sample autocovariance function $\widehat{\gamma}(\tau)$ by $\widehat{\gamma}(0)$ (the sample variance of the process), i.e.

$$\widehat{\rho}(\tau) = \frac{\widehat{\gamma}(\tau)}{\widehat{\gamma}(0)}.$$

Let's consider a sample of 500 independent realizations of a standard normal random variable:

```
randomize(0)
y = normal(500)
```
                                                          Q times02.xpl

Using acf we can evaluate the sample autocorrelations of the generated series. In the following example the result is stored in the vector x and the first five autocorrelations are displayed:

```
x = acf(y)
x[1:5]
```

The shape of the autocorrelation function may be easily analyzed using the
function acfplot which displays the **correlogram**, i.e. the graph of the autocorrelation function $\widehat{\rho}(\tau)$. Type

```
acfplot(y)
```

to get the following graph in Figure 9.2. Confidence levels $\pm 2/\sqrt{T}$ are plotted
to easily check the assumption that the series is a white noise.

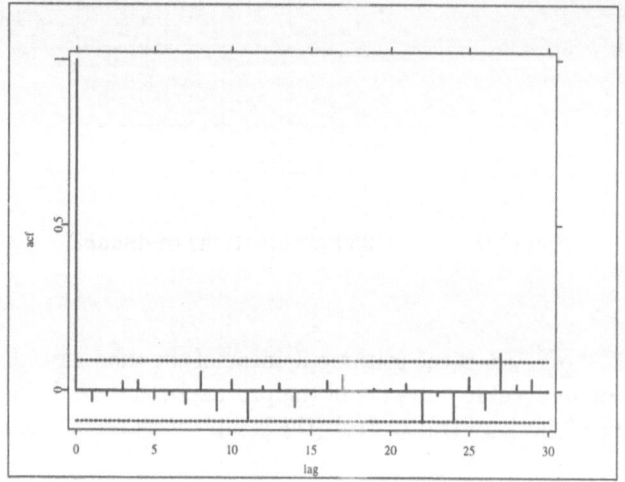

Figure 9.2. Autocorrelation function.

As another useful measure, the **partial autocorrelations** are implemented in
pacf. The partial autocorrelation of order $\tau \geq 2$ is calculated as the correlation
of the two residuals obtained after regressing $Y_{\tau+1}$ and Y_1 on the intermediate
observations $Y_2, \ldots, Y_\tau$, the partial autocorrelation at lag 1 being defined as
the correlation between Y_1 and Y_2.

XploRe also provides the function pacfplot as a partial autocorrelation equivalent to acfplot. The usage of these two quantlets is similar to the previously
mentioned autocorrelation function. To evaluate the sample partial autocorrelations of the generated series and plot the evaluated values type the following
instructions:

```
x = pacf(y,5)
```

```
x[1:5]
pacfplot(y)
```

The resulting plot is shown in Figure 9.3.

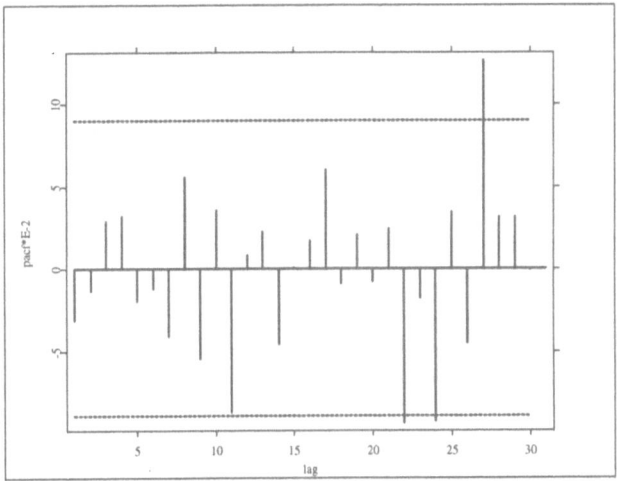

Figure 9.3. Partial autocorrelation function.

9.1.2   The Periodogram and the Spectrum of a Series

The frequency domain analysis is concerned with the decomposition of the observed series into periodic components. The main analytical tool for spectral analysis is the **spectrum** of a series defined as

$$f(\lambda) = (2\pi)^{-1} \left[\gamma(0) + 2 \sum_{\tau=1}^{\infty} \gamma(\tau) \cos(\lambda\tau) \right],$$

with λ the angular frequencies in $[-\pi, \pi]$ and $\gamma(\tau)$ the theoretical autocovariances. Since the spectrum is symmetric around zero, the analysis is restricted to the range of frequencies in $[0, \pi]$. For a sample of T observations, we consider the harmonic frequencies, or Fourier frequencies $\lambda_j = 2\pi j/T$, $j = 1, \ldots, [T/2]$.

The sample counterpart of the spectrum is the periodogram, defined as

$$I(\lambda) = (2\pi)^{-1} \left[\widehat{\gamma}(0) + 2 \sum_{\tau=1}^{T-1} \widehat{\gamma}(\tau) \cos(\lambda\tau) \right].$$

The periodogram is not a consistent estimator of the spectrum. When the sample size tends to infinity, more frequencies are considered without adding more precision on a particular one. The function pgram computes and plots the periodogram of the series. XploRe computes the periodogram for the frequencies $k_j \in [0, 0.5]$, which are linked to the angular frequencies by the relationship $k_j = \lambda_j/(2\pi)$. The periodogram of our generated series of independent normal random variables Y_t can be obtained as follows:

```
z = pgram(y)
```

Q times03.xpl

The periodogram for the series y is computed, the result is stored in the variable z and the periodogram is displayed on the screen (Figure 9.4.)

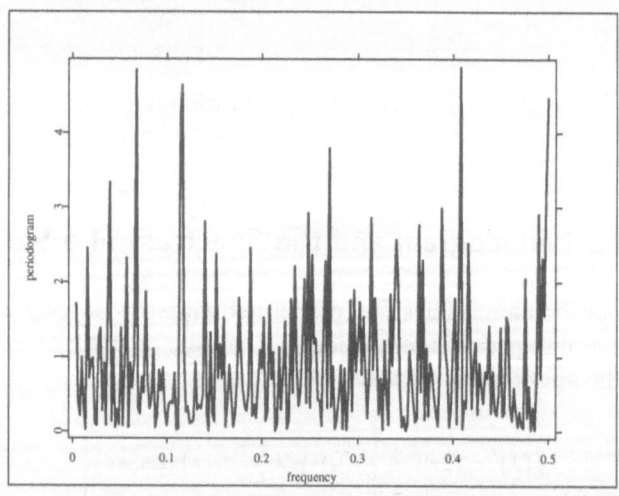

Figure 9.4. Periodogram.

The jagged shape of the periodogram illustrates the typical feature of a white noise series, i.e. a series of independent and identically distributed random variables. The lack of smoothness of the periodogram makes it difficult to interpret. In order to estimate the spectrum, the periodogram can be smoothed using the function `spec` that will be introduced in more details in the following section.

9.2 Linear Models

```
est = armacls (y, p, q)
        estimates parameters of an ARMA process using the conditional
        sum of squares

est = armalik (y)
        estimates parameters of an ARMA(1,1) process using the maxi-
        mum likelihood

y = genar (eps, startval, phi)
        generates an AR process

y = genarma (a, b, eps)
        generates an ARMA process with zero mean
```

In this section we focus our attention on the class of **linear models**, i.e. models driven by a general dynamic relationship of the form

$$Y_t = g(Y_{t-1}, Y_{t-2}, \ldots, Y_{t-p}),$$

where the function $g(.)$ is assumed to be linear.

9.2.1 Autoregressive Models

A process Y_t is called an autoregressive process of order p, AR(p), if it is driven by the relation

$$Y_t = \phi_1 Y_{t-1} + \phi_2 Y_{t-2} + \ldots + \phi_p Y_{t-p} + \varepsilon_t,$$

where the ε_t form a white noise process, they are also called **innovations**.

A sample of n observations of an AR(p) process can be generated using the command **genar**, which has the following syntax:

```
y = genar(eps, starval, phi)
```

where

- eps is the n-dimensional vector of white noises,
- starval is the p-dimensional vector of initial values,
- phi is the p-dimensional vector of autoregressive parameters $(\phi_1, \ldots, \phi_p)$.

In the following example a sample of 250 observations of an AR(1) process is generated. In this example, x is a vector of 250 independent realizations of standardized normal random variable.

```
randomize (0)
eps = normal(250)
starval = 0
phi = 0.5
y = genar(eps,starval,phi)
```

🔍 times04.xpl

In the following example, using the function **spec**, the typical spectrum of an autoregressive process is displayed. Note that the lowest frequencies have the highest contribution to the variation of the process. spec also displays the periodogram of the series. Type the instruction

```
spec(y)
```

to obtain the plots in Figure 9.5.

9.2.2   Autoregressive Moving Average Models

A process Y_t is called a moving average process of order q, MA(q), if it is driven by the relation

$$Y_t = \varepsilon_t + \psi_1 \varepsilon_{t-1} + \psi_2 \varepsilon_{t-2} + \ldots + \psi_q \varepsilon_{t-q},$$

where ε_t is a white noise innovation process.

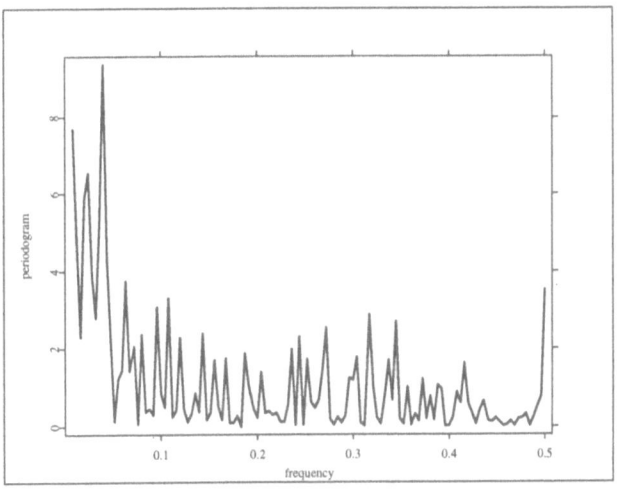

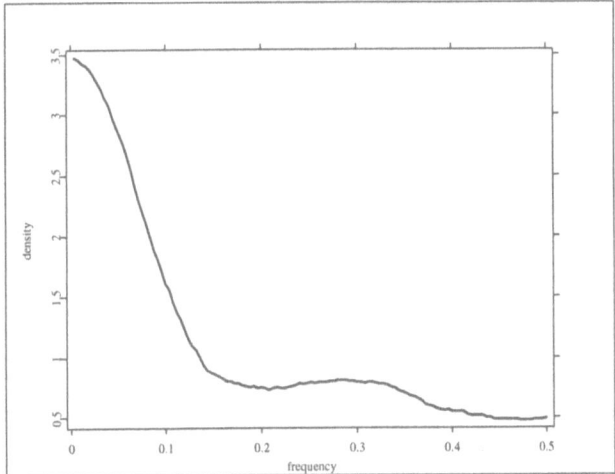

Figure 9.5. Periodogram and spectrum.

The structures of the autoregressive (AR) process and the moving average (MA) process may be combined into an autoregressive moving average process

$$Y_t = \phi_1 Y_{t-1} + \phi_2 Y_{t-2} + \ldots + \phi_p Y_{t-p} + \varepsilon_t + \psi_1 \varepsilon_{t-1} + \psi_2 \varepsilon_{t-2} + \ldots + \psi_q \varepsilon_{t-q}.$$

This process is denoted by ARMA(p, q), where p is the order of the autoregressive part and q is the order of the moving average part.

A sample of n observations of an ARMA(p, q) process can be generated using genarma, which has the following syntax:

```
y = genarma(a,b,eps)
```

where

- a is the p-dimensional vector of autoregressive parameters $(\phi_1, \ldots, \phi_p)$,

- b is the q-dimensional vector of moving average parameters $(\psi_1, \ldots, \psi_q)$,

- eps is the n-dimensional vector of white noises.

In the following example a sample of 250 observations of an ARMA$(1, 1)$ with Gaussian innovations process is generated:

```
randomize(0)
a = 0.5
b = 0.3
eps = normal(250)
y = genarma(a,b,eps)
```

🔍 times05.xpl

9.2.3   Estimating ARMA Processes

The parameters of an ARMA$(1, 1)$ process may be estimated using armalik, which has the following syntax:

```
est = armalik(y)
```

where

- y is the observed process,

- `est` is a list containing the estimated parameters, the corresponding asymptotic standard deviations, the asymptotic covariance and the estimate of the white noise variance.

To estimate the parameters of our generated ARMA(1,1) process, type:

```
est1 = armalik(y)
est1{1}
est1{2}
```

⟳times06.xpl

As output, XploRe returns

```
Contents of a
[1,]   0.49957
[2,]   0.25991
Contents of stderr
[1,]   0.057633
[2,]   0.064244
```

The parameters of a general ARMA(p, q) process can be estimated by **armacls**. This quantlet minimizes the conditional sum of squares and has the following syntax:

```
est = armacls(y,p,q)
```

where

- y is the sample of observations,

- p is the order of the autoregressive part,

- q is the order of the moving average part,

- est is a list containing the estimated parameters and information about the convergence of the method.

Since the generated sample y is the realization of an ARMA(1,1) process, we may estimate the parameters of our sample with the following instructions:

```
est2 = armacls(y,1,1)
est2
```

Q times07.xpl

As a result XploRe shows

```
Contents of est2.y.minimum
[1,]   0.49504
[2,]   0.28265
Contents of est2.y.iter
[1,]        20
Contents of est2.y.converged
[1,]         1
Contents of est2.wnv
[1,]   0.90943
```

9.3 Nonlinear Models

---

z = archest(y, q, p)
    estimates the parameters of a GARCH model

h = archtest (y {,lags {,testform}})
    test for ARCH effects

h = annarchtest (y {,nlags {,nodes {,testform}}})
    test for ARCH effects based on neural networks

y = genarch (a, b, n)
    generates an GARCH process with Gaussian innovations

y = genbil (phi, psi, gamma, noise)
    generates a bilinear process

y = genexpar (thrlag, gamma, phi0, phi1, noise)
    generates an exponential AR process

y = gentar (nr, thrlag, thr, phi, noise)
    generates a threshold AR process

gpplot(x, m, k)
    returns the Grassberger–Procaccia plot for time series

Nonlinear time series models have been recently explored by many authors. These models became important especially in analyzing financial and economic time series with underlying theoretical models that contain nonlinear relations. Several nonlinear models are implemented in XploRe with special focus on estimating and testing in ARCH and GARCH models which are often used in financial applications.

9.3.1 Several Examples of Nonlinear Models

We speak about a **threshold model** if the parameters of the model depend on the state of the observed system (random process). In building these models, the real line (please note that we are concerned only with univariate time series here) is divided into k parts by the set of ordered values $r_1 < \cdots < r_{k-1}$

(these values are called the threshold parameters). Depending on which interval $(r_j, r_{j+1}]$ contains the value Y_t the jth set of parameters is used to generate Y_{t+d}. The parameter $d < k$ is then called a delay parameter. The zero mean threshold AR(p) process (TAR(p) process) then may be introduced by the equation

$$Y_t = \phi_1^j Y_{t-1} + \phi_2^j Y_{t-2} + \ldots + \phi_p^j Y_{t-p} + \varepsilon_t,$$

where ε_t is a white noise process and $j \in \{1, \ldots, k\}$ is an indicator of the set of parameters to be used, i.e. j is determined by the condition $Y_{t-d} \in (r_j, r_{j+1}]$.

The threshold AR(p) may be generated in XploRe using **gentar**, which has the following syntax:

```
y = gentar(nr,thrlag,thr,phi,noise)
```

where

- **nr** is the number of threshold regions,

- **thrlag** is the threshold lag (the delay parameter),

- **thr** is a (nr-1)-dimensional vector of the threshold parameters that separates the regions,

- **phi** is a (nr*p)-dimensional vector of the AR parameters for all regions sorted as follows
$$(\phi_1^1, \ldots, \phi_p^1, \ldots, \phi_1^{nr}, \ldots, \phi_p^{nr}),$$

- **noise** is an n-dimensional vector of the noise (n is the number of observations to be generated).

The following example generates and displays 250 observations of the process

$$
\begin{aligned}
Y_t &= 0.6\,Y_{t-1} + \varepsilon_t \quad \text{for} \quad Y_{t-1} \leq 0 \\
&= 0.4\,Y_{t-1} + \varepsilon_t \quad \text{for} \quad Y_{t-1} > 0
\end{aligned}
$$

with $\varepsilon_t \sim N(0,1)$.

```
y=gentar(2,1,0,#(0.6,0.4), normal(250))
timeplot(y)
```

🔍 times08.xpl

The class of **exponential autoregressive** (EAR(p)) models with the lag d is characterized by the relation

$$Y_t = \sum_{j=1}^{p} \left[\phi_j^0 Y_{t-j} + (\phi_j^1 - \phi_j^0) \exp(-\gamma Y_{t-d}) \right] + \varepsilon_t,$$

where ε_t is a white noise process.

The EAR(p) process may be generated in XploRe using **genexpar**, which has the following syntax:

```
y = genexpar(thrlag,gamma,phi0,phi1,noise)
```

where

- **thrlag** is the threshold lag (the delay parameter),

- **gamma** is a positive parameter of the exponential function,

- **phi0** is a p-dimensional vector of AR parameters ($\phi_1^0, \ldots, \phi_p^0$),

- **phi1** is a p-dimensional vector of AR parameters ($\phi_1^1, \ldots, \phi_p^1$),

- **noise** is an n-dimensional vector of the noise (n is the number of observations to be generated).

The following example generates and displays 250 observations of an EAR(2) process:

```
y=genexpar(1,0.1,0.3|0.6, 2.2|-0.8,normal(250))
timeplot(y)
```

🔍 times09.xpl

The resulting time series is shown in Figure 9.6.

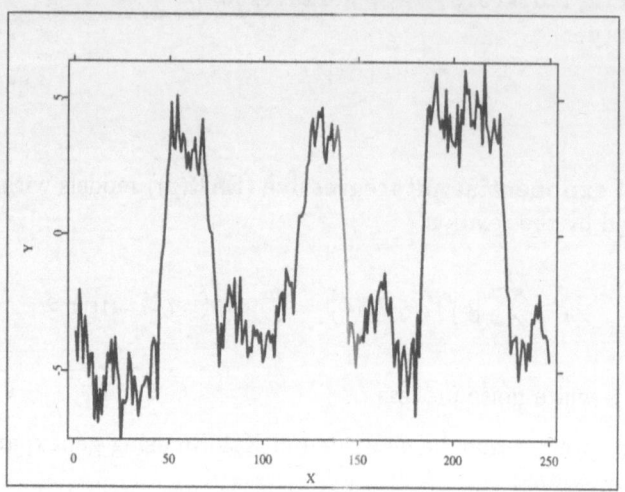

Figure 9.6. 250 observations of an EAR(2) process.

We speak about **bilinear model** if the process is driven by the relation

$$Y_t = \sum_{i=1}^{p} \phi_i Y_{t-i} + \sum_{j=1}^{q} \psi_j \varepsilon_{t-j} + \sum_{i=1}^{p} \sum_{j=1}^{q} \gamma_{i,j} Y_{t-i} \varepsilon_{t-j}$$

where ε_t is a white noise process.

The bilinear model may be generated in XploRe using `genbil`, which has the following syntax:

```
y = genbil(phi,psi,gamma,noise)
```

where

- `phi` is a p-dimensional vector of the AR parameters,

- `psi` is a q-dimensional vector of the MA parameters,

- `gamma` is a $p \cdot q$-dimensional vector of the bilinear parameters sorted as

$$(\gamma_{1,1}, \gamma_{1,2}, \ldots, \gamma_{1,q}, \gamma_{2,1}, \ldots, \gamma_{p,q}),$$

- **noise** is an n-dimensional vector of the noise (n is the number of observations to be generated).

The following example generates and displays 250 observations of a bilinear process, the resulting plot is shown in Figure 9.7.

```
y=genbil(0.5|0.2, 0.3|-0.3, 0.8|0|0|0.3,normal(250))
timeplot(y)
```

Q times10.xpl

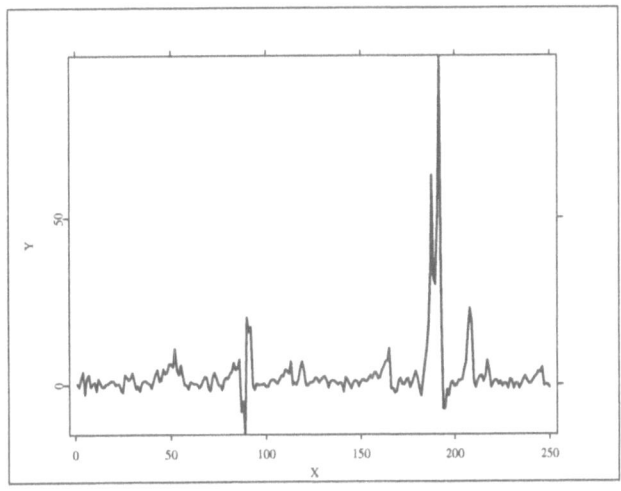

Figure 9.7. 250 observations of a bilinear process.

Based on the article Grassberger and Procaccia (1983), gpplot implements the Grassberger–Procaccia plot of time series. This function is applied as follows:

```
x=normal(100)
d=gpplot(x,2,10)
```

Q times11.xpl

Note that this example does not work in the Academic Edition of XploRe.

9.3.2   Nonlinearity in the Conditional Second Moments

A family of models with conditional heteroscedasticity is lately very popular especially among econometricians analyzing financial time series. This family may be introduced by a general formula

$$Y_t = f(Y_{t-1}, \ldots, Y_{t-\tau}) + \varepsilon_t, \quad t = 1, \ldots, T \quad \text{with} \quad \varepsilon_t | I_t \sim N(0, \sigma_t^2)$$

where I_t is the information set available at time t. The conditional variance σ_t^2 is a function of the information contained in the set I_t. We consider here the class of ARCH/GARCH models which restricts the information set I_t to the sequence of past squared disturbances ε_t^2.

ARCH models

The class of ARCH models represents the conditional variance σ_t^2 at time t as a function of the past squared disturbances

$$\sigma_t^2 = \omega + \sum_{i=1}^{q} \alpha_i \varepsilon_{t-i}^2.$$

This defines an ARCH(q) process where q is the order of the autoregressive lag polynomial $\alpha(L) = \alpha_1 L + \ldots + \alpha_q L^q$. This class of processes is generalized and nested in the class of generalized ARCH processes, denoted as GARCH and defined by the relation

$$\sigma_t^2 = \omega + \sum_{i=1}^{q} \alpha_i \varepsilon_{t-i}^2 + \sum_{j=1}^{p} \beta_j \sigma_{t-j}^2 = \omega + \alpha(L)\varepsilon_t^2 + \beta(L)\sigma_t^2.$$

This defines a GARCH(p, q) process, where p and q are the respective orders of the autoregressive and moving average lag polynomials. The moving average lag polynomial is defined as $\beta(L) = \beta_1 L + \ldots + \beta_p L^p$.

An GARCH process can be generated using genarch, which has the following syntax:

```
y = genarch(a,b,n)
```

where

- a is the vector which contains the constant and the coefficient of the autoregressive lag polynomial,

- b is the vector of parameters of the moving average lag polynomial,

- n is the size of the simulated series.

If the absolute value of the sum of coefficients of the autoregressive and moving average polynomials is greater than 1, i.e. if the generated process is not stationary, XploRe displays an error message.

In the following example, 250 observations of a GARCH(1,1) process are generated and displayed:

```
a = #(1,0.45)        ; AR part of GARCH errors
b = #(0.5)           ; MA part of GARCH errors
y = genarch(a,b,250)
```
                                                            ![Q] times12.xpl

Figure 9.8 shows the process.

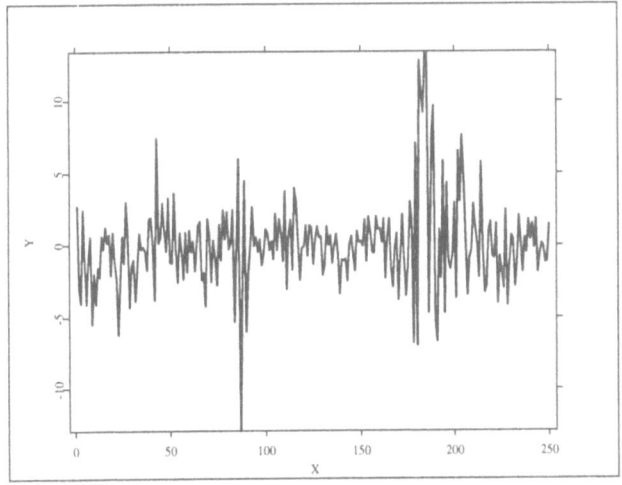

Figure 9.8. 250 observations of a GARCH(1,1) process.

The displayed time series shows that the GARCH process is characterized by the clustering of large deviating observations. This occurs typically in financial time series.

We can verify that the generated series is leptokurtic by evaluating its kurtosis with the command **kurtosis(y)** which yields the following output:

```
Contents of k
[1,]    7.1956
```

This value is greater than the kurtosis of the Gaussian distribution, i.e. 3. This feature is common to all ARCH and GARCH processes.

9.3.3   Estimating ARCH Models

ARCH/GARCH models are estimated in the time domain by the maximum likelihood method. This implies that we have to make an assumption about the distribution of the error terms. Usually the assumption of normality is sufficient and simplifies the estimation procedure. However, in some cases, this normality assumption is not sufficient for capturing the leptokurtosis of the sample under investigation. In that case we have to resort to other fat tailed distributions, such as the Student or the generalized exponential distribution.

Under the assumption of Gaussian innovations, we can estimate the parameters of a GARCH process using **archest**. Its syntax is as follows:

```
z = archest(y,q,p)
```

where

- y is the series of observations,
- p is the order of the $\beta(L)$ polynomial,
- q is the order of the $\alpha(L)$ polynomial.

To estimate the parameters of the previously generated process, type:

```
z = archest(y,1,1)
```

The quantlet **archest** estimates the parameters, their standard errors, and stores them in the form of a list. The first element of the list, **z{1}** contains the set of parameters, the second element of the list, **z{2}** contains the standard errors of estimates. Typing **z{1}** yields the parameter estimates

```
[1,]    1.9675
[2,]    0.40522
[3,]    0.42373
```

Typing z{2} returns the estimated standard errors

```
[1,]    0.57227
[2,]    0.07518
[3,]    0.077201
```

Note that **archest** returns additional values. For their description consult the APSS help file.

9.3.4 Testing for ARCH

We test for ARCH effects by investigating whether the components of the autoregressive lag polynomial are all equal to zero. Two formulations for a test for ARCH are usually considered:

- The Lagrange Multiplier test statistic, given by

$$\text{A-LM} = \frac{1}{2}\tilde{\varepsilon}^T\tilde{\varepsilon},$$

  where $\tilde{\varepsilon}$ is the T-dimensional vector of estimated endogenous variables in the auxiliary regression

$$\frac{\hat{\varepsilon}_t^2}{\hat{\sigma}^2} - 1 = \gamma_0 + \gamma_1\hat{\varepsilon}_{t-1}^2 + \ldots + \gamma_n\hat{\varepsilon}_{t-n}^2 + \zeta_t.$$

  Here $\hat{\varepsilon}_t$ are the residuals from the regression model, and $\sigma_t^2 = T^{-1}\sum_t \hat{\varepsilon}_t^2$.

- The R^2 form, with the test statistic equal to

$$TR^2,$$

  where R^2 is the squared multiple R^2 value of the regression of $\hat{\varepsilon}_t^2$ on an intercept and n lagged values of $\hat{\varepsilon}_t^2$.

Both tests are asymptotically equivalent, and asymptotically distributed as χ_n^2 under the null hypothesis.

The XploRe quantlet **archtest** performs these two tests. Its syntax is

```
h = archtest(y {,lags {,testform}})
```

where

- y is the vector of residuals.

- lags is the number of lags in the auxiliary regression. This argument
  may either be a scalar or a vector. In the latter case, the statistics is
  computed for all the order components of the vector. By default, the lag
  orders 2, 3, 4, and 5 are computed.

- testform is the form of the test. This argument is a string which can be
  either "LM" or "TR2". In the former case the Lagrange multiplier form is
  evaluated, while in the latter case the R^2 form is computed.

This function returns a table: The first column contains the order of the test,
the second column contains the value of the test, the third column contains the
95% critical value for the respective order, and the fourth column contains the
p-value of the test.

In our generated sample of observations, we test for ARCH with the command

```
archtest(y,"TR2")
```

This calculates the R^2 form of the ARCH test for the default lags 2, 3, 4 and
5. The results are displayed in the form of a table:

```
[1,] Lag order  Statistic  95% Critical Value  P-Value
[2,] ------------------------------------------------------
[3,]
[4,]     2       85.45238        5.97378          0.00000
[5,]     3      105.05328        7.80251          0.00000
[6,]     4      104.68014        9.47844          0.00000
[7,]     5      105.15906       11.06309          0.00000
```

We recommend to consult the APSS help file of **archest** and to play around
with the numerous variants of this quantlet.

Kamstra (1993), Caulet and Péguin–Feissolle (1997) and Péguin–Feissolle (1999)
consider a general nonparametric test for ARCH based on neural networks, in
the spirit of the Lee, White and Granger (1993) nonlinearity test presented

above. Caulet and Péguin–Feissolle (1997) consider the following parameterization of the conditional variance

$$\sigma_t^2 = \beta_0 + \sum_{j=1}^{q} \frac{\beta_j}{1 + \exp\{-(\gamma_{j0} + \gamma_{j1}\varepsilon_{t-1}^2 + \ldots + \gamma_{jn}\varepsilon_{t-n}^2)\}} \,. \qquad (9.1)$$

Péguin–Feissolle (1999) considers the more general case

$$\sigma_t^2 = \beta_0 + \sum_{j=1}^{q} \frac{\beta_j}{1 + \exp\{-(\gamma_{j0} + \gamma_{j1}\varepsilon_{t-1} + \ldots + \gamma_{jn}\varepsilon_{t-n})\}} \,,$$

i.e. extends the σ-field of the information set from the set of squared residuals to the set of residuals. This extended test appears to be more powerful than other tests for ARCH when the data generating process is a nonstandard one. The parameters $\gamma_{i,j}$ are randomly generated for solving the problem of parameter identification under the null hypothesis. All these neural network based test statistics are asymptotically χ_n^2 distributed under the null hypothesis.

The quantlet `annarchtest` evaluates the Lagrange multiplier form and the R^2 form for the specification (9.1). Its syntax is

```
h = annarchtest(y, {,nlags {,nodes {,testform}}})
```

where

- `y` is the vector of residuals.

- `nlags` is the number of lags in the auxiliary regression. This second argument may either be a vector or a scalar. If this argument is a vector, the test will be calculated for all order components of this vector. By default, the number of lags is set to 2, 3, 4 and 5.

- `nodes` is the number of hidden nodes in the neural network architecture. This argument may either be a vector or a scalar. By default, `nodes` is set to 3.

- If both the second and third arguments are vectors, the statistic will be calculated for all combinations of the second and third arguments.

- `testform` is the form of the test, which — as in the previous case — is either `"LM"` or `"TR2"` depending on the form of the test you wish to compute. By default, the `"LM"` form is calculated.

This function returns the results in the form of a table: The first column contains the number of lags in the auxiliary regression, the second column contains the number of hidden nodes, the third column contains the calculated statistic, the fourth column contains the 95% critical value for that order, and the last column contains the p-value of the test.

We calculate the test for ARCH based on neural networks with the command

```
annarchtest(y)
```

This returns

	Lag order	Nb of hidden units	Statistic	95% Critical Value	P-Value
[1,]					
[2,]					
[3,]	------	------	------	------	------
[4,]					
[5,]	2	3	26.56036	5.97378	0.00000
[6,]	3	3	60.80811	7.80251	0.00000
[7,]	4	3	14.44156	9.47844	0.00601
[8,]	5	3	27.32019	11.06309	0.00005

Bibliography

Andrews, D. W. K. (1991). Heteroskedasticity and Autocorrelation Consistent Covariance Matrix Estimation, *Econometrica* **59**: 817–858.

Brockwell, P. J. and Davis, R. A. (1991). *Time Series: Theory and Methods*, Springer-Verlag, New York.

Caulet, R. and Péguin–Feissolle, A. (1997). A Test for Conditional heteroskedasticity Based on Artificial Neural Networks, *GREQAM DT* **97A09**.

Grassberger and Procaccia (1983). Measuring the Strangeness of Strange Attractors, *Physica* **9D**: 189–208.

Harvey, A.C. (1993). *Time Series Models*, Harvester Wheatsheaf.

Kamstra, M. (1993). A Neural Network Test for Heteroskedasticity, *Simon Fraser Working Paper*.

Kwiatkowski, D., Phillips, P. C. B., Schmidt, P. and Shin, Y. (1992). Testing the Null Hypothesis of Stationarity Against the Alternative of a Unit Root: How Sure Are We that Economic Series Have a Unit Root, *Journal of Econometrics* **54**: 159–178.

Lee, T. H., White, H. and Granger, C. W. J. (1993). Testing for Neglected Nonlinearity in Time Series Models – A Comparison of Neural Network Methods and Alternative Tests, *Journal of Econometrics* **56**: 269–290.

Lee, D. and Schmidt, P. (1996). On the Power of the KPSS Test of Stationarity Against Fractionally-Integrated Alternatives, *Journal of Econometrics* **73**: 285–302.

Newey, W. K. and West, K. D. (1987). A Simple Positive Definite, Heteroskedasticity and Autocorrelation Consistent Covariance Matrix, *Econometrica* **55**: 703–705.

Péguin–Feissolle, A (1999). A Comparison of the Power of Some Tests for Conditional Heteroskedasticity, *Economics Letters*, forthcoming.

Siddiqui, M. (1976). The Asymptotic Distribution of the Range and Other Functions of Partial Sums of Stationary Processes, *Water Resources Research* **12**: 1271–1276.

Teräsvirta, T. (1998). Modeling Economic Relationships with Smooth Transition Regression Function, *Handbook of Applied Economic Statistics*.

Velasco, C. (1998). Gaussian Semiparametric Estimation of Non–Stationary Time Series, *Journal of Time Series Analysis*, forthcoming.

10 Kalman Filtering

Petr Franěk

In recursive methods the construction of an estimate at time t is based on an estimate from the previous time and the observations available in the time t. Exponential smoothing and Yule–Walker equations are examples of recursive algorithms but by defining a state–space model one can build a unifying theory of recursive methods with the Kalman filter as a general (linear) solution of filtering, smoothing and prediction problems.

All routines for Kalman filtering, which will be explained in the following, are part of the `times` library.

10.1   State–Space Models

`y = kemitor (T, x0, H, F, ErrY, ErrX)`
      simulates observations of a time-independent state–space model

A time-independent state–space model is defined by the two equations

$$y_t = Hx_t + v_t,$$
$$x_t = Fx_{t-1} + w_t, \quad t \in T,$$

where the first equation is called the **observation equation** and the second equation is called the **state equation**. The state x_t is assumed to be a $(n \times 1)$ vector and the observation y_t is assumed to be a $(m \times 1)$ vector. In practical applications, we usually work with a discrete-time model and therefore we usually take T as a set of integer numbers. The vectors v_t and w_t represent random effects and we assume that they are centered with a covariance structure given by

$$
\begin{aligned}
Ev_s v_t^T &= \delta_{st} Q, \\
Ew_s w_t^T &= \delta_{st} R, \\
Ev_s w_t^T &= 0,
\end{aligned}
$$

where the symbol δ_{st} stands for Kronecker delta, Q and R are some covariance matrices. To have the state–space model properly defined some assumptions about the initial state must also be made. We will denote the initial state as x_0 (supposing $t \in T = \{0, 1, \dots\}$). Standard assumptions are

$$
\begin{aligned}
Ex_0 &= \mu, \\
\operatorname{Var} x_0 &= \Sigma,
\end{aligned}
$$

where Σ is a $(n \times n)$ covariance matrix. Both μ and Σ may be known or unknown but we will assume here that Σ is known. Note that $\Sigma = 0$ corresponds to a situation when x_0 is a deterministic vector.

In order to keep the state–space model reasonably simple we assume the sequence $\{x_0, (v_t^T, w_t^T)^T\}_{t \in T}$ to be orthogonal. The matrices H, F, Q and R (sometimes they are referred to as system matrices) were originally (in engineering applications) assumed to be known but in econometric applications some of them may be unknown (but we still assume they are nonstochastic).

One can easily check that the sequence $\{x_t\}$ has a Markovian property if the vectors in the sequence $\{x_0, (v_t^T, w_t^T)^T\}_{t \in T}$ are independent. This is satisfied for example in a case when $\{v_t\}$ and $\{w_t\}$ are Gaussian errors (the model is then said to be Gaussian).

The state–space models include most of the well-known time series and linear models. In the following examples state representations of Holt–Winters method and ARMA(p, q) model are introduced.

10.1.1   Examples of State–Space Models

The state–space equivalent of the Holt–Winters method has the form

$$
\begin{aligned}
S_t &= S_{t-1} + T_{t-1} + \delta S_t, \\
T_t &= T_{t-1} + \delta T_t, \\
I_t &= -\sum_{i=1}^{p-1} I_{t-1} + \delta I_t, \\
y_t &= S_t + I_t + \varepsilon_t.
\end{aligned}
$$

The random variables δS_t, δT_t, δI_t and ε_t are assumed to be white noise and mutually uncorrelated. Note that the random components δS_t, δT_t and δI_t are not present in the classical form of the method and they represent wider abilities of state–space representation of the method.

ARMA(p, q) model

$$
X_t - \phi_1 X_{t-1} - \cdots - \phi_p X_{t-p} = \varepsilon_t + \theta_1 \varepsilon_{t-1} + \cdots + \theta_q \varepsilon_{t-q}, \quad t = 0, \pm 1, \ldots
$$

when $\{\varepsilon_t\}$ is a white noise and has also a state–space representation. Denoting

$$
m = \max\,(p, q + 1), \quad \phi_j = 0 \quad \text{for } j > p, \quad \theta_j = 0 \quad \text{for } j > q
$$

and

$$
x_t = \begin{bmatrix} X_t & X_{t-1} & \cdots & X_{t-m+1} \end{bmatrix},
$$

the state–space form of an ARMA(p, q) is

$$
\begin{aligned}
y_t &= \begin{bmatrix} 1 & 0 & \cdots & 0 \end{bmatrix} x_t, \\
x_t &= \begin{bmatrix}
\phi_1 & 1 & 0 & \cdots & 0 \\
\phi_2 & 0 & 1 & \cdots & 0 \\
\cdots & \cdots & \cdots & \cdots & \cdots \\
\phi_{m-1} & 0 & 0 & \cdots & 1 \\
\phi_m & 0 & 0 & 0 & 0
\end{bmatrix} x_{t-1} +
\begin{bmatrix}
1 \\
\theta_1 \\
\cdots \\
\theta_{m-1}
\end{bmatrix} \varepsilon_t, \quad t \in T.
\end{aligned}
$$

More detailed information about state–space models may be found for example in Harvey (1990).

10.1.2   Modeling State–Space Models in XploRe

Time-invariant state–space models can be easily simulated using the XploRe quantlet **kemitor**. One hundred observations of the state–space model

$$
\begin{aligned}
y_t &= \begin{bmatrix} 1 & 0 \end{bmatrix} x_t + v_t, \\
x_t &= \begin{bmatrix} 0.5 & -0.3 \\ 1 & 0 \end{bmatrix} x_{t-1} + \begin{bmatrix} w_t \\ 0 \end{bmatrix},
\end{aligned}
$$

with $x_0 = 0$, $v_t \sim N(0,4)$ and $w_t \sim N(0,1)$ may be simulated with the following set of instructions:

```
library("times")
T = 100
randomize(0)
ex = normal(T)~(vec(1:T).*0)
ey = normal(T).*2
H = 1~0
F = #(0.5,1)~#(-0.3,0)
x0 = #(0,0)
ar2 = kemitor(T,x0,H,F,ey,ex)
```

🔍 kalm01.xpl

Note that this time series corresponds to an AR(2) process

$$
y_t = 0.5\, y_{t-1} - 0.3\, y_{t-2} + \varepsilon_t
$$

with additive Gaussian noise.

The fact that the errors are pregenerated and supplied as parameters of the quantlet allows us to model errors with distributions different from the Gaussian one or error terms with some special interdependencies. In this framework, it is for example possible to model easily time series with different kinds of outliers.

10.2   Kalman Filtering and Smoothing

---

fy = kfilter (y, mu, Sig, H, F, Q, R)
    provides Kalman filtering of a (multivariate) time series

sy = ksmoother (y, mu, Sig, H, F, Q, R)
    provides Kalman smoothing of a (multivariate) time series

---

The state–space model consists of two processes — an observation process $\{y_t\}_{t\in T}$ and an unobservable state process $\{x_t\}_{t\in T}$. Having a sampling of observations made up to time k, denoted as $Y_k = \{y_1, \ldots, y_k\}$, we want to find the best estimate of the state x_t that we denote as $\widehat{x}_{t|k}$. The particular error covariance matrix is then denoted $P_{t|k} = (x_t - \widehat{x}_{t|k})(x_t - \widehat{x}_{t|k})^T$. Three different problems are recognized due to different relations between t and k:

- **Prediction**, i.e. estimating x_t in terms of $y_1, \ldots, y_{t-1}$ (i.e. $k = t - 1$).

- **Filtering**, i.e. estimating x_t in terms of $y_1, \ldots, y_t$ (i.e. $k = t$).

- **Smoothing**, i.e. estimating x_t in terms of all observations $y_1, \ldots, y_T$ (i.e. $k = T$).

To determine the best estimate, the Kalman filtering approach uses the mean squared error (MSE) criterion given as $E(x_t - \widehat{x}_{t|k})^T (x_t - \widehat{x}_{t|k})$. According to this criterion the best estimate is the conditional expectation $\widehat{x}_{t|k} = E(x_t \mid Y_k)$. This expectation is generally nonlinear (and usually difficult to find) and therefore we confine ourselves to linear filters. Since we assume the orthogonality of $(v_t^T, w_t^T)^T$ we are able to derive the desired best linear estimate using techniques of projections onto Hilbert space generated by the observations Y_k.

As a result we get the Kalman filter equations

$$\begin{aligned}
\widehat{x}_{t|t} &= \widehat{x}_{t|t-1} + P_{t|t-1} H^T \Delta_t^{-1}(y_t - H\widehat{x}_{t|t-1}), \\
P_{t|t} &= P_{t|t-1} - P_{t|t-1} H^T \Delta_t^{-1} H P_{t|t-1}^T,
\end{aligned}$$

where $\Delta_t = H P_{t|t-1} H^T + Q$, the Kalman prediction equations

$$\begin{aligned}
\widehat{x}_{t+1|t} &= F\widehat{x}_{t|t-1} + K_t(y_t - H\widehat{x}_{t|t-1}), \\
P_{t+1|t} &= F(P_{t|t-1} - P_{t|t-1} H^T \Delta_t^{-1} H P_{t|t-1})F^T + R,
\end{aligned}$$

and the Kalman smoothing equations

$$
\begin{aligned}
\widehat{x}_{t|T} &= \widehat{x}_{t|t} + P_t^*(\widehat{x}_{t+1|T} - F\widehat{x}_{t|t}), \\
P_{t|T} &= P_{t|t} - P_t^*(P_{t+1|T} - P_{t+1|t})P_t^{*T}, \\
P_t^* &= P_{t|t}F^T P_{t+1|t}^{-1}.
\end{aligned}
$$

The matrix $K_t = FP_{t|t-1}H^T\Delta_t^{-1}$ is called a gain matrix. Please note that the easy connection between the first two recursions is given by the equation

$$
\widehat{x}_{t|t-1} = F\widehat{x}_{t-1|t-1}
$$

and that the smoothing recursion consists of the backward recursion that uses the filtered values of x and P.

Several remarks should be made about the Kalman filter equations. First the inversion of Δ_t may be replaced with a pseudoinversion. This is generally the case in models with Q singular. Next, the Kalman filter is a minimum square error estimator among all linear estimators but in the case of a Gaussian model it is the minimum square error estimator among all estimators and $\mathcal{L}(x_t \mid Y_k) = N(\widehat{x}_{t|k}, P_{t|k})$, i.e. the Kalman filter yields the whole information about the conditional distribution of x_t (this, however, does not hold in a general non-Gaussian model where the Kalman filter yields only information about the first two moments of the conditional distribution of x_t).

The Kalman filtration equations are implemented in the quantlet kfilter. The input parameters of this quantlet are the time series to be filtered (possibly multivariate), and the system matrices of the underlying state–space model. To filtrate the time series **ar2** simulated in the first example type the following instructions.

```
x0 = #(0,0)
Sig = #(0,0)~#(0,0)
H = 1~0
F = #(0.5,1)~#(-0.3,0)
Q = 4
R = #(1,0)~#(0,0)
filtered = kfilter(ar2,x0,Sig,H,F,Q,R)
```

Q kalm02.xpl

The filtered series is in the variable **filtered** and the result may be visualized as follows:

```
library("plot")
orig = vec(1:T)~ar2
filt = vec(1:T)~filtered
orig = setmask(orig, "line", "red", "thin")
filt = setmask(filt, "line", "blue", "medium")
disp = createdisplay(1,1)
show(disp,1,1, orig, filt)
setgopt(disp,1,1, "title", "AR(2) with noise - filtered")
```
Q kalm02.xpl

The generated series is plotted as a thin red line while the filtered series is plotted as a medium blue line.

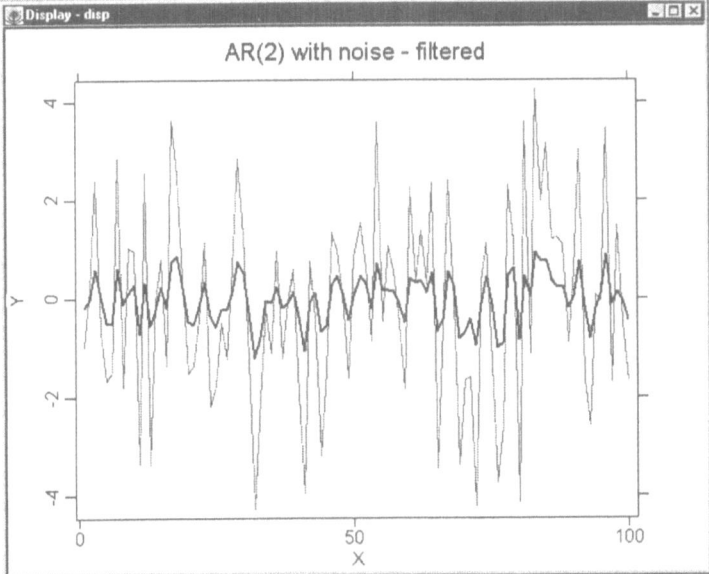

The quantlet kfilter saves the file KFOutPut.dat into the XploRe working directory. This file contains all filtered state estimates $x_{t|t}$ and their error covariance matrices $P_{t|t}$ and might be used to track the development of the estimations' errors or for prediction purposes.

The matrices $P_{t|t}$ are vectorized and appended to the state estimation but each of them may be easily recovered. The values of $x_{50|50}$ and $P_{50|50}$ might be recovered in the following way:

```
KFOutPut = read("KFOutPut.dat")
dimX = rows(x0)
x50 = (KFOutPut[50,1:dimX])'
P50 = reshape(KFOutPut[50,dimX+1:dimX+dimX^2],#(dimX,dimX))
```

Kalman smoothing equations are implemented by the quantlet `ksmoother`. Its usage is similar to the quantlet `kfilter`. Input parameters consist of the time series to be smoothed (possibly multivariate) and the system matrices of the underlying state–space model. In the following sample code the time series `ar2` is smoothed and the result is visualized.

```
smoothed = ksmoother(ar2,x0,Sig,H,F,Q,R)
smoot = vec(1:T)~smoothed
smoot = setmask(smoot, "line", "blue", "medium")
disp = createdisplay(1,1)
show(disp,1,1, orig, smoot)
setgopt(disp,1,1, "title", "AR(2) with noise - smoothed")
```

🔍 kalm03.xpl

The quantlet `ksmoother` uses the file `KFOutPut.dat` during the backward recursions. It generates the file `KSOutPut` and saves it into the XploRe working directory. Again, this file might be used for analytical purposes.

10.3   Parameter Estimation in State–Space Models

{Estmu,EstF,EstQ,EstR} = kem (ar2,x0,Sig,H,F,Q,R,Limit)
     provides estimations of μ, F, Q and R in a state–space model
     using the EM-algorithm

As stated in the previous paragraphs, some of the matrices may be unknown. In a usual application, the matrix H and the covariance matrix Σ of the initial state may be assumed known. Shumway and Stoffer (1982) have proposed an **EM-algorithm** for the iterative estimation of the unknown parameters μ, F, Q and R.

The algorithm is derived under the assumption of Gaussian error terms. Then the logarithm of the likelihood function is expressed as

$$
\begin{aligned}
\log L \;=\; & -\frac{1}{2}\log|\,\Sigma\,| - \frac{1}{2}(x_0 - \mu)^T \Sigma^{-1}(x_0 - \mu) \\
& -\frac{T}{2}\log|\,R\,| - \frac{1}{2}\sum_{t=1}^{T}(x_t - F x_{t-1})^T R^{-1}(x_t - F x_{t-1}) \\
& -\frac{T}{2}\log|\,Q\,| - \frac{1}{2}\sum_{t=1}^{T}(y_t - H_t x_t)^T Q^{-1}(y_t - H_t x_t).
\end{aligned}
$$

In the rth **E-step** of the algorithm, the conditional expectation

$$
G_r(\mu, F, R, Q) = E(\log L \mid Y_T)
$$

is derived using $\hat{x}_{t|T}$ and $P_{t|T}$ taken from Kalman recursions for smoothing used with parameters estimated in the previous step of the algorithm. In the rth **M-step** of the algorithm, the parameter estimates that maximize the function $G_r(\mu, F, R, Q)$ are taken. These values are used in the next E-step. The EM-algorithm provides estimates that converge to stable solutions. However, depending on initial values used in the first step of the algorithm, this solution may be a local maximum of the likelihood function. It is therefore recommended to try to run the algorithm with several different initial values. Please note that the algorithm sometimes converges to the stable solution quite slowly.

This algorithm is implemented in XploRe by the quantlet **kem**. The input parameters are almost the same as in the functions for Kalman filtering and smoothing, although the values of μ, F, Q and R are used as initial values for the first step of the EM-algorithm. The last parameter of the quantlet is the maximum number of iterations allowed. This quantlet returns estimated parameters μ, F, Q and R (in this order).

To run this quantlet on the **ar2** series and view the estimates, you can use the following sample code. Please note that depending on your system the processing may take a while.

```
x0 = #(0,0)
Sig = #(0,0)~#(0,0)
H = 1~0
F = #(5,1)~#(1,5)
Q = 10
```

```
R = #(1,0)~#(0,1)
{Estmu, EstF, EstQ, EstR} = kem(ar2,x0,Sig,H,F,Q,R,20)
Estmu
EstF
EstQ
EstR
```

Q kalm04.xpl

The state–space model proposed by the EM-algorithm should be similar to the
following one (the numbers may differ for your particular ar2 series).

```
Contents of Estmu
[1,] -0.049532
[2,]  0.011073

Contents of EstF
[1,]  0.59788  0.060844
[2,]  0.73477  3.7628

Contents of EstQ
[1,]  3.7002

Contents of EstR
[1,]  0.20202 -0.031133
[2,] -0.031133  0.76948
```

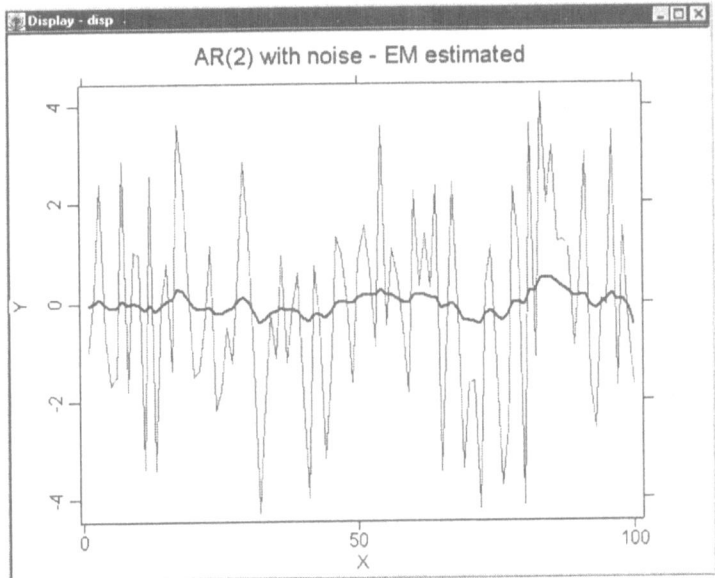

Note that not knowing the particular shape of the system matrices (i.e. the zero element in F, for example) the quantlet proposed its own state–space model. There are several works that generalize the EM-algorithm so that it allows us to have some elements of the unknown matrices fixed or to apply some linear constraints on them. You may, for example, refer to the article Hosking, Pai and Wu (1996). The implementation of these modifications in XploRe, based on the quantlet kem, should be straightforward.

Bibliography

Anderson, B. D. O. and Moore, J. B. (1979). *Optimal filtering*, Prentice–Hall, Englewood Cliffs.

Brockwell, P. J. and Davis, R. A. (1991). *Time Series: Theory and Methods*, Springer–Verlag, New York.

Harvey, A. C. (1990). *Forecasting, Structural Time Series Models and the Kalman Filter*, Cambridge University Press, Cambridge.

Hosking, J. R. M., Pai, J. S. and Wu, L. S. Y. (1996). An Algorithm for Estimating Parameters of State–Space Models, *Statistics and Probability Letters* **28**: 99–106.

Shumway, R. H. and Stoffer, D. S. (1982). An Approach to Time Series Smoothing and Forecasting Using the EM Algorithm, *Journal of Time Series Analysis* **4**: 253–263.

11 Finance

Stefan Sperlich and Wolfgang Härdle

There is growing interest in quantifying and simulating economic processes, particularly in the statistical analysis of the behavior of financial markets. The library **finance** is designed for this purpose. This chapter explains and illustrates the use of XploRe for theory and practice in this setting.

The **finance** library offers functions to predict, to simulate and to estimate time series processes as for example stock returns, to determine option prices and to evaluate different scenarios (e.g. for portfolio strategies). To give a survey of the library we will present the principal procedures implemented in the finance library and illustrate their use with several graphics.

Before starting to work with the finance library in XploRe you have to load all the functions contained in the library by typing the command

```
library("finance")
```

11.1   Outline of the Theory

In the last decades there has been a growing interest in the behavior of financial markets. Due to the increasing globalization of markets, they began to play a central role in international business and economic decision making. Thus, the meaning of "risk" became the central theme in this context.

Risk management is essential in a modern market economy. Financial markets enable firms and households to select an appropriate level of risk in their transactions, by redistributing risks towards other agents who are willing and able to assume them. Markets for options, futures and other so-called derivative securities — derivatives, for short — have a particular status. Futures allow agents to hedge against upcoming risks; such contracts promise the future delivery of a certain item at a certain price. As an example, a firm might decide to engage in copper mining after determining that the metal to be extracted can be sold in advance at the futures market for copper. The risk of future movements in the copper price is thereby transferred from the owner of the mine to the buyer of the contract. Due to their design, options allow agents to hedge against one-sided risks; options give the right, but not the obligation, to buy or sell something at a prespecified price in the future.

In avoiding the risk of long positions one could for example try to hedge the risk by going short in options on the corresponding asset and adapting the proportion held in assets and short-selled options according to the underlying price process of that asset. Therefore, formulas for the pricing of those derivative securities generated a lot of practical and theoretical interest.

Already in the year 1900, Bachelier introduced Brownian motion as a model for price fluctuations on a speculative market. In 1973, Black and Scholes founded their famous option pricing formula which calculates the "fair price" of an option (which means that there is no arbitrage). This has generated a lot of theoretical work relying on that basic model.

11.1.1   Some History

The valuation of derivatives has a long history. One of the earliest endeavors was undertaken by Louis Bachelier (thesis at Sorbonne, 1900). But his formula was based on such assumptions as zero interest rate, and a process that allowed for a negative share price.

This formula was improved by Case Sprenkle, James Boness and Paul Samuelson. They assumed that stock prices are log-normally distributed, guaranteeing that share prices are positive, and allowed for a nonzero interest rate. They also assumed that investors are risk averse and demand a risk premium additionally to the interest rate. In 1964, Boness suggested a formula that came close to the Black–Scholes formula, but still relied on an unknown interest rate, which included compensation for the risk associated with the stock.

Further attempts at valuation (before 1973) basically determined the expected value of a stock option at expiration and discounted its value back to the time of evaluation. Unfortunately, those approaches require taking a stance on which risk premium to use in the discounting. But assigning a risk premium is not straightforward, since it should reflect not only the risk for changes in the stock price, but also the investors attitude towards risk. The latter is hard or impossible to observe in reality.

11.1.2   The Black–Scholes Formula

A commonly used model for the description of fluctuations of asset prices is the following. $X(.)$ denotes the price process which is assumed to be the solution of the stochastic differential equation

$$dX(t) = s(t, x) \, dW(t) + m(t, x) \, dt.$$

Here $W(.)$ denotes Brownian motion, $s(., X)$ is the volatility process and $m(., X)$ is the trend or drift. Classical models suppose that $s(t, X) = \sigma X(t)$ and $m(t, X) = \mu X(t)$ which results in geometric Brownian motion.

Fischer Black, Robert Merton and Myron Scholes developed a new method of determining the value of derivatives. Their work (in the early 1970s) solved a longstanding problem in financial economics and has provided ways of dealing with financial risk, both in theory and in practice. Further, their methodology has proven general enough for a wide range of applications. It can thus be used to value not only the flexibility of physical investment projects but also insurance contracts and guarantees.

In the press release, when Scholes and Merton were awarded the Nobel Prize in 1997, was given the following example: Consider a European call option at a strike price of $100 in three months. (A European option gives the right to buy or sell only at a certain date, whereas a so-called American option gives the same right at any point in time up to a certain date.) Clearly, the value of this call option depends on the current share price; the higher the share price today the greater the probability that it will exceed $100 in three months, in which case it will pay to exercise the option. A formula for option valuation should thus determine exactly how the value of the option depends on the current share price. How much the value of the option is altered by a change in the current share price is called the "delta" of the option — see also the **greeks**.

Assume that the value of the option increases by $1 when the current share price goes up $2 and decreases by $1 when the stock goes down $2. Assume also that an investor holds a portfolio of the underlying stock and wants to hedge against the risk of changes in the share price. He can then construct a risk-free portfolio by selling twice as many options as the number of shares he owns. For reasonably small increases in the share price, the profit the investor makes on the shares will be the same as the loss he incurs on the options, and vice versa for decreases in the share price. As the portfolio thus constructed is risk free, it must yield exactly the same return as a risk-free three-month treasury bill. If it did not, arbitrage trading would begin to eliminate the possibility of making risk-free profits. As the share price is altered over time and as the time to maturity draws nearer, the delta of the option changes. In order to maintain a risk-free stock-option portfolio, the investor has to change its composition.

Black, Merton and Scholes assumed that such trading can take place continuously without any transaction costs. The condition that the return on a risk-free stock-option portfolio yields the risk-free rate, at each point in time, implies a partial differential equation, the solution of which is the Black–Scholes formula for a call option:

$$C = SN(d) - Le^{-rt}N\left(d - \sigma\sqrt{t}\right)$$

where $N()$ is the standard normal distribution, S, t, r, L see below and d is defined by

$$d = \frac{\log(S/L) + (r + \sigma^2/2)t}{\sigma\sqrt{t}}.$$

According to this formula, the value of the call option C is given by the difference between the expected share price — the first term on the righthand side — and the expected cost — the second term — if the option is exercised. The higher the option value, the higher the current share price S, the higher the volatility of the share price σ, the higher the risk-free interest rate r, the longer the time to maturity t, the lower the strike price L, and the higher the probability that the option will be exercised — see also the quantlet influence.

All the parameters in the equation can be observed except sigma, which has to be estimated from market data. Alternatively, if the price of the call option is known, the formula can be used to solve for the market-implied volatility. Market equilibrium is not necessary for option valuation; it is sufficient that there are no arbitrage opportunities. The method described in the example above is based precisely on the absence of arbitrage. It generalizes to valuation of other types of derivatives. Mertons 1973 article included the Black–Scholes formula and some generalizations, for instance, he allowed the interest rate to be stochastic. The theory of Merton, Black and Scholes can also be used for many other or related fields such as:

- **Corporate liabilities**
  Black, Merton and Scholes realized already in 1973 that a share can be interpreted as an option on the whole firm. When loans mature and the value of the firm is lower than the nominal value of debt, the shareholders have the right, but not the obligation, to repay the loans. The method can thus be used for determining the value of shares, which can be important if the shares are not traded. Since other corporate liabilities are also derivative instruments (whose value, too, depends on the value of the firm), they can be valued using the same method.

- **Investment evaluation**

- **Guarantees and insurance contracts**

- **Complete markets**

11.2 Assets

There exist several quantlets for the simulation and estimation of asset prices. Implemented in the finance library are

```
stocksim ()
    simulates random processes for stock prices
stockest (data)
    estimates a diffusion model for stock price data
stockestsim (data)
    simulates and estimates a Wiener process with Poisson jump
```

11.2.1   Stock Simulation

The quantlet `stocksim` simulates random processes for a stock price in three different ways:

- using a Wiener process,

- using a Poisson jump process,

- using a mixture of both.

It is invoked by typing `stocksim()` An interactive window appears which asks for the values of the process to be simulated.

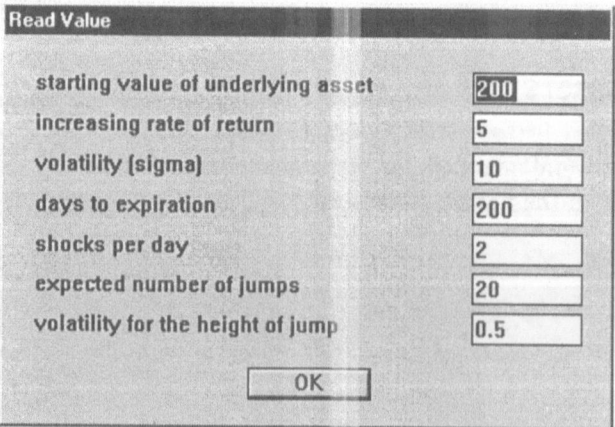

The function returns as output a display plotting the three processes and asks if one wants to repeat the simulation. In the interactive window one is asked for the starting values of the underlying asset, the increasing rate of return which corresponds to $m(t, x)$ in the underlying diffusion process. The volatility parameter σ corresponds to a constant $s(t, x)$. The expected number of jumps is the parameter for the underlying jump Poisson process. More precisely the geometric Brownian motion

$$\frac{dX(t)}{X(t)} = \mu dt + \sigma dW(t) \tag{11.1}$$

is simulated with an overlayed Poisson jump process. If we set the last two parameters (intensity and height of the jump) equal to zero we exactly simulate from (11.1) at discrete points. The first parameter equals $X(0)$ and the second is the drift μ. The volatility is given by b, the third parameter. If T^* denotes the days to expiration and nd is the observation frequency per day, the process is on the time interval $T^*/365 \subset [0, 1]$ on exactly $T = ndT^*$ discretization points. This may be checked by variation of these parameters. The process is recursively calculated as $X(t + 1) = X(t) \exp(\mu/nd + \sigma W(t)/\sqrt{nd})$ at the points $t = 1, 2, \ldots, T$.

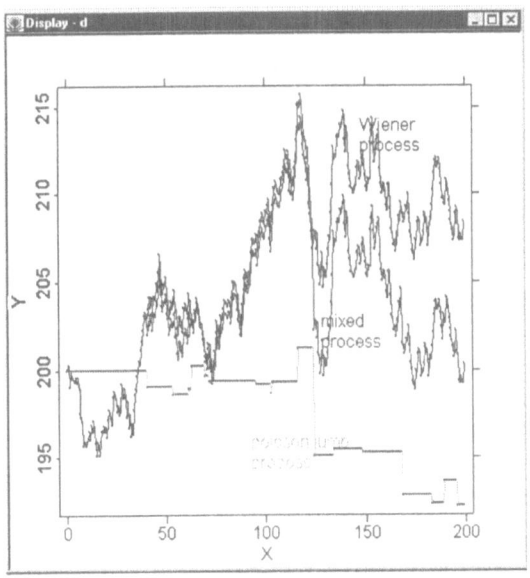

11.2.2   Stock Estimation

The stockest quantlet assumes that the underlying diffusion processes models are the same as under 11.2.1, i.e. a mixture of a Poisson jump and a Wiener process with drift. For the estimation of such a process, we have to choose a dataset that we want to examine. Let's estimate the parameters for the price process of the Motorola stock. The data is loaded into XploRe by typing

```
data=read("motorola")
```

in the command line of the XploRe input window. The data consists of 591 observations. It has 6 columns. We choose the second column — which simply contains the price notations of the stock — with the command data=data[,2] Estimation now takes place by executing stockest. This quantlet is executed by typing the name of the variable representing the dataset in parentheses:

```
stockest(data)
```

Now the corresponding parameters of the model are displayed in the XploRe output window. As an example, take the estimation of the volatility: In the output window you find the following information:

```
Content of object _tmp.sigma2
[1,]     38.819
```

The other estimated parameters are mue, the increasing rate of return, sigma the volatility of returns, lambda the number of jumps in the Poisson model and jump the volatility of the height of the jump.

11.2.3   Stock Estimation and Simulation

The quantlet stockestsim is a combination of the quantlets described in Subsections 11.2.1 and 11.2.2. At first it estimates with the first part of a given dataset the parameters of a random process. This is done for two kinds of models: a Wiener process and a combination of a Wiener and a Poisson jump process. Then both models are compared by a simulation with the rest of the real dataset.

As in the quantlet stockest, you need to choose the dataset first and then execute the function by putting the dataset as input parameter. This is done

in XploRe by typing the following sequence of commands in the command line
of the input window:

```
data=read("motorola")
data=data[,2]
stockestsim(data)
```

The result is a graphical display showing the three processes:

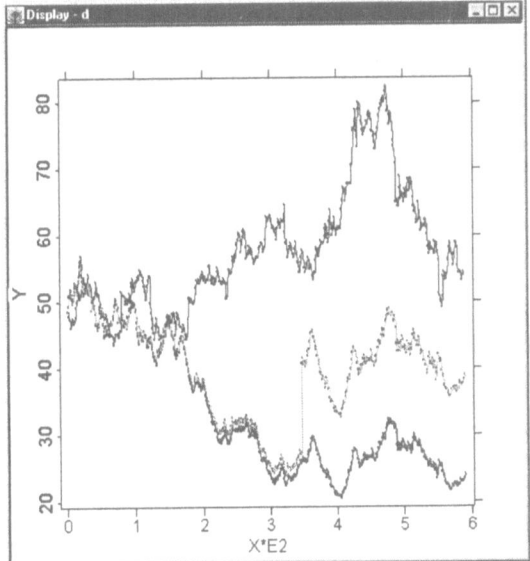

11.3   Options

11.3.1   Calculation of Option Prices and Implied Volatilities

A calculation of option prices is possible by using one of the following functions:

---

optstart ()
> starting program to calculate option prices or implied volatilities

bitree (vers, task)
> calculates option prices using the Binomial tree

{opvv,sel,ingred} = bs1 (task)
> calculates option prices using the Black–Scholes formula

mcmillan (eopv, sel, task, ingred)
> calculates option prices using the McMillan formula

american ()
> starting program to calculate option prices for american options

european ()
> starting program to calculate option prices or implied volatilities for european options

asset (vers)
> auxiliary quantlet to calculate option prices for american options

---

The interactive option pricing quantlet optstart is simply invoked by typing

    optstart()

in the XploRe command line. A selection box appears which starts the interactive option pricing procedure.

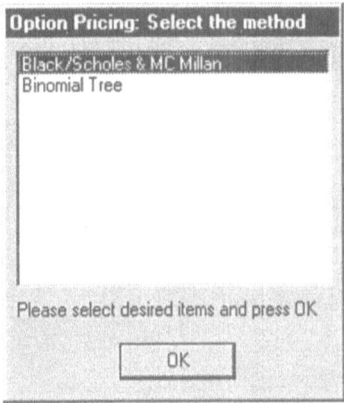

Simply select the method you want to use. If you wish to calculate the option price analytically, choose Black/Scholes & MC Millan; if you want XploRe to calculate it numerically, choose Binomial Tree. Let's choose Black/Scholes & MC Millan. In any case you will be asked whether you want to compute the price of an European (an option which can be executed only at a given date) or an American option (that can be executed anytime).

In this example we have chosen American. This is the kind of option that is usually traded e.g. in the USA or in Germany. The next decision is about the underlying asset (stock or exchange rate).

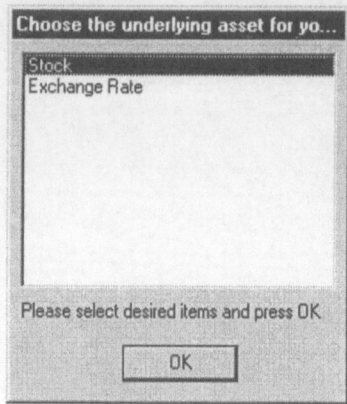

In our example we are regarding a stock as the underlying asset. In this setting you are questioned if you like to have dividends included in the stock and of what kind you want them to be. (If you choose **Exchange Rate** here, the next two menu items will be skipped.) Then you are asked whether you like to compute the price of an option or the implied volatility. Now we are ready to enter the parameters needed for the computation of the option prices. These are **Price of the Underlying Asset, Exercise Price, Domestic Interest Rate per Year** and **Volatility per Year** in percent as well as the **Time to Expiration** in years.

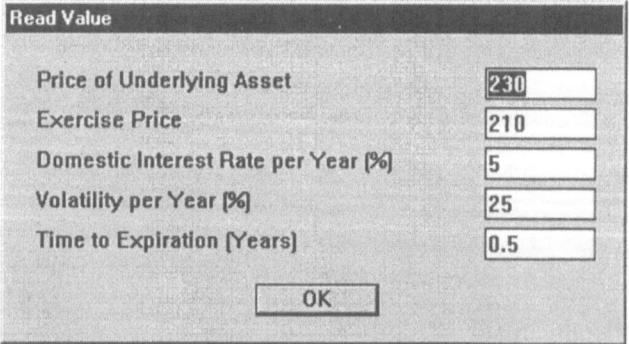

In case you have chosen a dividend payment, one more window will appear where you are asked to put the amount of the dividend.

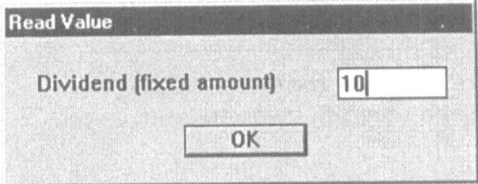

Finally you can choose the kind of option you like to calculate. Let's say we wanted to know the price of a call option:

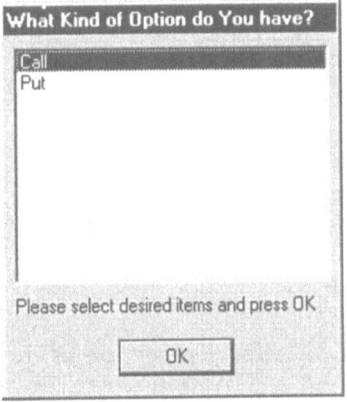

The price of our American call option on the given stock in the scenario (chosen through the corresponding parameters) with fixed dividend is now displayed in the XploRe output window. In case you have chosen a stock as underlying asset even the price of the European call option is displayed (in case you have not chosen a dividend, the price of a European call option equals that of an American call option):

```
[1,]
[2,] ------------------------------------
[3,]  The Price of Your European Call-Option
[4,]  on Given Stock with fixed Dividend is
[5,]  27.3669
[6,] ------------------------------------
[7,]
```

```
[1,]
[2,] ------------------------------------
[3,]  The Price of Your American Call-Option
[4,]  on Given Stock with fixed Dividend is
[5,]  27.4641
[6,] ------------------------------------
[7,]
```

11.3.2   Option Price Determining Factors

influence ()
    displays the influence of price determining parameters on options

The quantlet **influence** measures and visualizes the influence of different factors on the prices of options. It is simply started by typing

influence()

in the command line of XploRe. The option prices are calculated with the Black–Scholes formula. After starting the quantlet the following window appears:

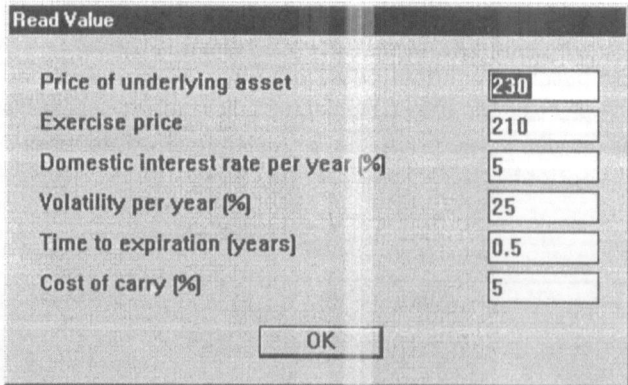

You may enter the different parameters needed to simulate the diffusion process. Next select the influence variables — you may select up to two variables. The following example demonstrates the use of just one variable:

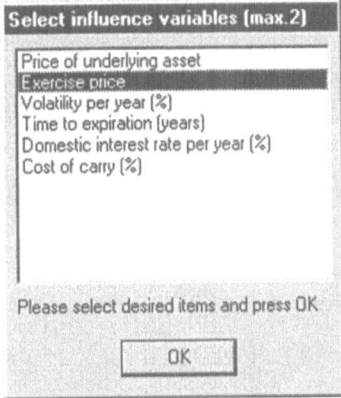

In this example we would like to calculate the influence of the exercise price on the option price. You must set the lower and upper bound for your chosen variable.

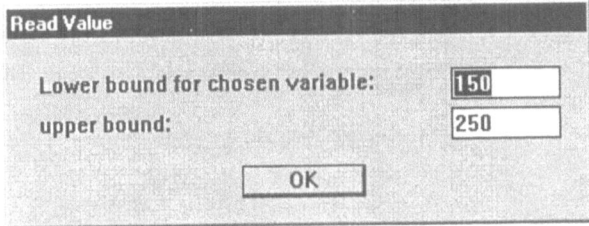

After pushing the OK button you will be asked for what kind of option the influence is to be calculated.

If you choose for example a Put option you will obtain the following graph which shows the influence of the factor (exercise price) on the price of the option:

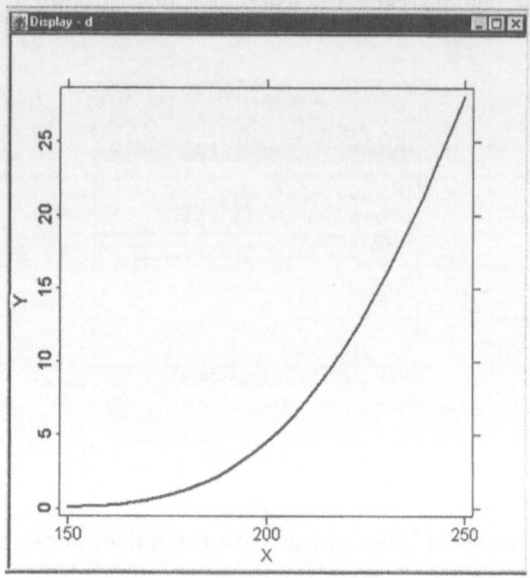

Using `influence` you can also select two variables as the following example demonstrates:

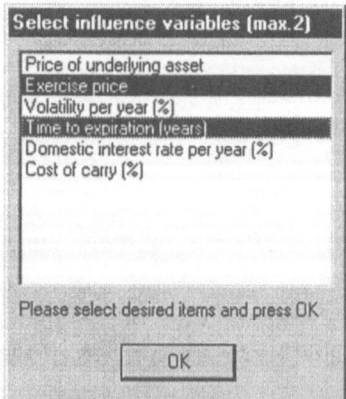

After selecting the two variables you wish to compute, XploRe asks you to set the lower and upper bound for both variables.

If you choose e.g. Put you will obtain a three-dimensional graphic with the two selected influence factors (exercise price and time to expiration) and the price of the option. You may turn the graphic around by using the cursor buttons.

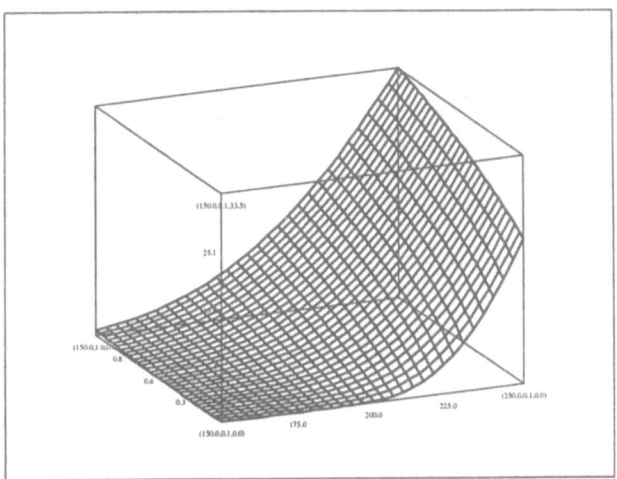

11.3.3   Greeks

> greeks ()
> > calculates and displays the different indices which are used for
> > trading with options

The interactive function **greeks** calculates and displays the different indices
used for analyzing and trading with options. You start it by

```
data=greeks()
```

The first step is to enter the asset's basic data:

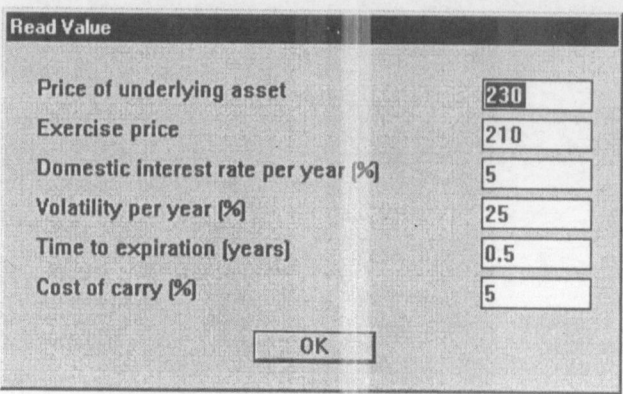

Next, you have to select the variables you want to analyze (at most two), e.g.

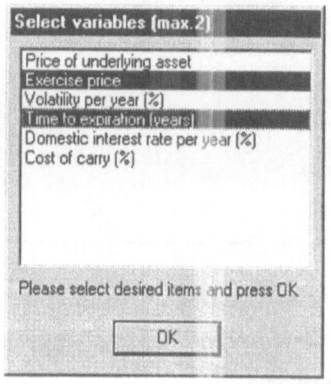

Select the ranges for the values of the chosen variables:

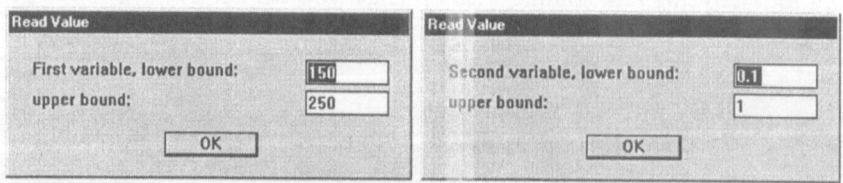

Now you can choose the index you are interested in:

After telling the program the kind of option you want, the quantlet **greeks** will produce a graphical output window for your result:

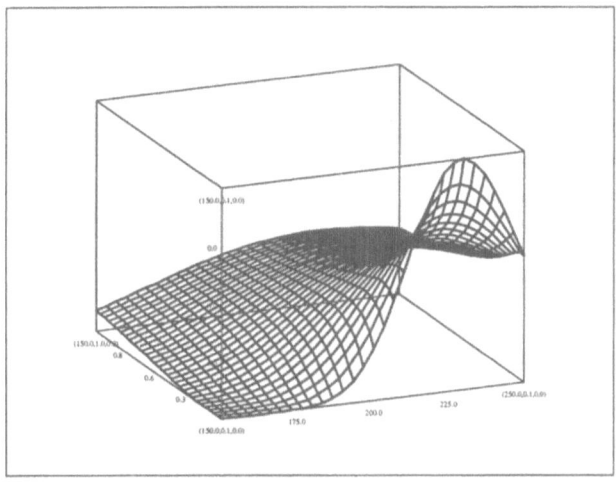

11.4   Portfolios and Hedging

11.4.1   Calculation of Arbitrage

```
arbitrage ()
       calculates an arbitrage table
```

The function **arbitrage** calculates an arbitrage table considering puts and calls with the same strike price. It is simply started by typing

```
arbitrage()
```

in the command line of XploRe. After starting arbitrage the following window will appear:

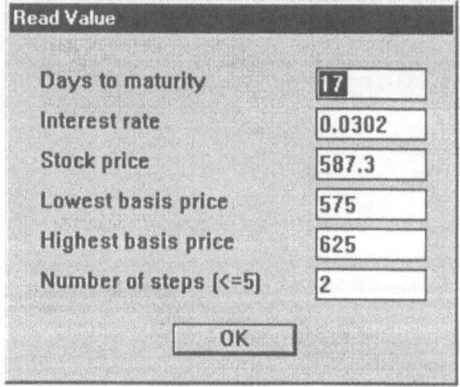

Here you are asked to put in your given data — Days to maturity, Interest rate, Stock price, Lowest basis price, Highest basis price and the Number of steps.

After pushing the OK button you can first put in the call prices and right afterwards the put prices:

```
Read Value                                Read Value

  Please input call prices.   [3]           Please input put prices.   [15.9]
                              [3]                                      [15.9]
                              [3]                                      [15.9]

            [   OK   ]                               [   OK   ]
```

As a result XploRe presents you the following table of arbitrage in the output
window, where

- **Call_price** is the vector of call prices

- **Put_price** is the vector of put prices

- **Basis** is the vector of basis prices

- **Stock_flow** is the amount we pay/get for buying/selling stock

- **Call_flow** is the amount we pay/get for buying/selling call option

- **Put_flow** is the amount we pay/get for buying/selling put option

- **Bank_flow** is the investment to/loan from a bank

- **Arbitrage** is the vector of arbitrage gains/losses

```
[ 1,] Stock price:    587.30
[ 2,] Interest rate:    0.0302
[ 3,] Days to maturity: 17.00
[ 4,]
[ 5,] Call_price Put_price Basis  Stock_flow Call_flow Put_flow  Bank_flow  Arbitrage
[ 6,] ---------------------------------------------------------------------------------
[ 7,]     3.00     15.90   575.00   587.30     -3.00     15.90   -574.18      26.02
[ 8,]     3.00     15.90   600.00   587.30     -3.00     15.90   -599.15       1.05
[ 9,]     3.00     15.90   625.00   587.30     -3.00     15.90   -624.11     -23.91
[10,]
```

11.4.2   Bull-Call Spreads

```
callbull ()
        calculates the results of a Bull-Call Spread for the context of
        option pricing
```

The function `callbull` calculates the results of a Bull-Call Spread for the context of option pricing. It is simply started by typing

```
callbull()
```

in the command line of XploRe. After starting `callbull()` the following window will appear:

Read Value

lowest quotation	540
highest quotation	610
Strike price of long call (C1)	550
Price for C1	35
Strike price of short call (C2)	600
Price for C2	15
Number of contracts	100

OK

After putting in all the basic data required just push the OK button. XploRe will calculate the results and present them in the XploRe output window in the following way:

```
[ 1,]
[ 2,]    Stock price     long Call    short Call    gain/loss
[ 3,] ---------------------------------------------------------
[ 4,]        540.00       -3500.00      1500.00      -2000.00
[ 5,]        550.00       -3500.00      1500.00      -2000.00
[ 6,]        560.00       -2500.00      1500.00      -1000.00
[ 7,]        570.00       -1500.00      1500.00          0.00
[ 8,]        580.00        -500.00      1500.00       1000.00
[ 9,]        590.00         500.00      1500.00       2000.00
[10,]        600.00        1500.00      1500.00       3000.00
[11,]
```

12 Microeconometrics and Panel Data

Jörg Breitung and Axel Werwatz

This chapter introduces the tools available in XploRe for analyzing microdata, i.e. data sets consisting of observations on N individual units, such as persons, households or firms.

Why does analyzing microdata require specific techniques? Which techniques have been collected under the heading "microeconometrics"?

"Micro" here is used as in microeconomics. Microeconomics provides theories of individual behavior. Microeconometrics provides the statistical tools for analyzing observed individual behavior.

Individual behavior and individual decision making often have discrete outcomes: students choose among majors, firms choose whether or not to launch a new product, etc. Consequently, several of the quantlets that we will describe are designed to deal with models where the dependent variable is not free to take on any value, i.e. models with **limited-dependent** or **qualitative dependent** variables.

Another feature of microdata stems from the fact that the observed units are individuals that pursue their own best interest. Those observed as lawyers pursued a career in law because they are probably talented for this line of work. Hence, the average earnings of observed lawyers are a too optimistic indicator for what a nonlawyer could earn if she were working as a lawyer. Observed lawyers self-selected into their profession. We will present several procedures that deal with **self-selection**.

When microdata first became available it usually consisted of observations on N individuals at a given point in time (cross-section data). Many microdata sets analyzed these days provide richer information: N individuals are observed repeatedly at (usually) equally spaced points in time. That is, contemporary microdata sets are often **panel data** sets. We will introduce the quantlets available in XploRe that take advantage of the panel structure of the data.

Summing up, we will (in this order) cover the XploRe quantlets for dealing with limited-dependent or qualitative dependent variables, self-selection and panel data.

12.1   Limited-Dependent and Qualitative Dependent Variables

The title of this section is taken from Maddala's (1983) well-known book of the same name which is still a very good reference for parametric models of this kind. XploRe's `metrics` library covers some of the parametric models. Its comparative strength, however, is in the semiparametric models for data with limited-dependent and qualitative dependent variables, which have been developed more recently. We will first discuss the parametric models and then turn to their semiparametric competitors.

12.1.1   Probit, Logit and Tobit

```
{b,s,cv} = tobit (x, y)
     two-step estimation of the Tobit model
```

Probit, Logit and Tobit are among the three most widely used parametric models for analyzing data with limited-dependent or qualitative dependent variables. Probit and Logit can be viewed as special cases of the generalized linear model (GLM). This is the perspective taken in XploRe. Hence we ask you to consult Chapter 7 for a description of XploRe's Probit and Logit (standard Logit, conditional Logit, multinomial Logit) quantlets.

The Tobit model is a parametric censored regression model. Formally, the model is given by

$$Y = \begin{cases} x^T\beta + \varepsilon & \text{if } Y^* = x^T\beta + \varepsilon \geq c \\ 0 & \text{otherwise} \end{cases}, \qquad (12.1)$$

where ε is an unobservable error term, assumed to be normally distributed with mean zero and variance σ^2, and c is a known constant. In the pioneering study of Tobin, Y is a consumer's expenditure on a durable good and $Y^* = x^T\beta + \varepsilon$ the consumer's willingness to pay for the good. Y is equal to Y^* only if the willingness to pay exceeds c, the minimum amount necessary to purchase the good. Otherwise, Y is equal to zero.

It is well known that an OLS regression of the nonzero Ys on the explanatory variables x will not produce consistent estimates of the regression coefficients in this situation. The model implies the following conditional mean function :

$$E(Y|x, Y^* > 0) = x^T\beta + \sigma\frac{\varphi(x^T\beta/\sigma)}{\Phi(x^T\beta/\sigma)}, \qquad (12.2)$$

where $\varphi(\bullet)$ and $\Phi(\bullet)$ are the probability density (*pdf*) and cumulative distribution functions (cdf) of the standard normal distribution.

XploRe offers a two-step estimation of β : In the first step, a pilot estimate of $\gamma = \beta/\sigma$ is obtained by estimating the model $P(I(Y^* > 0)) = \Phi(x^T\gamma)$ by Probit analysis (here, $I(\bullet)$ denotes the indicator function). Using this pilot estimate $\widehat{\gamma}$, we can compute $\varphi(x^T\widehat{\gamma})$ and $\Phi(x^T\widehat{\gamma})$ and use them as regressors when we estimate (12.2) on the part of the sample where $Y > 0$.

The tobit quantlet takes the observed x and y as inputs and returns the vector of estimated coefficients b, the estimated standard deviation of the error term s and the estimated covariance matrix cv of b and s:

```
{b, s, cv} = tobit(x, y)
```

The dependent variable must be equal to 0 for the censored observations. The known constant c in (12.1) is subsumed in the constant term of β. That is, the estimated constant term is an estimate of the original constant term minus c.

In the following example, simulated data is used to illustrate the use of tobit:

```
library("metrics")
randomize(241200)
n     =   500
k     =     2
x     =     matrix(n)~aseq(1, n ,0.25)
s     =     8
u     =     s*normal(n)
b     =     #(-9, 1)
```

```
ystar  =      x*b+u
y      =      ystar.*(ystar.>=0)
tstep  =      tobit(x,y)
dg     =      matrix(rows(tstep.cv),1)
dig    =      diag(dg)
stm    =      dig.*tstep.cv
std    =      sqrt(sum(stm,2))
coef   =      tstep.b|tstep.s
coef~(coef./std)      ; t-ratios
```

🔍metric01.xpl

We have generated the data in accordance with the assumptions of the Tobit model and chose values of -9 and 1 for the components of β and 8 for σ. The results in the XploRe output window show the estimated coefficients (first column) and their t-ratios:

```
Contents of _tmp
[1,]   -9.7023   -11.201
[2,]    1.0092    92.13
[3,]    6.8732    4.5859
```

12.1.2   Single Index Models

XploRe offers several quantlets to estimate semiparametric regression models where the conditional mean of Y is assumed to depend on the explanatory variables x only via a function of the single (linear) index $x^T\beta$:

$$E(Y|x) = g(x^T\beta), \qquad (12.3)$$

where $g(\bullet)$ is an unknown function. The Probit and Logit models of the previous sections are special cases of such single index models (SIMs) where $g(\bullet)$ is assumed to be a known function (the cdf of the standard normal and logistic distribution, respectively). XploRe offers several alternative procedures to estimate β in (12.3) that only require that $g(\bullet)$ is a smooth (but otherwise unspecified) function. All procedures are noniterative, easy to compute and have very desirable large-sample properties.

12.1.3   Average Derivatives

Define the vector of average derivatives of y with respect to x as

$$\delta = E\left\{\frac{\partial E(Y|x)}{\partial x}\right\}, \qquad (12.4)$$

i.e. δ is the vector of partial derivatives of the regression function $E(Y|x)$, averaged over the support of x. In a SIM, it turns out that

$$\delta = E\left\{g'(x^T\beta)\right\}\beta, \qquad (12.5)$$

where g' is the derivative of $g(\bullet)$. From (12.5) we can see that in a SIM δ is proportional to the vector of regression coefficients β. This implies that if we find a way to estimate δ, then we can estimate β up to scale. Estimating β up to scale is as good as we can do under the assumptions of the SIM, even if we had an infinite number of observations. That is, the scale of β is not identified in a SIM of the form (12.3). Hence we can focus on estimating δ. That is the approach followed by the XploRe quantlets adeind, dwade and adeslp which return estimates of the average derivative δ. Asymptotically, these estimators are equivalent, see Stoker (1991).

Two comments are in order:

- The average derivative is only defined for the continuous components of x. For the discrete components of x we have to find a different way to estimate their coefficients. The metrics library features the quantlet adedis which estimates the coefficients of discrete components of x by a method suggested in Horowitz and Härdle (1996).

- You can use adeind, dwade and adeslp to estimate the average derivative in any regression model where it exists. See Stoker (1991) and Härdle and Stoker (1989) for discussions of the interpretation of the average derivative in regression models that are not of the single index form.

Here is an overview of the XploRe commands for estimating the vector of average derivatives δ (and thereby β up to scale in a SIM):

{delta, dvar} = adeind (x, y, d, h)
> estimates the average derivative of Y with respect to x, which is proportional to β in SIMs

delta = dwade (x, y, h)
> an alternative to adeind that estimates the density-weighted average derivative

{delta, dvar} = adeslp(x, y, d, m)
> another alternative to adeind that estimates the average derivative by an instrumental variables regression

{delta, alpha, lim, hd} = adedis (z, x, y, h, hfac,c0,c1)
> estimates the coefficients of discrete components of x, in addition to dwade estimation of the coefficients of the continuous components

We will cover all quantlets, except for adeslp, which is discussed in Stoker (1991).

12.1.4   Average Derivative Estimation

It can be shown that

$$\delta = E\{s(x)E(Y|x)\} = E\left\{\frac{1}{f(x)}\frac{\partial f(x)}{\partial x}E(Y|x)\right\},\qquad(12.6)$$

where $f(x)$ is the density of x. Equation (12.6) says that δ is equal to the expectation of the product of the score vector $s(x)$ and the conditional expectation function of Y. Hence, we can estimate δ using the sample analogs of the quantities on the right-hand side of (12.6):

$$\widehat{\delta} = \frac{1}{N}\sum_{i=1}^{N}\widehat{s}_h(x_i)Y_i\,I(\widehat{f}_h(x_i) > b_N) = \frac{1}{N}\sum_{i=1}^{N}\frac{1}{\widehat{f}_h(x_i)}\frac{\partial\widehat{f}_h(x_i)}{\partial x}Y_i\,I(\widehat{f}_h(x_i) > b_N)$$

$$(12.7)$$

where $\widehat{f}_h(x)$ denotes the kernel density estimator with bandwidth h, $\widehat{f}_h'(x)$ the vector of kernel estimators of the partial derivatives of $f(x)$ and $I(\widehat{f}_h(x_i) > b_N)$

is an indicator that trims out observations at which the estimated density is very small.

Equation (12.7) defines the average derivative estimator (ADE) of Härdle and Stoker (1989) which is computed by the XploRe quantlet adeind:

```
{delta, dvar} = adeind(x, y, d, h)
```

adeind takes the data (x and y) as well as the bandwidth h and the binwidth d as inputs.

The bandwidth is visible in (12.7) but the binwidth d is not. This is because binning the data is merely a computational device for speeding up the computation of the estimator. Larger binwidth will speed up computation but imply a loss in information (due to using binned data rather than the actual data) that may be prohibitively large.

You may wonder what happened to the trimming bounds b_n in (12.7). Trimming is necessary for working out the theoretical properties of the estimator (to control the random denominator of $\widehat{s}_h(x)$). In adeind, the trimming bounds are implicitly set such that 5 % of the data are always trimmed off at the boundaries (this is done by calling trimper within adeind; trimper is an auxiliary quantlet, not covered here in more detail).

In the following example, simulated data are used to illustrate the use of adeind:

```
library("metrics")
randomize(333)
n       =  500
x       =  normal(n,3)
beta    =  #(0.2 , -0.7 , 1)
index   =  x*beta
eps     =  normal(n,1) * sqrt(0.5)
y       =  2 * index^3 + eps
d       =  0.2
m       =  5
{delta,dvar} = adeind(x,y,d,m)
(delta/delta[1])~(beta/beta[1])
```

Q metric02.xpl

We have generated the data such that $g(x^T\beta) = 2(x^T\beta)^3$. Recall that the estimate of δ is an estimate of β up to scale. Since the scale of β is not identified you may normalize the estimated coefficients by dividing each of them by the first coefficient and then interpret the effects of the explanatory variables relative to the effect of the first explanatory variable (which is normalized to 1). The estimated and true coefficient vectors, both normalized, are shown in the output window:

```
Contents of _tmp
[1,]          1              1
[2,]    -4.2739         -3.5
[3,]     5.8963            5
```

Note that this example does not work in the Academic Edition of XploRe.

12.1.5 Weighted Average Derivative Estimation

The need for trimming in ADE is a consequence of its random denominator. This and other difficulties associated with a random denominator are overcome by the Density Weighted Average Derivative Estimator (DWADE) of Powell, Stock and Stoker (1989). It is based on the density weighted average derivative of Y with respect to x:

$$\tilde{\delta} = E\left\{\frac{\partial g(x^T\beta)}{\partial x}f(x)\right\} = E\{g'(x^T\beta)f(x)\}\beta. \tag{12.8}$$

Obviously, $\tilde{\delta}$ shares the property of the (unweighted) average derivative of being proportional to the coefficient vector β in single index models. The estimation strategy is therefore basically the same: find a way to estimate $\tilde{\delta}$ and you have an estimator of β up to a scaling factor.

It can be shown that

$$\tilde{\delta} = -2E\{E(Y|x)f'(x)\}. \tag{12.9}$$

Thus we may estimate $\tilde{\delta}$ (and thereby β) by

$$\widehat{\tilde{\delta}} = -\frac{2}{N}\sum_{i=1}^{N}Y_i\widehat{f'_h}(x_i), \tag{12.10}$$

where $\widehat{f'_h}(x_i)$ again denotes the vector of kernel estimators of the partial derivatives of $f(x)$. The DWADE estimator defined in (12.10) shares the desirable distributional features of the ADE estimator ($\sqrt{N}$-consistency, asymptotic normality). It is computed in XploRe by the `dwade` quantlet:

```
d = dwade(x, y, h)
```

`dwade` needs the data (x and y and the bandwidth h as inputs and returns the vector of estimated density weighted average derivatives.

We illustrate `dwade` with simulated data:

```
library("metrics")
randomize(333)
n      =  500
x      =  normal(n,3)
beta   =  #(0.2 , -0.7 , 1)
index  =  x*beta
eps    =  normal(n,1) * sqrt(0.5)
y      =  2 * index^3 + eps
h      =  0.3
d      =  dwade(x,y,h)
(d/d[1])~(beta/beta[1])
```

metric03.xpl

Note that we used the same data generating process as in the previous section and that we again show estimated and true coefficient vectors, both normalized by dividing through by their first element:

```
Contents of _tmp
[1,]      1         1
[2,]  -3.4727      -3.5
[3,]   4.9435       5
```

12.1.6   Average Derivatives and Discrete Variables

By definition, derivatives can only be calculated for continuous variables. Thus, `adeind` and `dwade` will not produce estimates of the components of β that belong to discrete explanatory variables.

Most discrete explanatory variables are 0/1 dummy variables. How do they enter into SIMs ? To give an answer, let us assume that x consists of several continuous and a single dummy variable and let us split x and β accordingly into x_1 (continuous component) and x_2 (dummy), and β_1 and β_2, respectively. Then we have

$$
\begin{aligned}
E(Y|x_1, x_2) &= g(x_1^T\beta_1), & \text{if } x_2 = 0, \\
E(Y|x_1, x_2) &= g(x_1^T\beta_1 + \beta_2), & \text{if } x_2 = 1.
\end{aligned}
$$

Graphically, changing x_2 from 0 to 1 means shifting $g(\bullet)$ (as a function of $x_1^T\beta_1$) by β_2. This is depicted in Figure 12.1, where β_2 is labeled b_2 for technical reasons.

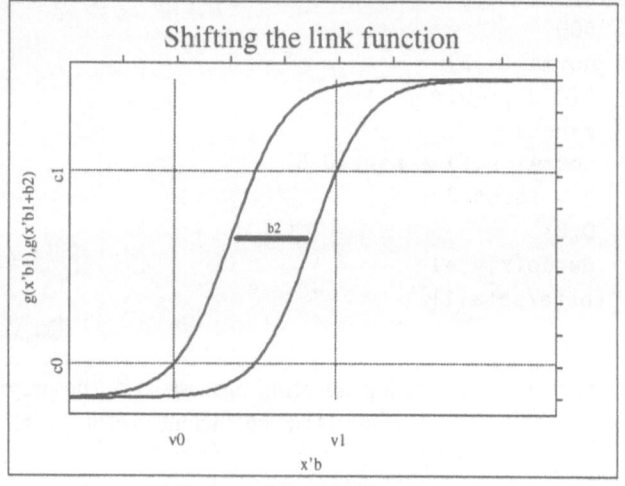

Figure 12.1. Shifting $g(\bullet)$ by β_2.

This is the basic idea underlying the estimator of β_2 proposed by Horowitz and Härdle (1996): given an estimate of β_1 (which can be obtained by using dwade, for instance) we can estimate both curves in Figure 12.1 by running separate kernel regression of Y on $x_1^T\widehat{\beta}$ for the data points for which $x_2 = 0$ (to get an estimate of $g(x_1^T\beta_1)$) and for which $x_2 = 1$ (to get an estimate of $g(x_1^T\beta_1 + \beta_2)$). Then we can compute the horizontal differences between the two estimated curves to get an estimate of β_2. This procedure is implemented in the XploRe quantlet adedis:

```
{d, a, lim, h} = adedis(x2, x1, y, hd, hfac,c0,c1)
```

It takes as inputs the data, consisting of x2 (discrete explanatory variables), x1 (continuous explanatory variables) and y, and several parameters that are needed in the three steps of adedis estimation:

- The (vector) of bandwidth(s) hd
  used in estimating the coefficient(s) of the continuous explanatory variable(s) x_1. These coefficients are implicitly estimated by dwade. Separate estimates of β_1 are computed for groups of observations that share the same value of x_2. That is, if x_2 is a dummy variable then separate estimates of β_1 are computed for observations with $x_2 = 0$ and $x_2 = 1$. These separate estimates are subsequently averaged to get a unique estimate of β_1. This is the first step of the estimation algorithm.

- The bandwidth scaling factor hfac
  for estimating the $g(\bullet)$ functions by kernel regression of Y on $x_1^T \widehat{\beta}_1$, where $\widehat{\beta}_1$ is the estimate of β_1 from the first step. The scaling factor hfac is used in calculating the bandwidth for the kernel regression in the following way:

$$h = \text{hfac} \sqrt{\text{Var}(x_1^T \widehat{\beta}_1)} \, N^{-1/7.5}, \tag{12.11}$$

where N is the sample size and $\text{Var}(x_1^T \widehat{\beta})$ is the sample variance of $x_1^T \widehat{\beta}_1$. This rule for calculating the bandwidth is suggested by simulation results in Horowitz and Härdle (1996). Separate kernel regressions are performed for each group of observations with a common value of x_2. This is the second step of the estimation procedure.

- The monotonicity constants c0 and c1
  which are part of the following monotonicity conditions that $g(\bullet)$ is assumed to satisfy: There are finite numbers v_0, v_1, c_0, and c_1 such that $v_0 < v_1$, $c_0 < c_1$ and

$$g(x_1^T \beta_1 + \beta_2 x_2) < c_0 \quad \text{for each value of } x_2 \text{ if } x_1^T \beta_1 < v_0,$$
$$g(x_1^T \beta_1 + \beta_2 x_2) > c_1 \quad \text{for each value of } x_2 \text{ if } x_1^T \beta_1 > v_1.$$

Figure 12.1 shows an example where $g(\bullet)$ is the cdf of the standard normal. You may verify that the values of v_0, v_1, c_0 and c_1 in the figure satisfy the conditions but many other choices for these constants would have worked too. adedis essentially calculates the horizontal differences between the estimated $g(\bullet)$ functions (corresponding to the different values of x_2) over the range of function values which satisfy $c_0 \leq \widehat{g}(\bullet) \leq c_1$.

The above conditions imply that in this range $\widehat{g}(\bullet)$ is monotonous. You may want to check this by graphing the estimated $g(\bullet)$ functions against $x_1^T \widehat{\beta}_1$.

Whereas you have to specify c_0 and c_1, the constants v_0 and v_1 are implicitly set to the minimum and maximum of $x^T \widehat{\beta}_1$, plus or minus the bandwidth used in the kernel estimation of $g(\bullet)$. The values of v_0 and v_1 are returned by adedis in the vector lim, along with the bandwidth h, calculated according to (12.11.) The most important outputs are, of course, the estimates of β_1 and β_2 which are stored in d and a, respectively.

12.1.7   Parametric versus Semiparametric Single Index Models

```
{t, p} = hhtest (vhat, y, yhat, h {,c {,m}}))
    tests a parametric against a semiparametric SIM
```

In the previous sections, we have seen that even the noniterative estimators of SIMs implemented in XploRe require quite a bit of care and computational effort. More importantly, semiparametric estimators are less efficient than parametric estimators if the assumptions underlying the latter are satisfied. It is therefore desirable to know whether the distributional flexibility of these models justifies the loss in efficiency and the extra computational cost. That is, we would like to statistically test semiparametric SIMs against easily estimable parametric SIMs.

Horowitz and Härdle (1994) have developed a suitable test procedure that is implemented in the hhtest quantlet. Formally, the HH-test considers the following hypotheses:

$$H_0 \; : \; E(Y|x) = f(x^T \beta),$$
$$H_1 \; : \; E(Y|x) = g(x^T \beta),$$

where $f(\bullet)$ is a known function (such as the cdf of the standard normal distribution) and $g(\bullet)$ is an unknown, smooth function. Hence, the parametric model is in the null whereas the semiparametric model is in the alternative hypothesis.

Here is the main idea underlying the HH-test: if the model under the null is true (and given an estimate of β), then a nonparametric regression of Y on $x^T\widehat{\beta} = v$ will give a consistent estimate of the parametric function $f(\bullet)$. If, however, the parametric model is wrong, then the nonparametric regression of Y on $x^T\widehat{\beta} = v$ will deviate systematically from $f(\bullet)$.

This insight is reflected in the HH-test statistic:

$$T = \sqrt{h}\sum_{i=1}^{N} w(x_i^T\widehat{\beta})\{Y_i - f(x_i^T\widehat{\beta})\}\{\widetilde{f}_i(x_i^T\widehat{\beta}) - f(x_i^T\widehat{\beta})\} \qquad (12.12)$$

where h is the bandwidth used in the nonparametric (kernel) regression, and $w(\bullet)$ is a weight function that downweighs extreme observations. In practice $w(\bullet)$ is defined to be identically equal to one for 90% or 95% of the central values of $x_i^T\widehat{\beta}$ and zero otherwise. The term $\{\widetilde{f}_i(x_i^T\widehat{\beta}) - f(x_i^T\widehat{\beta})\}$ compares the kernel regression of Y on $x_i^T\beta$ (denoted $\widetilde{f}_i(x_i^T\widehat{\beta})$) with the parametric model implied by the null hypothesis.

The HH-test statistic is computed by the XploRe quantlet `hhtest`:

```
{t, p} = hhtest(vhat, y, yhat, h {,c {,m}}})
```

The function `hhtest` takes as inputs

vhat the vector with the estimated index $x_i^T\widehat{\beta}$. Horowitz and Härdle (1994) suggest that the index can be estimated under the null hypothesis. Hence, if your model in H_0 is the Probit model, then you get **vhat** by running **glmest** with the `"bipro"` option on your data. See Chapter 7.

y the observations on the dependent variable.

yhat the parametric estimate of $E(Y|x)$, i.e. $f(x_i^T\widehat{\beta})$.

h the bandwidth for the kernel regression of **y** on **vhat**.

c optional parameter that must be lying in the interval $0 =< c < 1$. Proportion of the sample to be cut at each extreme. Default is 0.05.

m optional $n\mathrm{x}1$ vector or scalar. m should be given only if Y is a binomial or binary dependent variable. If it is binomial, then m should be the vector of binomial coefficients. If Y is binary, then set m equal to 1. This will improve estimation of the variance of the test statistic.

`hhtest` returns the test statistic t and the corresponding p-value p. Under H_0 the test statistic defined in (12.12) is asymptotically normally distributed with zero mean and finite variance. The test, however, is a one-sided test because deviations of H_0 of the semiparametric kind considered in (12.12) will lead to large positive values of the test statistic.

We illustrate `hhtest` using the `kyphosis` data:

```
library("metrics")
x      =  read("kyphosis")
y      =  x[,4]
x      =  x[,1:3]
x      =  matrix(rows(x))~x
h      =  2
g      =  glmbilo(x,y)
eta    =  x*g.b
mu     =  g.mu
{t,p}  =  hhtest(eta,y,mu,h,0.05,1)
t~p
```

Q metric04.xpl

The value of the test statistic and the p-value are displayed in the XploRe output window:

```
Contents of _tmp
[1,] -0.79444  0.21346
```

The null hypothesis is the Logit model that was used to fit the data. We cannot reject H_0 at conventional significance levels because the p-value is greater than 0.2.

12.2   Multiple Index Models

The single index model of the previous sections has been extended to multiple index models in various ways. For instance, popular parametric models for data with multicategorical response variables (representing the choice of individuals among more than two alternatives) are of the multi-index form. In the Multinomial Logit model, the probability that an individual will choose alternative j depends on characteristics x of the individual through the indices $x^T\beta_1, \ldots, x^T\beta_J$:

$$P(Y = j | x = x) = \frac{\exp(x^T\beta_j)}{1 + \sum\limits_{k=1}^{J} \exp(x^T\beta_k)} . \qquad (12.13)$$

Quantlets to estimate this and related models can be found in the glm library.

12.2.1   Sliced Inverse Regression

```
{edr, eigen} = sir (x, y, h)
        fits a multiple index model by the method of Sliced Inverse Re-
        gression
```

A semiparametric multiple index model is the (SIR) model considered in Li (1991). Given a response variable Y and a (random) vector $x \in \mathbb{R}^p$ of explanatory variables, SIR is based on the model:

$$Y = m(x^T\beta_1, \ldots, x^T\beta_k, \varepsilon), \qquad (12.14)$$

where $\beta_1, \ldots, \beta_k$ are unknown projection vectors, k is unknown and assumed to be less than p, $m : \mathbb{R}^{k+1} \to \mathbb{R}$ is an unknown function, and ε is the noise random variable with $E(\varepsilon|X) = 0$.

Model (12.14) describes the situation where the response variable Y depends on the p-dimensional variable x only through the indices $x^T\beta_1, \ldots, x^T\beta_k$. The smaller k is relative to p, the better able is SIR to represent the p-dimensional regression of Y on x by a parsimonious k-dimensional subspace. The unknown $\beta_1, \ldots, \beta_k$, which span this space, are called *effective dimension reduction directions* (EDR directions). The span is referred to as the *effective dimension reduction space* (EDR space).

SIR tries to find this k-dimensional subspace of $\mathbb{R}^p$ by considering the inverse regression (IR) curve, i.e. $E(x|Y)$. Under some weak assumptions on the joint distribution of the elements of x, it can be shown that the centered inverse regression $E(x|Y) - E(x)$ lies in the subspace formed by the β_js. The β_js are found by an eigenvalue/eigenvector decomposition of the estimated covariance matrix of the vector $E(z|y)$, where z is a standardized version of x. In XploRe, this is achieved by using the sir quantlet:

```
{edr, eigen} = sir(x, y, h)
```

It takes as inputs the data (x and y) and the parameter h. h is related to the "slicing" part of sliced inverse regression. The algorithm actually works on nonoverlapping intervals (slices) of the data. There are different ways to divide the data into slices and the value of h has to be set accordingly. Three cases are distinguished:

- $h <= -2$
  In this case the absolute value of h is computed and interpreted as giving the number of elements in each slice. The number of slices is calculated as floor(N/h).

- $h >= 2$
  In this case h is interpreted as the number of slices. The h slices are constructed such that they have equal width.

- $0 < h < 1$
  In this case h is understood to give the percentage of the range of Y to be allocated to each slice. The number of slices is calculated as floor($1/h$).

SIR shares with other semiparametric procedures in that there is no clear-cut way of choosing the parameter h (which is not a bandwidth here). It is a good idea to try different values of h and see how the results are affected.

The outputs of sir are edr, the matrix containing estimates of the effective dimension reduction directions (i.e. the $\widehat{\beta}_j$s) and the eigenvalues of the estimated covariance matrix of the vector $E(z|Y)$, $\widehat{\text{Cov}}\{E(z|Y)\}$.

sir2 provides an alternative to sir. Usage of sir2 is very similar to sir but the details of the algorithm as well as the underlying theory are different. For further details on SIR II, see Li (1991).

12.2.2   Testing Parametric Multiple Index Models

Parametric multiple index models like the multinomial logit model (12.13) are based on rather strong functional form assumptions. If these assumptions are not valid, then maximum-likelihood estimators of the parameters of these models are inconsistent.

XploRe provides a procedure to test the functional form assumptions of parametric multiple index models:

{t, p} = hhmult (vhat, y, yhat, h)
    tests a parametric polychotomous response model against a semi-parametric alternative

---

The function hhmult is a generalization of the HH-test from Subsection 12.1.7. See Werwatz (1997) for details. hhmult can only be applied for models with a multicategorical dependent variable. The null hypothesis corresponds to a parametric model such as the multinomial logit model (12.13). You first have to estimate this model (in the case of the multinomial logit model you can use glmmultlo) and save the estimated indices in the matrix vhat and the predicted probabilities in the matrix yhat. Along with vhat and yhat, you provide the observations on the dependent variable and a vector of bandwidths h.

hhmult expects that you convert the observations on the multicategorical dependent variable into a set of dummy variables, one dummy variable per category of Y. The bandwidths are used for the nonparametric regressions of the dummy dependent variables on the estimated indices vhat. These nonparametric regressions provide an estimate of the parametric link functions, which take the following form in the multinomial logit model:

$$F_j(v_1, \ldots, v_J) = \frac{\exp(v_j)}{1 + \sum_{k=1}^{J} \exp(v_k)}. \qquad (12.15)$$

If the parametric model is correct, then the nonparametric estimate should differ from the link function implied by the parametric model only by random sampling error. If, however, the parametric model is misspecified, then there should be significant differences between the hypothesized and estimated link functions.

All these differences are summarized into one test statistic which, asymptotically, follows a χ^2 distribution. hhmult returns the value of the test statistic as well as the associated p-value.

12.3   Self-Selection Models

Self-selection or sample selection models are applied when the individuals in the sample are not randomly chosen from the population from which one would like to draw inferences. In the prototypical self-selection model, interest centers on estimating the parameters of a regression equation (often labeled "outcome equation") from observations on individuals who self-selected into the sample on the basis of a criterion that is correlated with the dependent variable of the outcome equation.

To illustrate, suppose we are interested in estimating the expected income of a randomly chosen individual if she were working as a lawyer. Computing the average income of those who actually work as lawyers is likely to be an upward biased estimate since those observed as lawyers probably chose their profession because they are talented for this line of work and expect to earn a relatively high income.

The solution to the self-selection problem proposed by Heckman (1979) is to propose and estimate a model of the self-selection decision. That is, Heckman's solution adds a "decision equation" to the outcome equation. Formally, the model consists of the following two equations:

$$
\begin{aligned}
I^* &= z^T\gamma + \delta & \text{decision equation,} \\
Y^* &= x^T\beta + \varepsilon & \text{outcome equation.}
\end{aligned}
\tag{12.16}
$$

Here I^* is the unobserved propensity to select into the sample, z a vector of observable explanatory variables and δ an unobservable error term. In the outcome equation, all quantities have the usual meaning. Note, however, that we denote the dependent variable as Y^* because it is not the observed dependent variable but rather the "potential" dependent variable. In our lawyer example, Y^* is the potential income as a lawyer for a randomly chosen individual. Y^* is observed only for those who actually choose to be lawyers. Formally,

$$
Y = Y^* \cdot I(I^* > 0) = \begin{cases} 0 & \text{if } I^* \leq 0, \\ Y^* & \text{if } I^* > 0. \end{cases}
\tag{12.17}
$$

That is, potential income Y^* and observed income Y are equal only if the propensity to select into the sample (e.g. to become a lawyer) is positive ($I^* > 0$). For those not selecting into the sample ($I^* \leq 0$), Y^* is not observed and set equal to 0. Regarding the propensity to select into the sample, we are only able to observe whether it is positive (in which case $I(I^* > 0|z) = 1$) or negative ($I(I^* > 0|z) = 0$).

The self-selection problem arises if δ and ε are correlated, i.e. the (unobservable part of the) decision to select into the sample is correlated with the (unobservable part of the) outcome of interest.

12.3.1 Parametric Model

```
heckit = heckman (vhat, y, yhat, h)
      two-step estimation of a parametric self-selection model
```

In Heckman (1979)'s classical solution to the problem it is assumed that δ and ε are jointly normally distributed:

$$\begin{pmatrix} \varepsilon \\ \delta \end{pmatrix} = N\left(\begin{pmatrix} 0 \\ 0 \end{pmatrix}, \begin{pmatrix} \sigma_\varepsilon & \sigma_{\delta,\varepsilon} \\ \sigma_{\delta,\varepsilon} & 1 \end{pmatrix} \right). \tag{12.18}$$

The variance of δ is not identifiable and set to 1. Under this assumption, the regression function for the observed dependent variable Y can be written as

$$\begin{aligned} E(Y|x) = E(Y^*|x, I^* > 0) &= E(x^T\beta|x, I^* > 0) + E(\varepsilon|x, I^* > 0) \\ &= x^T\beta + E(\varepsilon|x, z^T\gamma > -\delta) \\ &= x^T\beta + \sigma_{\delta,\varepsilon}\,\phi(z^T\gamma)/\Phi(z^T\gamma), \end{aligned}$$

where $\sigma_{\delta,\varepsilon}$ is the covariance between δ and ε. The parameters of the model (β, γ, $\sigma_{\delta,\varepsilon}$) may be estimated by the following two-step procedure, implemented in heckman:

1. **Probit step**
   Estimate γ by fitting the probit model
   $$P(\,I(I^* > 0|z) = 1) = \Phi(z^T\gamma) \tag{12.19}$$

   using all observations, i.e. those with $I(I^* > 0|z) = 1$ (the lawyers, in the example) and $I(I^* > 0|z) = 0$ (the nonlawyers).

2. **OLS step**
   Using only observations with $I(I^* > 0|z) = 1$ to estimate the regression function
   $$E(Y|x) = x^T\beta + \sigma_{\delta,\varepsilon}\,\phi(z^T\gamma)/\Phi(z^T\gamma)$$

   by an OLS regression of the observed Ys on x and $\phi(z^T\widehat{\gamma})/\Phi(z^T\widehat{\gamma})$, where $\widehat{\gamma}$ is the first-step estimate of γ.

heckman takes as inputs the data, i.e. observations on Y, x, z, and $I(I^* > 0|z)$ (the latter are labeled q) and returns estimates of β (stored in heckit.b), $\sigma_{\delta,\varepsilon}$ (heckit.s) and γ (heckit.g).

We illustrate heckman with simulated data where the error terms in the decision and outcome equations are strongly correlated:

```
library("metrics")
randomize(10178)
n       =  500
s1      =  1
s2      =  1
s12     =  0.7
ss      =  #(s1,s12)~#(s12,s2)
ev      =  eigsm(ss)
va      =  ev.values
ve      =  ev.vectors
ll      =  diag(va)
ll      =  sqrt(ll)
sh      =  ve*ll*ve'
u       =  normal(n,2)*sh'
z       =  2*normal(n,2)
g       =  #(1,2)
q       =  (z*g+u[,1].>=0)
x       =  matrix(n)~aseq(1, n ,0.25)
b       =  #(-9, 1)
y       =  x*b+u[,2]
y       =  y.*(q.>0)
heckit = heckman(x,y,z,q)
heckit.b
heckit.s
heckit.g
```

metric05.xpl

The estimates of β, $\sigma_{\delta,\varepsilon}$ and γ are displayed in the XploRe output window:

```
Contents of b
[1,]  -8.9835
[2,]   0.99883
Contents of s
[1,]   0.77814
Contents of g
[1,]   1.0759
[2,]   2.1245
```

Since the data generation process satisfies the assumptions of the parametric self-selection model it is not surprising that the **heckman** coefficient estimates are quite close to the true coefficients.

12.3.2 Semiparametric Model

```
{a, b} = select (x, y, id, h)
        three-step estimation of a semiparametric self-selection model
```

The distributional assumption (12.18) of the parametric self-selection model is, more than anything else, made for convenience. If, however, (12.18) is violated then the **heckman** estimator is not consistent. Hence, there is ample reason to develop consistent estimators for self-selection models with weaker distributional assumptions.

Powell (1987) considers a semiparametric self-selection model that combines the two-equation structure of (12.16) with the following weak assumption about the joint distribution of the error terms:

$$f(\delta, \varepsilon | z) = f(\delta, \varepsilon | z^T \gamma). \tag{12.20}$$

That is, all we assume about the joint density of δ, ε (conditional on z) is that it is a smooth but unknown function $f(\bullet)$ that depends on z only through the linear index $z^T \gamma$. Under these assumptions the regression function for the observed outcomes Y takes the following form:

$$
\begin{aligned}
E(Y|x) = E(Y^*|x, I^* > 0) &= x^T \beta + E(\varepsilon | x, z^T \gamma > -\delta) \\
&= x^T \beta + \lambda(z^T \gamma),
\end{aligned}
$$

where $\lambda(\bullet)$ is an unknown smooth function.

Note that for any two observations i and j with $x_i \neq x_j$ but $z_i^T \gamma = z_j^T \gamma$ we can difference out the unknown function $\lambda(\bullet)$ by subtracting the regression functions for i and j :

$$
\begin{aligned}
E(Y_i | x = x_i) - E(Y_j | x = x_j) &= (x_i - x_j)^T \beta + \lambda(z_i^T \gamma) - \lambda(z_j^T \gamma) \\
&= (x_i - x_j)^T \beta.
\end{aligned}
$$

This is the basic idea underlying the estimator of β proposed by Powell (1987): regress differences in Y on differences in x, weighting differences strongly for which $z_i^T \widehat{\gamma}$ and $z_j^T \widehat{\gamma}$ are close together (and hence, $\lambda(z_i^T \widehat{\gamma}) - \lambda(z_j^T \widehat{\gamma}) \approx 0$). That is, β is estimated by the following weighted least squares estimator:

$$
\begin{aligned}
\widehat{\beta} = &\left[\binom{n}{2} \sum_{i=1}^{N} \sum_{j=i+1}^{N} \widehat{w}_{ijN} (x_i - x_j)(x_i - x_j)' \right]^{-1} \\
&\left[\binom{n}{2} \sum_{i=1}^{N} \sum_{j=i+1}^{N} \widehat{w}_{ijN} (x_i - x_j)(y_i - y_j) \right],
\end{aligned} \qquad (12.21)
$$

where $\widehat{w}_{ijN} = 1/h \, K \left[(z_i^T \widehat{\gamma} - z_i^T \widehat{\gamma})/h \right]$ with symmetric kernel function $K(\bullet)$ and bandwidth h.

In (12.21), we tacitly assume that we have already obtained $\widehat{\gamma}$, an estimate of γ. Under assumption (12.20), we get a single index model for the decision equation in place of the Probit model (12.19) in the parametric case:

$$
P(I(I^* > 0|z) = 1) = g(z^T \gamma). \qquad (12.22)
$$

Here, $g(\bullet)$ is an unknown, smooth function. Estimators for γ in this model have been discussed in Subsections 12.1.4–12.1.6. Any of these methods may be applied to get an estimate of γ. This is the first step of the semiparametric procedure. Given $\widehat{\gamma}$, the second step consists of estimating β using (12.21).

It should be pointed out that (12.21) does not produce an estimate of the constant term because the constant term is eliminated by taking differences in (12.21). An estimator of the constant term can be obtained in a third step by a procedure suggested by Andrews and Schafgans (1997). Their estimator is defined as

$$
\widehat{\beta}_0 = \frac{\sum_{i=1}^{N} (Y_i - x_i^T \widehat{\beta}) K(z_i^T \widehat{\gamma} - k)}{K(z_i^T \widehat{\gamma} - k)}, \qquad (12.23)
$$

where $K(\bullet)$ is a nondecreasing $[0, 1]$-valued weight function that is set equal to zero for all negative values of its argument, and k is a threshold parameter that has to increase with increasing n for the estimator to be consistent. The basic idea behind (12.23) is that for large values of $z^T \gamma$, $E(\varepsilon | x, z^T \gamma > -\delta) = \lambda(z^T \gamma) \approx 0$ and the intercept of the $\lambda(\bullet)$ function can be separated from the "classical" constant term of the regression equation.

The quantlet `select` combines the second and third steps of the semiparametric estimation procedure. Both steps are also separately available in the quantlets `powell` and `andrews`.

`select` takes as inputs the data for the outcome equation (x and y, where x may not contain a vector of ones), the vector `id` containing the estimated first step index $z^T \hat{\gamma}$, and the bandwidth vector h. The first element of h is the threshold parameter k used for estimating the intercept coefficient while the second element is the bandwidth h used for estimating the slope coefficients according to (12.21).

We illustrate `hhtest` using the `kyphosis` data:

```
library("metrics")
randomize(66666)
n     =  200
ss1   =  #(1,0.9)~#(0.9,1)
g     =  #(1)
b     =  #(-9, 1)
u     =  gennorm(n, #(0,0), ss1)
ss2   =  #(1,0.4)~#(0.4,1)
xz    =  gennorm(n, #(0,0), ss2)
z     =  xz[,2]
q     =  (z*g+u[,1].>=0)
hd    =  0.1*(max(z) - min(z))
d     =  dwade(z,q,hd)*(2*sqrt(3)*pi)
id    =  z*d
h     =  (quantile(id, 0.7))|(0.2*(max(id)-min(id)))
x     =  matrix(n)~xz[,1]
y     =  x*b+u[,2]
zz    =  paf(y~x~id, q)
y     =  zz[,1]
x     =  zz[,3:(cols(zz)-1)]
id    =  zz[,cols(zz)]
```

```
{a,b} = select(x,y,id,h)
d~a~b
```

metric06.xpl

`select` returns the first-, second- and third-step coefficient estimates. For the data at hand, they turn out to be equal to

```
Contents of _tmp
[1,]  0.81368  -8.7852  1.0471
```

Note that these estimates are quite close to the value of the corresponding population parameter. Yet, since the data generation process satisfies the assumptions of the parametric model, one could have obtained more efficient estimates using the parametric `heckman` estimator.

12.4   Panel Data Analysis

```
z = pansort (z0 {,N})
     arranges the data set appropriately

z = panstats (z {,T})
     computes summary statistics for the variables

{output, siga, sige} = panfix (z, m {,T})
     estimates a fixed effects (or mixed) model

output = panrand (z, siga, sige, m {,T})
     estimates a random effects (or mixed) model

z = pantime (z0 {,T})
     removes fixed time effects

output = panhaus (z, siga, sige, m {,T})
     performs a Hausman specification test
```

In a panel data set, a given sample of N individuals is observed at different time periods and thus provides multiple observations on each individual in the sample. As a simple example, assume that the relationship between the earnings and the experience of an employee is given by the linear model

$$y = \alpha + x\beta + u,$$

where y denotes log of total earnings and x is a measure of experience. Furthermore, assume that there are data for $i = 1, \ldots, N$ individuals at $t = 1, \ldots, T_i$ time periods. If $T_i = T$ is constant for all cross-section units, then the data set is called a balanced panel. Otherwise the panel is unbalanced.

We expect of course that a skilled worker earns more than an unskilled worker. However, if there is no information on the educational background of the employee, we might represent the impact of unobserved education by introducing an individual specific constant α_i. Therefore, the model is written as

$$y_{it} = \alpha_i + x_{it}\beta + u_{it},$$

where α_i depends on the skills (or ability) of the employee. One may also expect that the parameter β varies across individuals but this would amount to a separate analysis of the N cross section units. In typical panel data models, individual heterogeneity is represented by an individual specific intercept alone.

We may also include a measure of heterogeneity with respect to time. For example there may be a common time trend of an unknown form. Hence the model is augmented by a time specific constant λ_t:

$$y_{it} = \alpha_i + \lambda_t + x_{it}\beta + u_{it} .$$

Comprehensive overviews of this type of models are given by Hsiao (1986) and Baltagi (1995).

The effects α_i and λ_t are assumed to be either deterministic (fixed effects model) or random (random effects model). The crucial assumption associated with these different models is that the random effects model assumes that

$$E(x_{it}\alpha_i) = 0 \quad \text{and} \quad E(x_{it}\lambda_t) = 0 \quad \text{for all } i, t ,$$

whereas in a fixed effects model, the regressors may be correlated with the individual and time effects. For the error u_{it} it is usually assumed that $E(u_{it}) = 0$, $E(u_{it}^2) = \sigma^2$ and that u_{it} is uncorrelated across i and t. However, the panfix quantlet does not require such restrictive assumptions.

We may also mix the assumptions of a random and fixed effects model. For example, we may assume that α_i is random and λ_t is fixed. Moreover, it may be assumed that some regressors are correlated with the random effects, while others are uncorrelated. Such a specification is in particular attractive if the set of regressors contains some variables that are constant in time. We therefore write the mixed model as

$$y_{it} = x_{it}^T \beta + z_{it}^T \gamma + \alpha_i + \lambda_t + u_{it} \qquad (12.24)$$

where

- x_{it} is a $m \times 1$ vector of explanatory variables assumed to be time varying and correlated with α_i ,

- z_{it} is a $k \times 1$ vector of explanatory variables assumed to be uncorrelated with α_i,

- λ_t is assumed to be fixed.

We will neglect random time effects because it is practically of minor importance and we prefer a dynamic model to represent random effects in time.

A typical (static) panel data analysis usually proceeds as follows:

1. **Model specification.** First, an F-test or an LM-test is used to test the hypothesis that individual effects are significant. Of course, if these tests do not provide any evidence for individual effects, then the model can simply be estimated by ordinary least squares (OLS). Second, the Hausman test is used to decide whether the regressors are correlated with the individual effect. In XploRe these tests can be performed by using the `panfix` and `panhaus` quantlets.

2. **Variance decomposition.** If at least one variable is assumed to be un-correlated with the random effects, the generalized least squares (GLS) estimator requires an estimate of the variance of α_i and u_{it}. These estimates are obtained in XploRe by using the `panfix` quantlet. If all re-gressors are assumed to be correlated with the individual effects, `panfix` gives the final estimates.

3. **GLS and GIV estimation.** Using the estimates of the error variances, Generalized Least Squares (GLS) or Generalized Instrumental Variable (GIV) estimates are computed by the `panrand` quantlet.

12.4.1   The Data Set

The data set is assumed to be ordered by the cross-section units. That is, the complete data of the first individual are given in the first T_1 rows, then the data of the second individual in the rows $T_1 + 1, \ldots, T_1 + T_2$ and so on. If the data are an unbalanced data set, the first two columns must provide the identification number of the cross-section unit and the time period. If the data are in the form of a balanced panel, it is sufficient to provide the common number of time periods T to assign the data to the cross-section units. Most procedures do not use the time index in the second column. Exceptions are the `pantime` and the `pandyn` quantlets. Thus, the user may insert some arbitrary column (for example, a column generated by the XploRe command `matrix(rows(z),1)`) in all other cases.

The data matrix must be organized in the following form:

{1}	{2}	3	4	$\cdots$	$3+m$	$m+4$	$\cdots$	$3+m+k$
1	1	y_{11}	$x_{11,1}$	$\cdots$	$x_{11,m}$	$z_{11,1}$	$\cdots$	$z_{11,k}$
$\vdots$								$\vdots$
1	T_1	y_{1T_1}	$x_{1T_1,1}$	$\cdots$	$x_{1T_1,m}$	$z_{1T_1,1}$	$\cdots$	$z_{1T_1,k}$
2	1	y_{21}	$x_{21,1}$	$\cdots$	$x_{21,m}$	$z_{21,1}$	$\cdots$	$z_{21,k}$
$\vdots$								$\vdots$

An example for such a data set is the file **earnings.dat** (see Appendix B.8). The first 10 rows look like:

```
[   1,]      1      1      2600      20      9      2
[   2,]      1      2      2500      21      9      2
[   3,]      1      3      2700      22      9      2
[   4,]      1      4      3400      23      9      2
[   5,]      1      5      3354      24      9      2
[   6,]      2      1      2850      19     13      1
[   7,]      2      2      4000      20     13      1
[   8,]      2      3      4000      21     13      1
[   9,]      2      4      4200      22     13      1
[  10,]      2      5      5720      23     13      1
```

The first column is the individual index, the second column is the time index and the third column gives the monthly earnings. The fourth column shows the experience in years and the fifth column presents the years of schooling. The elements in the final column are 2 for a female and 1 for a male participant. All cross sectional units are observed at $T_i = T$ periods and, therefore, the first two columns can be skipped. In this case we specify the joint number of time periods $T = 5$ when calling the quantlets.

Summing up, the columns of the matrix are defined as follows:

first column: <u>Identification number for cross section units</u>
> This column is skipped if a balanced panel is specified by the variable T. The quantlets `pantime` and `pandyn` do not use this column for an unbalanced data set and, thus, the entries of this column may be set to arbitrary values.

second column: <u>Time index</u>
> This column is skipped if a balanced panel is specified by the variable T. Only the quantlets `pantime` and `pandyn` make use of the time index so that in these cases the time index may be set to arbitrary values for all other quantlets involving panel data.

third column: <u>Dependent variable</u>

next m columns: <u>First set of explanatory variables</u>
> If a mixed specification is used (as in `panfix, panrand, panhaus`) these columns give the values for those variables, which are assumed to be (i) time varying and (ii) uncorrelated with the individual effects.

next k columns: <u>Second set of explanatory variables</u>
> For a mixed specification the second set of explanatory variables are assumed to be uncorrelated with the individual effects. If no distinction between the explanatory variables is made, all explanatory variables are listed in the first set.

If it is the case that a balanced data set is (inappropriately) ordered by time so that all observations for period 1 are given first, then the observations of period 2 and so on, then the quantlet `pansort` can be used to rearrange the data set:

```
z = pansort(z0 {,N})
```

where N indicates the number of cross section units in the original data set z0. If the parameter N is not given, the **pansort** quantlet will sort the unbalanced data set with respect to the cross sectional units and the time periods.

The **panstats** quantlet provides some summary statistics for the variables in the data set. It is highly recommended to first compute the summary statistics to find out possible problems with the data. The column **Within Var.%** gives the fraction of variance due to the within-group deviations. If this fraction is zero, the respective variable is constant over time. This is an important information for the estimation of fixed effects models. For the earnings data we type

```
panstats(z)
```

The output is

```
[ 1,]
[ 2,]  N*T:    500,  N:     100,  Min T(i):   5,  Max T(i):   5
[ 3,]  -----------------------------------------------------------
[ 4,]   Minimum    Maximum     Mean    Within Var.% Std.Error
[ 5,]  -----------------------------------------------------------
[ 6,]      1350    1.3e+04     3938      17.87        1513
[ 7,]         5         45    24.25       2.27        9.351
[ 8,]         9         13     9.94      0.2764       1.319
[ 9,]         1          2     1.34         0         0.4742
[10,]
```

This table indicates that the second variable (g) is constant over time (**Within Var.%** is zero) and there are only a few individuals with a changing schooling variable (**Within Var.%** is only 0.03 percent of the total variance). Accordingly, the schooling variable can be treated as (roughly) constant as well.

12.4.2   Time Effects

When specifying a panel data model, we first have to decide whether fixed time effects are to be included in the regression. Since in most panel data sets the number of time periods is small compared to the number of cross-section units, including time effects does not imply a severe loss in efficiency even if there are no time effects at all. So it is recommended to include time effects regularly. After the final model is specified, we may also test for a presence of time effects.

If the data set is a **balanced panel**, no information about the individuals and time periods is needed. The program simply assumes that the observations are in a consecutive order. Therefore, only the common number of time periods has to be indicated. To remove the time effects the following command can be used:

```
z1 = pantime(z {,T})
```

This will produce a new data set z1 that results from subtracting the time mean and adding the overall mean of the variables. If the input parameter T is given, the data is assumed to be a balanced panel with a common number of time periods, and the columns containing the individual number and time periods are skipped.

Applying such a data transformation and estimating with individual effects will give the same coefficient estimates as if the models are estimated with time dummies. However, the variance estimates differ due to the different degrees of freedom correction. Whenever the (average) number of time series is much smaller than the number of individuals N, the difference is negligible. More importantly, the R^2 measure is lower than in the case of an explicit inclusion of time effects because the variance of the dependent variable becomes smaller after subtracting the time mean. If these differences matter, time dummies should be included in the variable list instead of using the pantime quantlet.

A test for the existence of time effects can be computed as follows. First the model is estimated with fixed individual effects by using the original data set z. Then, the pantime quantlet is applied to remove the time effects. The resulting data set z1 is again estimated using the panfix quantlet. The likelihood-ratio statistic is obtained as $\mathrm{LR}_\lambda = 2(\ell_1 - \ell_0)$, where ℓ_0 is the log-likelihood value for the model without time effects and ℓ_1 is the log-likelihood value of the model after removing the time effects. The LR statistic has an asymptotic χ^2 distribution with $\max(T_i) - 1$ degrees of freedom.

12.4.3 Model Specification

To test for the existence of individual effects, two different test statistics can be used. The F-statistic and the LM-statistic are computed by the quantlet panfix. With respect to the power of the test it is important to indicate a possible correlation of the regressors with the individual effect. If it is assumed that all regressors are correlated with the individual effects (the fixed effects model), then the F-test is optimal. In the case of the random effects model, the LM-statistic is preferable. In a mixed specification the LM-statistic is computed by using the residuals from the instrumental variable (IV) estimation instead of the ordinary least squares residuals.

The quantlet `panfix` is called as

```
{output, siga, sige} = panfix(z, m {,T})
```

The string `output` yields the output table of an estimation assuming the first given m explanatory variables as time varying and correlated with the individual effects. The remaining k variables are assumed to be uncorrelated with the individual effect. The common time period T is included in the list of input parameters if the data are a balanced panel. An example for the use of this quantlet is given in the next section.

Besides the test statistic(s) for the existence of individual effects, the `panfix` quantlet computes estimates for the error variances σ_α^2 (`siga`) and σ_ε^2 (`sige`) which are used for the Hausman test called by

```
output = panhaus(z, siga, sige, m {,T})
```

This quantlet computes Hausman's test statistics for the hypothesis that the first m explanatory variables are correlated with the individual effect α_i (e.g. Hsiao, 1986). The test is based on the difference of the between- and the within-group estimator. This version of the Hausman test is numerically identical to the usual version which is based on a comparison of the within-group and the GLS estimator. The advantage of the former version of the Hausman test is that the differences between the coefficient estimates can be seen as an estimate for the parameter vector δ in the "Mundlak model":

$$\alpha_i = \bar{x}_i^T \delta + \eta_i + \varepsilon_{it} \ ,$$

where $\bar{x}_i$ denotes the time mean for the individual i. If the jth element of the vector δ is zero, then there is no correlation between the (mean of the) jth variable in x_{it}. Therefore, the t-statistics for the elements of δ can be seen as tests of the correlation with the respective explanatory variables. Accordingly, the test procedure can be used to select mixed specifications.

12.4.4 Estimation

If it is assumed that all variables are correlated with the individual effect α_i the "within-group estimator" can be applied. This estimator requires that all variables are time varying at least for some cross-section units. The respective command is

```
{output, siga, sige} = panfix(z, m {,T})
```

where the parameters are the same as before. If $m = 0$, then the OLS estimator is computed and for $m > 0$ an Instrumental Variable (IV) estimator is applied. The standard errors of this estimator are estimated in a robust fashion, that is, the standard errors are valid for quite general forms of autocorrelation and heteroscedasticity (Arrelano, 1987). The usual standard deviations and t-statistics for the within-group estimator can be obtained from the quantlet panrand by setting m equal to the number of all explanatory variables.

If it is assumed that at least one variable is uncorrelated with the individual effect α_i and that the error u_{it} is homoscedastic and serially uncorrelated, a more efficient estimate can be obtained using the panrand quantlet. This estimator employs the error variances σ_α^2 (siga) and σ_ε^2 (sige) from the panfix quantlet and computes the Generalized Least Squares (GLS) (for $m = 0$) or Generalized Instrumental Variable (GIV) (for $m > 0$) estimates:

```
{outfix, siga, sige} = panfix(z, m {,T})
panrand(z ,siga, sige, m {,T})
```

The panfix quantlet stores the estimates of the variances in the variables siga and sige. These variance estimates are provided as an input for the panrand quantlet. This estimation procedure is similar to the one suggested by Hausman and Taylor (1981). However, the only difference between the original estimator of Hausman and Taylor and the one used in XploRe is that the former uses $\bar{z}_i$ and $z_{it} - \bar{z}_i$ as separate instruments, whereas the panrand quantlet uses $z_{it} = (z_{it} - \bar{z}_i) + \bar{z}_i$ as a joint instrument. The reason for using the latter estimator is that it is easier to compute and usually no efficiency gain results from splitting z_{it} into $\bar{z}_i$ and $z_{it} - \bar{z}_i$.

12.4.5   An Example

Assume that we want to estimate an earnings function of the form

$$y_{it} = b_0 + b_1 x_{it} + b_2 x_{it}^2 + b_3 g_i + b_4 s_i + \alpha_i + \varepsilon_{it}$$

where

> y_{it} is the log of earnings,
>
> x_{it} is an experience variable,
>
> g_i is a dummy variable coding the gender and
>
> s_i is a measure of schooling attainment.

See Berndt (1991) for an introduction to the estimation of an earnings function.

The following code can be found in the example quantlet **Q metric07.xpl**. First we run the library and load the data set:

```
library("metrics")
z = read("earnings")
```

To estimate a fixed effects specification we only include the variables with a substantial variation in time. Here only the experience variables x and x^2 are included. The new data set is constructed as follows:

```
z1=z[,1:2]~log(z[,3])~z[,4]~(z[,4]^2)
```

To estimate the fixed effects model the **panfix** quantlet is used:

```
{output,siga,sige} = panfix(z1,2)
output
```

Since we have set $m = 2$, all regressors are allowed to be correlated with the individual effect. The output is

```
[ 1,] =========================================================
[ 2,] Fixed-Effect Model: y(i,t)=x(i,t)'beta+ a(i) + e(i,t)
[ 3,] =========================================================
[ 4,] PARAMETERS          Estimate        robust SE      t-value
[ 5,] =========================================================
[ 6,] beta[ 1 ]=          0.085408        0.01362          6.271
[ 7,] beta[ 2 ]=         -0.00038595      0.000245        -1.575
[ 8,] CONSTANT =          6.4021          0.1929          33.185
[ 9,] =========================================================
[10,] Var. of a(i):       0.48222         e(i,t):        0.015594
[11,] AR(1)-test     p-val: 0.9967        Autocorr.:       0.1628
[12,] F(no eff.)     p-val: 0.0000        R-square:        0.9050
[13,] LM(siga=0)     p-val: 0.0000        Log-Like:    -2080.423
[14,] =========================================================
```

Comparing the estimate of σ_α^2 with the estimate of σ_ε^2 it turns out that the individual effect clearly dominates the remaining error.

A problem with this estimate is that we have no estimates of the coefficients attached to g and s. Such estimates are available using the more restrictive framework of a random effects model. This specification is estimated by using the commands

```
z1 = z1~z[,5:6]
panrand(z1,siga,sige,0)
```

Setting $m = 0$ implies that no variable is allowed to be correlated with the individual effect. The resulting output table is as follows:

```
[ 1,] =========================================================
[ 2,] Random-Effect Model: y(i,t)=x(i,t)'beta+ a(i) +e(i,t)
[ 3,] =========================================================
[ 4,] PARAMETERS          Estimate        SE             t-value
[ 5,] =========================================================
[ 6,] beta[ 1 ]=          0.070265        0.01067          6.586
[ 7,] beta[ 2 ]=         -0.00035168      0.0002077       -1.693
[ 8,] beta[ 3 ]=          0.18934         0.04433          4.271
[ 9,] beta[ 4 ]=         -0.32376         0.1471          -2.201
[10,] constant =          5.298           0.5076          10.436
[11,] =========================================================
```

```
[12,] R-square:  0.9979 ,     N =    100 ,   N*T =     500
[13,] =====================================================
```

A problem with this specification is, however, that the regressors may be correlated with the individual effect. To test this hypothesis, the Hausman test is
performed:

```
panhaus(z1,siga,sige,2)
```

The resulting output table is

```
[ 1,] =====================================================
[ 2,] Hausman Specification Test for the null hypothesis:
[ 3,]  2 Variables x(i,t) are uncorrelated with a(i)
[ 4,] =====================================================
[ 5,]   d = beta(between)-beta(within)     SE        t-value
[ 6,] =====================================================
[ 7,]   d [ 1 ] =         -0.0630        0.0398       -1.585
[ 8,]   d [ 2 ] =          0.0000        0.0008        0.047
[ 9,] =====================================================
[10,] P-Value (RANDOM against FIXED effects):       0.0000
[11,] =====================================================
```

The test yields some indication for a correlation of the individual effects with
the first variable (x) but only a weak evidence for the second variable (x^2). The
joint test clearly rejects the random effects specification.

For the remaining variables which are assumed to be constant in time, no test
statistic is available. Accordingly, we may specify a mixed specification with
the first variable (or the first two variables are allowed to be correlated with the
individual effect and all other variables are assumed to be uncorrelated with
the individual effect). This specification is estimated in two steps.

First we ignore the random effects structure and perform a simple instrumental
variable estimation of the model:

```
{output,siga,sige} = panfix(z1,2)
output
```

```
[ 1,] =========================================================
[ 2,] Mixed Specification: y=x(i,t)'b1+z(i,t)'b2+a(i)+e(i,t)
[ 3,] =========================================================
[ 4,] PARAMETERS        Estimate       robust SE       t-value
[ 5,] =========================================================
[ 6,] beta[ 1 ]=         0.083721        0.01378         6.077
[ 7,] beta[ 2 ]=        -0.00036598      0.0002477      -1.478
[ 8,] beta[ 3 ]=         0.19717         0.0449          4.391
[ 9,] beta[ 4 ]=        -0.32609         0.1347         -2.421
[10,] CONSTANT =         4.9066          0.5434          9.030
[11,] =========================================================
[12,] Var. of a(i):      0.38122        e(i,t):      0.015434
[13,] AR(1)-test   p-val: 0.9939        Autocorr.:      0.1535
[14,] LM(siga=0)   p-val: 0.0000
[15,] R2:                0.5274
[16,] =========================================================
[17,] The first  2 Variables x(i,t) are assumed to be
[18,] correlated with the individual effects
[19,] ---> IV estimate to compute siga and sige
[20,] =========================================================
```

Since we ignore the covariance matrix of the errors at this stage, the estimation may be inefficient. However, the standard errors of the estimate are estimated robustly and thus are valid under very general assumptions. To improve the efficiency of the estimator, a GIV estimator is applied at the second step:

```
panrand(z1,siga,sige,2)
```

```
[ 1,] =========================================================
[ 2,] GLS-IV estimation of the Mixed specification
[ 3,] =========================================================
[ 4,]      y = x(i,t)'b1 + z(i,t)'b2 + a(i) + e(i,t)
[ 5,] =========================================================
[ 6,] PARAMETERS        Estimate         SE            t-value
[ 7,] =========================================================
[ 8,] beta[ 1 ]=         0.083717        0.01107         7.565
[ 9,] beta[ 2 ]=        -0.00036593      0.0002128      -1.719
[10,] beta[ 3 ]=         0.19769         0.04057         4.872
[11,] beta[ 4 ]=        -0.32608         0.1309         -2.492
```

```
[12,] constant =              4.9015         0.4428        11.069
[13,] ===========================================================
[14,] R-square:  0.9970 ,    N =    100 ,   N*T =        500
[15,] ===========================================================
[16,] The first  2 Variables x(i,t) are assumed to be
[17,] correlated with the individual effects
[18,] ===========================================================
```

The *t*-statistics of the GIV estimation is substantially larger than the (robust) *t*-statistics from the first stage estimation. This reflects the improved efficiency of the estimation.

12.5   Dynamic Panel Data Models

{output, beta} = pandyn (z, p, IVmeth {,T})
> computes 1st-stage GMM estimate of a dynamic linear model with p lags of the dependent variables

output = pandyn2 (z, p, IVmeth, beta {,T})
> computes 2nd-stage (robust) GMM estimate of a dynamic linear model

The dynamic model is given by

$$y_{it} = \gamma_1 y_{i,t-1} + \cdots + \gamma_p y_{i,t-p} + x_{it}^T \beta + \alpha_i + \varepsilon_{it}$$

For such a model the within-group estimator (for the fixed effects models) and the GLS estimator (for the random effects model) are not applicable. Therefore, Arrelano and Bond (1991) suggest to estimate the model using a GMM estimation procedure. The idea is to estimate the differenced model

$$\Delta y_{it} = \gamma_1 \Delta y_{i,t-1} + \cdots + \gamma_p \Delta y_{i,t-p} + x_{it}^T \beta + \Delta \varepsilon_{it},$$

where Δ is the difference operator such that $\Delta y_{it} = y_{it} - y_{i,t-1}$ by using the instruments

$$y_{i,t-2}, y_{i,t-3}, \ldots, Y_{i1}, \Delta x_t.$$

Using these instruments, different GMM estimators can be constructed. Let $z_{it} = [y_{it}, y_{i,t-1}, \ldots, y_{i1}]^T$. Then, Arrelano and Bond (1991) suggest to use the instrumental variable (IV) matrix

Method 4:
$$\begin{bmatrix} z_{ip} & 0 & 0 & \cdots & 0 \\ 0 & z_{i,p+1} & 0 & \cdot & 0 \\ 0 & 0 & z_{i,p+2} & \cdot & 0 \\ \cdot & \cdot & \cdot & \cdot & \\ 0 & 0 & 0 & \cdot & z_{i,T-2} \\ \Delta x_{i,p+2} & \Delta x_{i,p+3} & \Delta x_{i,p+4} & \cdot & \Delta x_{iT} \end{bmatrix}.$$

A problem with this IV matrix is that the number of rows grows with T^2 so that for large T the number of instruments may be larger as N and the GMM estimator cannot be computed. Therefore, in many cases a more parsimonious arrangement of the instruments is required:

Method 3:
$$\begin{bmatrix} y_{ip} & 0 & 0 & \cdots & 0 \\ y_{ip} & y_{i,p+1} & 0 & \cdot & 0 \\ y_{ip} & y_{i,p+1} & y_{i,p+2} & \cdot & 0 \\ \cdot & \cdot & \cdot & \cdot & \cdot \\ y_{ip} & y_{i,p+1} & y_{i,p+2} & \cdot & y_{i,T-2} \\ \Delta x_{i,p+2} & 0 & 0 & \cdots & 0 \\ 0 & \Delta x_{i,p+3} & 0 & \cdot & 0 \\ 0 & 0 & \Delta x_{i,p+4} & \cdot & 0 \\ \cdot & \cdot & \cdot & \cdot & \cdot \\ 0 & 0 & 0 & \cdot & \Delta x_{iT} \end{bmatrix}.$$

In this IV matrix the number of moments grows with $k \cdot T$. However, the number of instruments is still rather large so that a further reduction may be necessary. In these cases the following IV matrix is used:

Method 2:
$$\begin{bmatrix} y_{ip} & 0 & 0 & \cdots & 0 \\ y_{ip} & y_{i,p+1} & 0 & \cdot & 0 \\ \cdot & \cdot & \cdot & \cdot & \cdot \\ y_{ip} & y_{i,p+1} & y_{i,p+2} & \cdot & y_{i,T-2} \\ \Delta x_{i,p+2} & \Delta x_{i,p+3} & \Delta x_{i,p+4} & \cdot & \Delta x_{iT} \end{bmatrix}.$$

Finally we may have a panel with $T > N$. In this case these three approaches (methods 2–4) are not applicable. Therefore another method is implemented using only $y_{i,t-2}$ and $x_{i,t-1}$ as instruments for the lagged differences:

$$
\text{Method 1:} \quad
\begin{bmatrix}
y_{ip} & y_{i,p+1} & y_{i,p+2} & \cdot & y_{i,T-2} \\
\Delta x_{i,p+1} & \Delta x_{i,p+2} & \Delta x_{i,p+3} & \cdot & \Delta x_{i,T-1} \\
\Delta x_{i,p+2} & \Delta x_{i,p+3} & \Delta x_{i,p+4} & \cdot & \Delta x_{iT}
\end{bmatrix}.
$$

Accordingly, only k over-identifying moment conditions are used and thus the resulting estimator should be applicable in almost all cases.

The computation of the GMM estimator may be computationally burdensome when the number of instruments gets large. Therefore, it is highly recommended to start with the simplest GMM estimator (i.e. method 1) and then to try the more computer intensive methods 2–4. For a more efficient estimator, the standard errors of the coefficients should be smaller than for less efficient estimators. Therefore, it is expected that the standard errors tend to decrease with a more efficient GMM method. However in small samples the difference may be small or one may even encounter situations where the standard errors increase with an (asymptotically) more efficient estimator. This may occur by chance in a limited sample size or it may indicate a serious misspecification of the model.

To estimate the optimal weight matrix of the GMM estimate, two different approaches can be used. First, under the standard assumptions of the errors, the weight matrix may be estimated as

$$
W_N = \left[\sum_{i=1}^{N} \widehat{\sigma}_\varepsilon^2 Z_i D Z_i^T \right]^{-1} , \quad \text{where } D =
\begin{bmatrix}
2 & -1 & 0 & 0 & . & 0 \\
-1 & 2 & -1 & 0 & . & 0 \\
0 & -1 & 2 & -1 & . & 0 \\
. & . & . & . & . & . \\
0 & 0 & 0 & 0 & . & 2
\end{bmatrix},
$$

where Z_i is the IV matrix. If, however, the error are heteroskedastic or if $\Delta \varepsilon_{it}$ is uncorrelated but not independent of z_{t-2}, then the weight matrix is estimated as

$$
W_N = \left[\sum_{i=1}^{N} \sigma_\varepsilon^2 Z_i \Delta \widehat{\varepsilon}_i \Delta \widehat{\varepsilon}_i^T Z_i^T \right]^{-1} ,
$$

where $\Delta \widehat{\varepsilon}_i$ is the residual vector $\widehat{\varepsilon}_i = [\Delta \widehat{\varepsilon}_{i,p+2}, \dots, \Delta \widehat{\varepsilon}_{iT}]^T$ obtained from a consistent first-step estimation of the model. Therefore, the estimation is more cumbersome and may have poor small sample properties if the number of instruments is relatively large compared to the sample size N.

To assess the validity of the model specification, Hansen's misspecification statistic is used. This statistic tests the validity of the overidentifying restrictions resulting from the difference of the number of conditional moments and the number of estimated coefficients. If the model is correctly specified, the statistic is χ^2 distributed and the p-value of the statistic is given in the output string of the `pandyn` quantlet. Furthermore, a Hausman test (computed as a conditional moments test) can be used to test the hypothesis that the individual effects are correlated with the explanatory variables.

The data set z is similarly arranged as in the case of a static panel data estimation. If the data set is unbalanced, the identification number of the cross-section unit and the time period are given in the first two columns. However, the quantlet `pandyn` only uses the time index so that the first column may have arbitrary values. The following columns are the dependent variable y_{it} and the explanatory variables $x_{1it}, \dots, x_{kit}$, where all explanatory variables must vary in time. This is necessary because if the variables are constant, they become zero after performing differences.

If the data set is in a balanced form, the first two columns can be dropped and the common number of time periods is given. Furthermore, the number of lagged dependent variables p must be indicated. Accordingly, the quantlet is called by

```
{output,beta} = pandyn(z,p,IVmeth {,T})
```

The output table is returned in the string output and the coefficient estimates are stored in the vector β. The variable IVmeth specifies the method for constructing the instrument matrix. For example, IVmeth=1 give the GMM estimator with the smallest set of instruments, while IVmeth=4 gives the Arellano–Bond estimator. If IVmeth=0, then the program will select an appropriate instrument matrix by choosing the (asymptotically) most efficient GMM procedure subject to the constraint that the number of instruments does not exceed $0.9N$.

For the two-stage GMM estimator the weight matrix is computed using a consistent first-step estimate of the model. This can be done using the following estimation stages:

```
{out1,beta} = pandyn(z,p,IVmeth {,T})
out2 = pandyn(z,p,IVmeth,beta {,T})
out2
```

The output for the second estimation stage is presented in the string out2. In small samples, the two-stage GMM estimator may have poor small sample properties. Thus, if the results of the two stages differ substantially, it is recommended to use the (more stable) first-stage estimator.

12.6   Unit Root Tests for Panel Data

```
output = panunit (z, m, p, d {,T})
        computes various panel unit root statistics for the mth variable
        in the data set z with p lagged and different deterministic term
        indicated by d
```

Using panel data, powerful tests for a unit root in the autoregressive representation of the series can be constructed. Following Dickey and Fuller (1979), the unit root hypothesis can be tested by performing the regression:

$$\Delta y_{it} = \mu_i + \beta_i t + \varrho_i y_{i,t-1} + \alpha_{i1} \Delta y_{i,t-1} + \cdots + \alpha_{ip} \Delta y_{i,t-p} + u_{it} \qquad (12.25)$$

and testing $\varrho_i = 0$ for all i. For the test procedure of Breitung and Meyer (1994) it is assumed that $\beta_1 = \cdots = \beta_N$ and $\alpha_{1j} = \cdots = \alpha_{Nj}$ for $j = 1, \ldots, N$. Thus, as in traditional panel data models, heterogeneity is represented by an individual specific intercept. A simple test is obtained by using $\widehat{\mu}_i = y_{i1}$ which is the best estimator given that the null hypothesis is true. The resulting test regression is

$$\Delta y_{it} = \beta t + \varrho(y_{i,t-1} - y_{i1}) + \alpha_1 \Delta y_{i,t-1} + \cdots + \alpha_p \Delta y_{i,t-p} + u_{it}^* \qquad (12.26)$$

and the OLS regression yields an asymptotically normally distributed t-statistic as $N \to \infty$. Note that this test procedure is asymptotically valid for fixed T.

Levin and Lin (1993) extend the test procedure to individual specific time trends and short run dynamics. At the first stage the individual specific parameters are "partialled out" by forming the residuals e_{it} and $v_{i,t-1}$ from a regression of Δy_{it} and $y_{i,t-1}$ on the deterministics and the lagged differences. To account for heteroscedasticity the residuals are adjusted for their standard deviations yielding $\tilde{e}_{it}$ and $\tilde{v}_{i,t-1}$. The final regression is of the form

$$\tilde{e}_{it} = \varrho \tilde{v}_{i,t-1} + \nu_{it}.$$

If there are no deterministics in the first-stage regressions, the resulting t-statistic for the hypothesis $\varrho = 0$ is asymptotically standard normally distributed as $T \to \infty$ and $N \to \infty$. However, if there is a constant or a time trend in the model, then second-stage t-statistic tends to infinity as $T \to \infty$, even if the null hypothesis is true. Levin and Lin (1993) suggest a correction of the t-statistic to remove the bias and to obtain an asymptotic standard normal distribution for the test statistic.

Another way to deal with the bias problem of the t-statistic is to adopt a different adjustment for the constant and the time trend. The resulting test statistics are called the modified Levin–Lin statistic. In the model with a constant term only, the constant can be removed by using $(y_{i,t-1} - y_{i1})$ instead of $y_{i,t-1}$. The first stage regression only uses the lagged differences as regressors. At the second stage, the regression is

$$\widetilde{e}_{it} = \varrho(\widetilde{v}_{i,t-1} - \widetilde{v}_{i1}) + \nu_{it}$$

and the resulting t-statistic for $\varrho = 0$ is asymptotically standard normal as $T \to \infty$ and $N \to \infty$. If there is a linear trend in the model the nuisance parameters are removed by estimating the current trend value by using past values of the process only. Accordingly, the series are adjusting according to

$$\widetilde{e}_{it}^{*} = \widetilde{e}_{it} - \widetilde{e}_{i1},$$

$$\widetilde{v}_{i,t-1}^{*} = \widetilde{v}_{i,t-1} - \widetilde{v}_{i1} - \frac{t+1}{t-2}\sum_{s=2}^{t-1}\Delta\widetilde{v}_{is}.$$

Again, the resulting modification yields a t-statistic with a standard normal limiting distribution.

Im, Pesaran, and Shin (1997) further extended the test procedure by allowing for different values of ϱ_i under the alternative. Accordingly, all parameters were estimated separately for the cross-section units. Let τ_i denote the individual t-statistic for the hypothesis $\varrho_i = 0$. As $T \to \infty$ and $N \to \infty$, we have

$$\sqrt{N}\bar{\tau} = N^{-1/2}\sum_{i=1}^{N}\tau_i \xrightarrow{d} \mathcal{N}(\mu_\tau, v_\tau),$$

where μ_τ and v_τ is the expectation and variance of the statistic τ_i. Im, Pesaran, and Shin (1997) present the mean and variances for a wide range of T and p. These values are interpolated by regression functions on $\sqrt{T}$, T, and T^2 and p.

The quantlet computing all these unit root statistics is called

```
output = panunit(z,m,p,d {,T})
```

The parameters necessary for computing the statistics are as follows. The parameter m indicates the number of variable in the data set which is to be tested for a unit root. The parameter p indicates the number of lagged differences in the model. The parameter d indicates the kind of deterministics used in the regressions. A value of d=0 implies that there is no deterministic term in the model. If d=1, a constant term is included and for d=2 a linear time trend is included. Finally, if a balanced panel data set is used, the common time period T is given. For example, assume that the second variable in a balanced data set with T=32 is to be tested including a constant and a lagged difference. Then the respective command is

```
output = panunit(z,2,1,1,32)
```

The string output first gives an output table of a pooled Dickey–Fuller regression. In a second table, the four unit root statistics are presented.

Bibliography

Andrews, D. and Schafgans, M. (1997). Semiparametric Estimation of the Intercept of a Sample Selection Model, *Review of Economic Studies*.

Arellano, M. (1987). Computing Robust Standard Errors for Within-groups Estimators, *Oxford Bulletin of Economics and Statistics*, 49, 431–434.

Arellano, M. and Bond, S.R. (1991). Some Tests of Specification for Panel Data: Monte Carlo Evidence and an Application to Employment Equations, *Review of Economic Studies*, 58, 277–297.

Baltagi, B.H. (1995). *Econometric Analysis of Panel Data*, Wiley, Chichester.

Berndt, E.R. (1991). *The Practice of Econometrics: Classic and contemporary*, Addison–Wesley, Reading.

Breitung, J. and Meyer, W. (1994). Testing for Unit Roots in Panel Data: Are wages on different bargaining levels cointegrated? *Applied Economics*, 26, 353–361.

Dickey, D.A. and Fuller, W.A. (1979). Distribution of the Estimates for Autoregressive Time Series With a Unit Root, *Journal of the American Statistical Association*, 74, 427–431.

Härdle, W. and Stoker, T. (1989). Investigating Smooth Multiple Regression by the Method of Average Derivatives, *Journal of the American Statistical Association*, 84, 986–995.

Horowitz, J.L. and Härdle, W. (1994). Testing a parametric model against a semiparametric alternative, *Econometric Theory*, 10, 821–848.

Horowitz, J.L. and Härdle, W. (1996). Direct Semiparametric Estimation of a Single-Index Model with Discrete Covariates, *Journal of the American Statistical Association*, 91, 1632–1640.

Hausman, J.A. and Taylor, W.E. (1979). Panel Data and Unobservable Individual Effects, *Econometrica*, 49, 1377–1399.

Heckman, J.J. (1979). Sample Selection Bias as a Specification Error, *Econometrica*, 47, 153–161.

Hsiao, C. (1986). *Analysis of Panel Data*, Cambridge University Press, Cambridge.

Im, K.S., Pesaran, M.H. and Shin, Y. (1997). Testing for Unit Roots in Heterogenous Panels, University of Cambridge, revised version of the DAE Working Paper No 9526.

Levin, A. and Lin, C.-F. (1993). Unit Root Tests in Panel Data: Asymptotic and Finite-Sample Properties, University of California San Diego, Unpublished Working Paper (revised).

Li, K.-C. (1991). Sliced inverse regression, *Journal of the American Statistical Association*, 86, 316–342.

Maddala, G.S. (1983). *Limited-dependent and qualitative variables in econometrics*, Cambridge University Press, Econometric Society Monographs No. 4.

Powell, J.L. (1987). Semiparametric Estimation of Bivariate Latent Variable Models, University of Wisconsin-Madison, Working Paper 8704.

Powell, J.L., Stock, J. H. and Stoker, T. M. (1989). Semiparametric Estimation of Index Coefficients, *Econometrica*, 57, 1403–1430.

Stoker, T.M. (1991). Equivalence of direct, indirect, and slope estimators of average derivatives, in: Barnett, Powell and Tauchen (eds.), *Nonparametric and Semiparametric Methods in Econometrics and Statistics*, Cambridge University Press.

Werwatz, A.K. (1997). A consistent test for misspecification in polychotomous response models, Discussion Paper No. 97–74, SFB 373, Humboldt University Berlin.

13 Extreme Value Analysis

Rolf-Dieter Reiss and Michael Thomas

The extreme (upper or lower) parts of a sample, such as

- flood discharges;

- high concentration of air pollutants;

- claim sizes over a higher priority in reinsurance business;

- larger losses on financial markets,

have exhibited an increasing risk potential during the last decades. This is the reason why the statistical analysis of extremes has become an important question in theory and practice.

We give an outline of the most important statistical models (extreme value (EV) and generalized Pareto (GP)) of tools for assessing the adequacy of the parametric models (mean excess functions) and of relevant statistical estimation procedures.

The parametric distributions are fitted to observed maxima, minima, intermediate data or peaks over thresholds and extrapolated to regions of interest, where only a few or no observations are available. Such an extrapolation is not possible when a nonparametric approach is utilized. It is evident that the choice of the parametric model is a crucial issue.

The statistical models and procedures are implemented as a part of the finance library, so one needs to load that library first by using the command

```
library ("finance")
```

13.1   Extreme Value Models

```
r = randx ("ev", n, gamma)
     returns a vector with n pseudorandom EV variables under the
     shape parameter gamma

r = pdfx ("ev", x, gamma)
     returns the values of the EV density with shape parameter gamma
     for all elements of the vector x

r = cdfx ("ev", x, gamma)
     returns the values of the EV distribution function with shape
     parameter gamma for all elements of the vector x

r = qfx ("ev", x, gamma)
     returns the value of the EV quantile function with shape param-
     eter gamma for all elements of the vector x
```

In this section, we introduce a parametric model for maxima $y_i = \max\{x_{i,1}, \ldots, x_{i,n}\}$, $i = 1, \ldots, m$, where the $x_{i,j}$ may not be observable. Such data occur, e.g. when annual flood maxima or monthly maxima of temperatures are recorded. In the first case, we have $n = 365$ while m is the number of observed years. We mention a limit theorem which suggests a parametric modeling of such maxima using extreme value (EV) distributions.

Assume that $X_1, \ldots, X_n$ are independent random variables with common distribution function (df) F, and let $Y := \max\{X_1, \ldots, X_n\}$. Then, $P\{Y \leq t\} = F^n(t)$ is the df of Y. If, for an appropriate choice of constants a_n and b_n, the dfs of the maxima converge to a continuous limiting distribution function, i.e.

$$F^n(a_n x + b_n) \to G(x), \quad n \to \infty,$$

then G is equal to one of the following types of distributions (a result due to Fisher and Tippett, 1928):

(i)   Gumbel (EV0) $G_0(x) = \exp(-e^{-x})$, $x \in \mathbb{R}$,
(ii) Fréchet (EV1) $G_{1,\alpha}(x) = \exp(-x^{-\alpha})$, $x \geq 0, \alpha > 0$,
(iii) Weibull (EV2) $G_{2,\alpha}(x) = \exp(-(-x)^{-\alpha})$, $x \leq 0, \alpha < 0$.

Moreover, the df F is said to belong to the domain of attraction of the EV distribution G (in short, $F \in \mathcal{D}(G)$). These EV distributions are also the limiting distributions of maxima of dependent random variables under weak conditions (see Leadbetter and Nandagopalan, 1989).

By employing the reparametrization $\gamma = 1/\alpha$ and appropriate scale and location parameters, one can unify these models in the von Mises parameterization by

$$
G_\gamma(x) = \begin{cases} \exp(-(1+\gamma x)^{-1/\gamma}), & 1+\gamma x > 0, \; \gamma \neq 0, \\ \exp(-e^{-x}), & x \in \mathbb{R}, \; \gamma = 0. \end{cases}
$$

Notice that $G_\gamma(x) \to G_0(x)$ for $\gamma \to 0$. Moreover, the relation $G_\gamma = G_{i,\alpha,-\alpha,|\alpha|}$ holds with $\alpha = 1/\gamma$, $i = 1$ for $\gamma > 0$, and $i = 2$ for $\gamma < 0$.

The quantlets concerning densities, distribution and quantile functions (qfs) of extreme value distributions as well as the generation of pseudorandom variables are listed at the beginning of this section. The routines belonging to the von Mises parameterization are merely displayed. One can address the three submodels by providing the names `"ev0"`, `"ev1"` and `"ev2"` in place of `"ev"`. Notice the shape parameter is not required within the EV0 model. For example, use

```
pdfx("ev2", x, alpha)
```

to calculate the Weibull density with shape parameter α. In addition, location and scale parameters μ and σ can easily be included, e.g. take

```
r = mu + sigma * randx("ev", n, gamma)
```

to generate a data set with n independent realizations under $G_{\gamma,\mu,\sigma}$.

13.2   Generalized Pareto Distributions

r = randx ("gp", n, gamma)
      returns a vector with n pseudorandom GP variables under the
      shape parameter gamma

r = pdfx ("gp", x, gamma)
      returns the value of the GP density with shape parameter gamma
      for all elements of the vector x

r = cdfx ("gp", x, gamma)
      returns the value of the GP distribution function with shape pa-
      rameter gamma for all elements of the vector x

r = qfx ("gp", x, gamma)
      returns the value of the GP quantile function with shape param-
      eter gamma for all elements of the vector x

In this section, we introduce a parametric model for those data $y_1, \ldots, y_k$ which exceed a threshold t, where $x_1, \ldots, x_n$ are the original data. Likewise, one may consider the k largest values $(y_1, \ldots, y_k) = (x_{n-k+1:n}, \ldots, x_{n:n})$ of the data x_i. Generalized Pareto (GP) distributions constitute adequate models for such data (Reiss and Thomas, 1997). They consist of the following three submodels:

(i)   Exponential (GP0) $W_0(x) = 1 - e^{-x}$, $x \geq 0$,
(ii)   Pareto (GP1) $W_{1,\alpha}(x) = 1 - x^{-\alpha}$, $x \geq 1, \alpha > 0$,
(iii   Beta (GP2) $W_{2,\alpha}(x) = 1 - (-x)^{-\alpha}$, $-1 \leq x \leq 0, \alpha < 0$.

Again, one can unify these distributions by using the parameterization with $\gamma = 1/\alpha$. We have

$$W_\gamma(x) = \begin{cases} 1 - (1 + \gamma x)^{-1/\gamma} & x > 0, \gamma > 0 \\ & 0 < x < -1/\gamma, \gamma < 0 \\ 1 - e^{-x} & x \geq 0, \gamma = 0 \end{cases} .$$

A mathematical justification of the modeling is obtained by a limit theorem. Assume that X is a random variable with df $F \in \mathcal{D}(G_\gamma)$, and consider the conditional distribution

$$F^{[t]}(x) := P(X \leq x | X > t), \quad x > t,$$

which is the common df of the exceedances y_j above the threshold t. For the sequence of thresholds $t_n = a_n t + b_n$ the convergence

$$F^{[a_n t + b_n]}(a_n(t + s) + b_n) \rightarrow 1 + \log G_\gamma(s/(1 + \gamma t))$$

holds (Falk, Hüsler and Reiss, 1994). Formally, the relation $W_\gamma = 1 + \log G_\gamma$ holds between EV and GP dfs G_γ and W_γ.

GP distributions are the only continuous dfs such that

$$F^{[t]}(a_t x + b_t) = F(x).$$

More precisely, the relation

$$W^{[t]}_{\gamma,\mu,\sigma} = W_{\gamma,t,\sigma+\gamma(t-\mu)}$$

holds. The location parameter of the truncated df is the truncation point t, while the shape parameter γ remains unchanged.

The quantlets concerning densities, distribution and quantile functions of generalized Pareto distributions as well as the generation of pseudorandom variables are shown at the beginning of this section. Again, the routines belonging to the von Mises parameterization are merely displayed. One can address the three submodels by providing the names "gp0", "gp1" and "gp2" instead of "gp". Within the GP0 model, the shape parameter is not required.

13.3   Assessing the Adequacy: Mean Excess Functions

```
r = empme (x, t)
      returns the value of the empirical mean excess function based on
      the real vector x at all elements of the vector t

r = gpme (gamma, t)
      returns the value of the GP mean excess function of a GP dis-
      tribution with the shape parameter gamma at all elements of the
      vector t

r = gp1me (alpha, t)
      returns the value of the mean excess function of a Pareto (GP1)
      distribution with shape parameter alpha at all elements of the
      vector t
```

Let X be a random variable with df F. Then, the mean excess function of F is

$$e_F(t) := E(X - t | X > t).$$

If F includes location and scale parameters μ and σ, then

$$e_{F_{\mu,\sigma}} = \sigma e_F \left(\frac{t - \mu}{\sigma} \right).$$

The mean excess function of a Pareto (GP1) distribution $W_{1,\alpha}$ is

$$e_{W_{1,\alpha}}(t) = \frac{t}{\alpha - 1}, \quad t > 1, \alpha > 1.$$

For the generalized Pareto distribution W_γ, the mean excess function is given by

$$e_{W_\gamma}(t) = \frac{1 + \gamma t}{1 - \gamma}$$

for $t > 0$, if $0 \le \gamma < 1$, and $0 < t < -1/\gamma$, if $\gamma < 0$. Notice that the mean excess function does not exist for $\gamma \ge 1$ ($\alpha \le 1$ in the Pareto (GP1) model).

Mean excess functions are linear if, and only if, F is a generalized Pareto distribution. Therefore, the empirical mean excess function

$$e_n(t) = \frac{\sum_{i=1}^n (x_i - t) I(t < x_i)}{\sum_{i=1}^n I(t < x_i)}, \quad x_{1:n} \le t < x_{n:n},$$

can be employed to check if a GP modeling of a given data set is plausible.

Moreover, by comparing the empirical mean excess function and a parametric one, fitted by an estimator, one obtains a visual tool to control the result of the estimation. A stronger deviation from the empirical mean excess function shows that an estimator may not be applicable. In the example provided in Section 13.8, we apply this tool to make a choice between two different parametric estimation procedures.

13.4   Estimation in EV Models

Different parametric estimation procedures are provided for estimating the shape, scale and location parameter of an extreme value distribution.

13.4.1   Linear Combination of Ratios of Spacings (LRS)

```
{gamma, mu, sigma} = lrseev (x)
        applies the LRS estimator to the vector x and returns the esti-
        mated shape, location and scale parameter of an EV distribution
```

Let
$$\widehat{\gamma}_i = \frac{\log(\widehat{r}_i)}{-\log(c)},$$

where
$$\widehat{r}_i = \frac{x_{[nq_2]:n} - x_{[nq_1]:n}}{x_{[nq_1]:n} - x_{[nq_0]:n}}$$

with
$$q_0 = i/n, \quad q_1 = q_0^c, \quad q_2 = (n-i)/n$$

and
$$c = (\log((n-i)/n)/\log(i/n))^{1/2}.$$

The previous construction becomes plausible by noting that $x_{[nq]:n}$ is an estimator of the q-quantile $G^{-1}_{\gamma,\mu,\sigma}(q)$, and therefore the relation

$$\widehat{r}_i \approx \frac{G^{-1}_\gamma(q_2) - G^{-1}_\gamma(q_1)}{G^{-1}_\gamma(q_1) - G^{-1}_\gamma(q_0)} = \left(\frac{1}{c}\right)^\gamma$$

holds. A natural estimator of γ is given by the linear combination

$$\widehat{\gamma} = \frac{1}{[n/4]} \sum_{i=1}^{[n/4]} \widehat{\gamma}_i.$$

Estimates of the location and scale parameters are derived from a least square line $x \to ax + b$ fitted to the QQ-plot

$$\left(G_{\widehat{\gamma}}^{-1} \left(\frac{i}{n-1} \right), x_{i:n} \right), \quad i = 1, \dots, n.$$

One obtains $\widehat{\mu} = b$ and $\widehat{\sigma} = a$ as estimates.

13.4.2   ML Estimator in the EV Model

{gamma, mu, sigma} = mleev (x, k)
    applies the ML estimator in the EV model to the vector x and
    returns the estimated shape, location and scale parameter of an
    EV distribution

The maximum likelihood estimator in the EV model is numerically evaluated by using an iteration procedure. The LRS estimator described in Subsection 13.4.1 serves as an initial value.

Note that the maximum likelihood estimator fails for $\gamma \leq -1$ because in that case no global maximum of the likelihood function exists. Yet it seems to be that a local maximum close to the initial value is attained.

13.4.3   ML Estimator in the Gumbel Model

{mu, sigma} = mleev0 (x)
    applies the ML estimator in the Gumbel (EV0) model to the
    vector x and returns the estimated location and scale parameters

---

The ML estimator in the Gumbel (EV0) model must be evaluated numerically. A certain moment estimator is utilized as the initial value in the iteration procedure.

13.5 Fitting GP Distributions to the Upper Tail

The GP approach concerns a local, parametric modeling and estimation of the underlying distribution in the upper tail. By using the estimation procedures in Section 13.6, one may fit a GP distribution in two steps.

Firstly, one estimates a GP distribution, say $W_{\tilde{\gamma},t,\tilde{\sigma}}$, within the GP submodel $\{W_{\gamma,t,\sigma} : \gamma \in \mathbb{R}, \sigma > 0\}$, based on the exceedances $y_1,\ldots,y_k$ over a selected threshold t. Notice that the location parameter is equal to the truncation point t, which is also the left endpoint of the estimated GP distribution. The estimated GP df, density, quantile function and mean excess function can be fitted to the empirical df, density, quantile function and mean excess function based on the exceedances y_j.

Secondly, a fit to the original data $x_1,\ldots,x_n$ is achieved by selecting location and scale parameters $\hat{\mu}$ and $\hat{\sigma}$ such that

$$W^{[t]}_{\hat{\gamma},\hat{\mu},\hat{\sigma}} = W_{\tilde{\gamma},t,\tilde{\sigma}}$$

and

$$W_{\hat{\gamma},\hat{\mu},\hat{\sigma}}(t) = \frac{n-k}{n},$$

where n is the total sample size and $k = \sum_{i=1}^{n} I(t < x_i)$ is the number of exceedances above t. One gets a GP distribution with the same shape as the first one, yet the reparametrized distribution possesses the mass $1 - k/n$ just as the empirical distribution function of the x_i above the threshold t. As a result, the reparametrized df, density, quantile function and mean excess function can be fitted to the upper part of the empirical df, density, quantile function and mean excess function based on the original data x_i. Explicitly, one gets the estimators

$$
\begin{aligned}
\hat{\gamma} &= \tilde{\gamma}, \\
\hat{\sigma} &= \tilde{\sigma}(k/n)^{\tilde{\gamma}}, \\
\hat{\mu} &= t - \frac{\tilde{\sigma} - \hat{\sigma}}{\tilde{\gamma}}.
\end{aligned}
$$

Figure 13.1 exemplifies this procedure. The left-hand plot shows the empirical df (solid line) based on the exceedances above the threshold 1.13 and a fitted GP df (dotted). The plot on the right-hand side shows the empirical df of the original data set and the reparametrized GP df that fits to the upper tail.

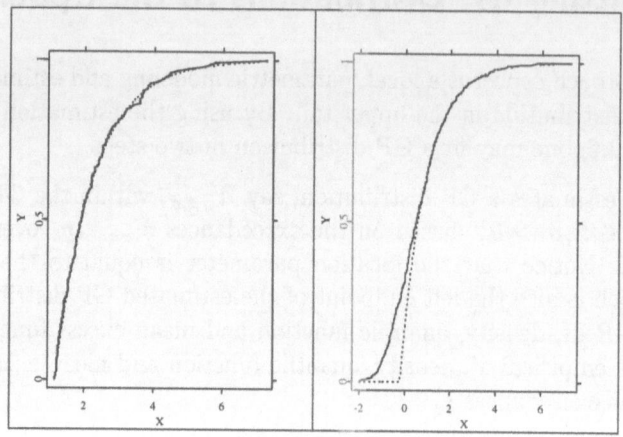

Figure 13.1. Fitting a GP df to the exceedance df above $t = 1.13$ (left) and fitting a reparametrized GP df to the upper tail of the empirical df of the complete sample (right).

Within the GP1-submodel of Pareto dfs $W_{1,\alpha,0,\sigma}$ with location parameter $\mu = 0$, we have

$$W^{[t]}_{1,\alpha,0,\sigma} = W_{1,\alpha,0,t}.$$

After having estimated the shape parameter α, based on the exceedances above the threshold t, one obtains the required scale parameter σ from the relation

$$W_{1,\alpha,0,\sigma}(t) = \frac{n-k}{n},$$

which yields the estimator

$$\widehat{\sigma} = t\left(\frac{k}{n}\right)^{1/\widehat{\alpha}}.$$

In our implementation, one has to select the number k of exceedances. Then, the threshold $t = x_{n-k+1:n}$ is utilized.

13.6 Parametric Estimators for GP Models

We describe the estimators for the GP models which are implemented in XploRe. The output always concerns the reparametrized GP or Pareto (GP1) distributions which were introduced in Section 13.5.

13.6.1   Moment Estimator

`{gamma, mu, sigma} = momentgp (x, k)`
    applies the moment estimator to the k largest order statistics of
    the vector x and returns the estimated shape, location and scale
    parameter of the GP model

The moment estimator (Dekkers, Einmal and de Haan, 1989) for the shape parameter γ in the von Mises parameterization, based on the k largest values of the sample, is given by

$$\widehat{\gamma}_{n,k} = \widehat{\gamma}_{1,n,k} + \widehat{\gamma}_{2,n,k},$$

where

$$\widehat{\gamma}_{1,n,k} = l_{1,k}$$

and

$$\widehat{\gamma}_{2,n,k} = 1 - \frac{1}{2}\left(1 - \frac{l_{1,k}^2}{l_{2,k}}\right)^{-1}$$

with

$$l_{j,k} = \frac{1}{k-1}\sum_{i=1}^{k-1}\log\left(\frac{x_{n-i+1:n}}{x_{n-k+1:n}}\right)^j, \quad j = 1, 2.$$

The scale parameter $\widetilde{\sigma}$ is estimated by fitting a least squares line to the GP QQ-plot under the estimated shape parameter $\widehat{\gamma}_{n,k}$, i.e. to the points

$$\left(W_{\widehat{\gamma}_{n,k}}^{-1}\left(\frac{i}{k+1}\right), y_{i:k}\right), \quad i = 1, \ldots, k,$$

where the y_i are the k largest values of the data set $x_1, \ldots, x_n$. The location parameter is equal to the threshold $t = x_{n-k+1:n}$, and the transformation to the tail described in Section 13.5 is applied. As an example, we simulate a sample with 500 data points under $W_{1,1,2}$ and apply the moment estimator, based on all values of the sample:

```
x = 1 + 2 * gpdata (1, 500)
momentgp (x, 500)
```

Then, XploRe displays in its output window

```
Contents of _tmp.gamma
[1,]   1.0326
Contents of _tmp.mu
[1,]   1.0024
Contents of _tmp.sigma
[1,]   1.9743
```

13.6.2   ML Estimator in the GP Model

{gamma, mu, sigma} = mlegp (x, k)
>   applies the ML estimator in the GP model to the k largest order
>   statistics of the vector x and returns the estimated shape, location
>   and scale parameter of the GP model

The maximum likelihood estimator in the GP model is numerically evaluated by using an iteration procedure. The moment estimator described in Subsection 13.6.1 serves as an initial value.

The remarks about the ML estimator in the EV model (see Subsection 13.4.2) also apply to this estimator.

13.6.3   Pickands Estimator

{gamma, mu, sigma} = pickandsgp (x, k)
>   applies the Pickands estimator to the k largest order statistics of
>   the vector x and returns the estimated shape, location and scale
>   parameter of the GP model

The Pickands estimator (Pickands, 1975) of the shape parameter γ is given by

$$\widehat{\gamma}_n(m) = \log\left(\frac{x_{n-m+1:n} - x_{n-2m+1:n}}{x_{n-2m+1:n} - x_{n-4m+1:m}}\right) / \log 2,$$

where $m = [k/4]$. This construction is similar to the one for the LRS estimator for the EV model described in Subsection 13.4.1. Scale and location parameter are estimated as described in Subsection 13.6.1.

13.6.4 Drees–Pickands Estimator

{gamma, mu, sigma} = dpgp (x, k)
  applies the Drees–Pickands estimator to the k largest order statistics of the vector x and returns the estimated shape, location and
  scale parameter of the GP model

A refinement of the Pickands estimator was introduced by Drees (1995). It uses a convex combination of Pickands estimates

$$\widehat{\gamma}_{n,k} = \sum_{i=0}^{\infty} \widehat{c}_{i,n}\widehat{\gamma}_n([m2^{-i}] + 1)$$

with $m = [k/4]$ and

$$\widehat{c}_{i,n} = c_{i,n}(\widehat{\gamma}_n(m)),$$

where

$$c_{i,n}(\gamma) = \begin{cases} \frac{2^{\gamma+1}-1}{2^{\gamma}-1}(1 - 2^{-(i+1)\gamma})2^{-(i+2)} & \gamma \neq 0 \\[2mm] (i+1)2^{-(i+2)} & \gamma = 0 \end{cases}$$

if $\gamma \geq -1/2$ and

$$c_{i,n}(\gamma) = c_{i,n}(-(1+\gamma))$$

if $\gamma < -1/2$.

13.6.5   Hill Estimator

{alpha, sigma} = hillgp1 (x, k)
      applies the Hill estimator to the k largest order statistics of the
      vector x and returns the estimated shape and scale parameter of
      the GP1 model

The celebrated Hill estimator is a maximum likelihood estimator for the GP1 submodel of Pareto dfs $W_{1,\alpha,0,t}$ with left endpoint t. It is given by

$$\widehat{\alpha}_{n,k} = \left(\frac{1}{k-1} \sum_{i=1}^{k-1} \log \frac{x_{n-i+1:n}}{x_{n-k+1:n}} \right)^{-1}.$$

Recall that $t = x_{n-k+1:n}$ is used as the threshold. Notice that $W_{1,\alpha,\mu,t-\mu}$ are further Pareto dfs with left endpoint equal to t. When $\mu = 0$, then we are in the above-mentioned submodel. When $\mu \neq 0$, then the Hill estimator can be inaccurate. Therefore, one should be cautious when the Hill estimator is applied to real data.

13.6.6   ML Estimator for Exponential Distributions

{mu, sigma} = mlegp0 (x, k)
      applies the ML estimator for the exponential distributions (GP0)
      to the k largest order statistics of the vector x and returns the
      estimated location and scale parameter of the GP0 model

The maximum likelihood estimator for the exponential distributions, based on the k largest values, is given by

$$\widehat{\sigma}_{n,k} = \frac{1}{k-1} \sum_{i=1}^{k-1} x_{n-i+1:n} - x_{n-k+1:n},$$

$$\widehat{\mu}_{n,k} = x_{n-k+1:n} - \widehat{\sigma}_{n,k} \log(n/k).$$

13.6.7 Selecting a Threshold by Means of a Diagram

```
r = momentgpdiag (x, k)
      evaluates the moment estimator for all number of extremes given
      in the vector k

r = mlegpdiag (x, k)
      evaluates the MLE of the GP model for all number of extremes
      given in the vector k

r = pickandsgpdiag (x, k)
      evaluates the Pickands estimator for all number of extremes given
      in the vector k

r = dpgpdiag (x, k)
      evaluates the Drees–Pickands estimator for all number of ex-
      tremes given in the vector k

r = hillgp1diag (x, k)
      evaluates the Hill estimator for all number of extremes given in
      the vector k
```

A visual tool to facilitate the selection of a threshold t (or, likewise, the number of upper extremes) is the diagram of estimates $k \to \widehat{\alpha}_{n,k}$ or $k \to \widehat{\gamma}_{n,k}$. For small values of k one recognizes a high random fluctuation of the estimator, while for large values of k, there is typically a bias due to a model deviation. Within an intermediate range, the estimates often stabilize around a value which gives a hint for the selection of the number of extremes. Of course, one should also apply QQ-plots and empirical mean excess functions to justify the choice of the threshold. In the statistical literature one can also find the advice to take the upper ten percent of a sample. Hydrologists take the 3–4 highest flood peaks in a water year. The automatic choice of a threshold is presently a hot research topic.

A diagram option is provided for each estimator of the shape parameter of a GP distribution. The corresponding calls are listed above. These quantlets return a vector with the estimates for each value of k. Thus, to plot a diagram of the Hill estimates based on a simulated Fréchet data set with shape parameter $\alpha = 1$, one can use the commands

```
x = ev1data (1, 500)
line (2:500~hillgp1diag (x,2:250))
```

The output of the above commands is shown in Figure 13.2. One can see that after a strong fluctuation for small values of k the estimates are close to the true parameter $\alpha = 1$. Moreover, a bias for large values of k is visible.

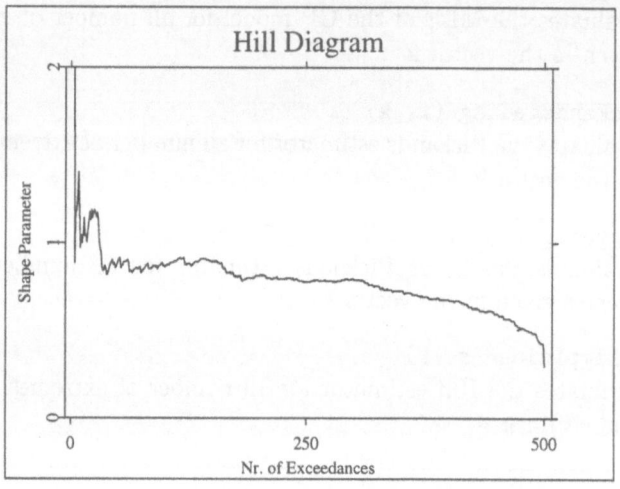

Figure 13.2. Diagram of Hill estimates applied to Fréchet data $G_{1,1}$.

13.7   Graphical User Interface

---

xtremes (x)
     starts menu interface for interactive analysis of data set provided
     in vector x

A graphical user interface allows the interactive analysis of a data set. After calling the quantlet

```
xtremes (x)
```

a menu box opens which lets the user select a parametric estimation procedure or a nonparametric visualization. Different curves of the same type (e.g. densities, dfs, etc.) are plotted into the same window. Therefore, one can easily compare empirical and parametric versions of a curve. The application of the menus and dialog boxes is self-explanatory.

13.8 Example

We analyze the daily returns of the exchange rate of the Yen related to the U.S. Dollar from Dec. 1, 1978 to Jan. 31, 1991. Our aim is to fit a generalized Pareto model to the lower tail of the returns to estimate the probability of extreme losses. We start the analysis by loading the finance library and the data set dyr.dat:

```
library("finance")
dyr=read("dyr.dat")
```

A scatter plot of the data set can be obtained using the command

```
plot(1:rows(dyr)~dyr)
```

In the Academic Version of XploRe the following examples can be executed with the smaller data set dyr1000.dat. Although slightly different results are obtained, one can still recognize that the Hill estimator is unsuited for that data set.

One recognizes from Figure 13.3 that the distribution of the returns possesses a fat tail. Because our estimators are defined for the upper tail, one must change the sign of the data set with the command

```
dyr = -dyr
```

A suitable threshold can be selected by plotting an estimator diagram. The call

```
r=momentgpdiag(dyr,5:500)
plot(5:500~r)
```

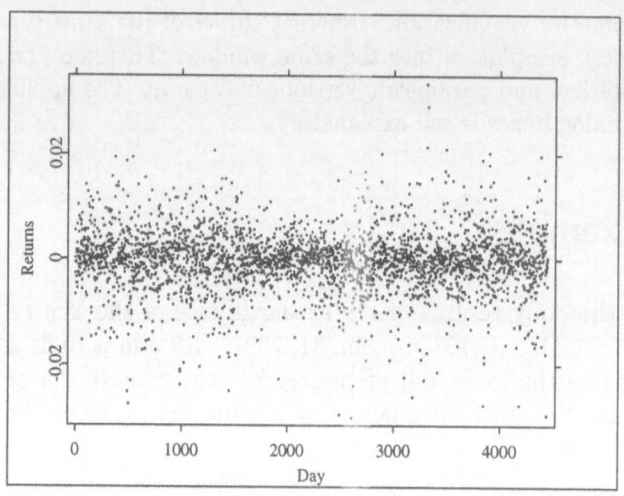

Figure 13.3. Scatter plot of daily returns of Yen related to U.S. Dollar from Dec. 1978 to Jan. 1991

produces the diagram in Figure 13.4. We select $k = 160$ extremes (yielding the threshold $t = 0.00966$) and plot the empirical quantile function as well as the estimated parametric one to check the quality of the fit.

This task is performed by the following code (calls to format the graphical output are not shown):

```
m=momentgp(dyr,160)

d=createdisplay(1,1)
t=aseq(0.965,350,0.0001)
qf=t~m.mu+m.sigma*qfx("gp",t,m.gamma)
show(d,1,1,qf)

empqf=(4284:4444)/4445~sort(dyr)[4284:4444]
adddata(d,1,1,empqf)
```

Q xtrm01.xpl

The resulting plot is shown in Figure 13.5.

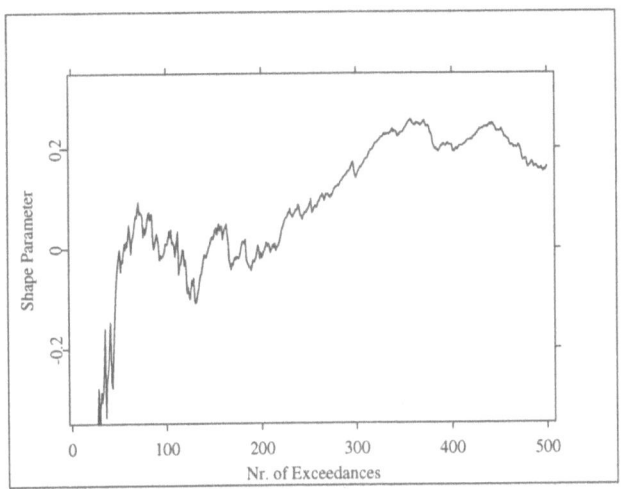

Figure 13.4. Diagram of moment estimator applied to daily returns.

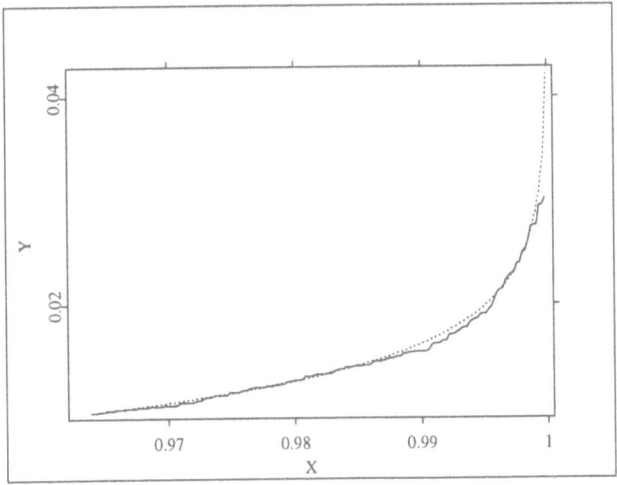

Figure 13.5. GP qf fitted to tail of returns.

The Hill estimator yields a similar picture (execute the following lines to add the pertaining Pareto quantile function to the plot).

```
h=hillgp1(dyr,160)
hqf=t~h.sigma*qfx("gp1",t,h.alpha)
adddata(d,1,1,hqf)
```

🔍 xtrm01.xpl

To decide which estimator should be preferred, we employ the mean excess function. Execute the next lines to create the plot of the empirical mean excess function as well as the parametric ones that is shown in Figure 13.6.

```
h=hillgp1(dyr,160)
m=momentgp(dyr,160)
d=createdisplay(1,1)
t=aseq(0.009,210,0.0001)
;
; plot empirical mean excess function
;
et=sort(dyr)[rows(dyr)-160:rows(dyr)-1]
eme=et~empme(dyr,et)
show(d,1,1,eme)
;
hme=t~gp1me(h.alpha,t/h.sigma)*h.sigma
adddata(d,1,1,hme)
;
mme=t~gpme(m.gamma,(t-m.mu)/m.sigma)*m.sigma
adddata(d,1,1,mme)
```

🔍 xtrm02.xpl

One can recognize that the empirical mean excess function is close to a straight line, which justifies the GP modeling. Yet, the GP mean excess function, based on the Hill estimator, strongly deviates from the empirical mean excess function. This indicates that the Hill estimator is not applicable.

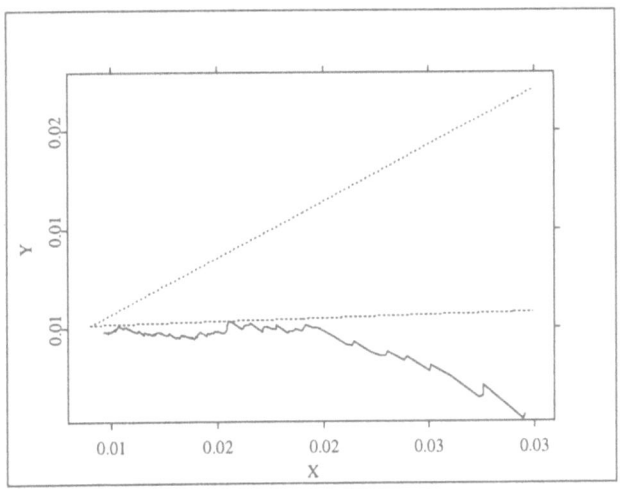

Figure 13.6. Empirical mean excess function (solid) and GP mean excess function fitted by Hill (dotted) and Moment estimator (dashed).

Bibliography

Dekkers, A.L.M., Einmahl, J.H.J. and de Haan, L. (1989). A moment estimator for the index of an extreme-value distribution, *Annals of Statistics* **17**: 1833–1855.

Drees, H. (1995). Refined Pickands estimators of the extreme value index, *Annals of Statistics* **23**: 2059–2080.

Falk, M., Hüsler, J. and Reiss, R.-D. (1994). *Laws of Small Numbers: Extremes and Rare Events*, DMV-Seminar, Birkhäuser, Basel.

Fisher, R.A. and Tippett, L.H.C. (1928). Limiting forms of the frequency distribution of the largest and smallest member of a sample, *Proc. Camb. Phil. Soc.* **24**: 180–190.

Leadbetter, M.R. and Nandagopalan, S. (1989). On exceedance point processes for stationary sequences under mild oscillation restrictions. In: *Extreme Value Theory*, J. Hüsler and R.-D. Reiss (eds.), *Lect. Notes in Statistics 51*, Springer, New York.

Pickands, J. (1975). Statistical inference using extreme order statistics, *Annals of Statistics* **3**: 119–131.

Reiss, R.-D. and Thomas, M. (1997). *Statistical Analysis of Extreme Values*, Birkhäuser, Basel.

14 Wavelets

Yuri Golubev, Wolfgang Härdle, Zdeněk Hlávka, Sigbert Klinke,
Michael H. Neumann and Stefan Sperlich

Wavelets are a powerful statistical tool which can be used for a wide range of applications, namely

- describing a signal, nonparametric estimation

- parsimonious (approximate) representation, data compression

- smoothing and image denoising

- jump detection and test procedures.

One of the main advantages of wavelets is that they offer a simultaneous localization in time and frequency domain. The second main advantage of wavelets is that, using fast wavelet transform, it is computationally very fast.

You can learn more about wavelets in the following overview. We advise you to consult e.g. Daubechies (1992), Kaiser (1995) or Härdle et al. (1998) if you wish to study wavelets in more detail.

Overview

Wavelet expansion of a certain function f is a special case of an orthonormal series expansion. Orthonormal series expansions of functions, or more generally transformations in the frequency domain, have several important applications. First, this is simply an alternative way of describing a signal. Such a description in the frequency domain often provides a more parsimonious representation than the usual one on a grid in the time domain. Orthonormal series expansions can also serve as a basis for nonparametric estimators and much more.

You can recall the classical Fourier series expansion of a periodic function. A nonperiodic function in $L_2(R)$ can be described by the Fourier integral transform which is represented by an appropriately weighted integral over the harmonic functions.

A first step towards a time-scale representation of a function with a time-varying behavior in the frequency domain is given by the windowed Fourier transform, which approximately amounts to a local Fourier expansion of f. To attain time localization, one multiplies f with a time window g and performs the usual Fourier integral transform. For the windowed Fourier transform,

$$\mathcal{F}(\omega, t) = \int g(u - t) f(u) e^{-2\pi i \omega u} du;$$

the inverse transform exists which means that no information is lost by this transformation. The main drawback of this transform is that the same time window is used over the whole frequency domain. Thus, for signals with a high power at high frequencies, this window will turn out to be unnecessarily large whereas it is too small for signals with dominating contributions from low frequencies.

In contrast, the wavelet transform provides a decomposition into components from different scales whose degree of localization is connected to the size of the scale window. This is achieved by translations and dilations of a single function, the so-called wavelet. It provides an amount of simultaneous localization in time and scale domain which is in accordance to the so-called uncertainty principle. This principle basically says that the best possible amount of localization in one domain is inversely proportional to the size of the localization window in the other domain. Note that we do not use the term "frequency" in connection with wavelet decompositions. The use of this term would be adequate in connection with functions that show an oscillating behavior. For wavelets it is more convenient to use the term "scale" to describe such phenomena.

A function f in $L_2(R)$ can be represented by its continuous wavelet transform

$$\mathcal{F}(\omega, t) = \int f(s) \psi \left(\frac{s - t}{\omega} \right) \omega^{-1/2} ds.$$

One can also obtain an invertible transform by a simple dyadic translation and dilation scheme which is based on functions $\psi_{j,k}(x) = 2^{j/2} \psi(s^j x - k)$, where $j, k \in Z$. There exist functions ψ such that $\{\psi_{j,k}; j, k \in Z\}$ forms a complete orthonormal basis of $L_2(R)$. A particular construction of such a function is described by Daubechies (1988). For a function $f \in L_2(R)$ we have the homogeneous wavelet expansion

$$f = \sum_{j=-\infty}^{\infty} \sum_{k=-\infty}^{\infty} \langle f, \psi_{j,k} \rangle \psi_{j,k},$$

where $\langle f, \psi_{j,k} \rangle = \int f(t) \psi_{j,k}(t) dt$. If we truncate the above expansion at a certain resolution scale $j = J - 1$, we obtain the projection of f on some linear space, say V_j. This space can be alternatively spanned by translates of an accordingly dilated version of a so-called scaling function or father wavelet φ. Let $\varphi_{l,k}(x) = 2^{l/2} \varphi(2^l x - k)$. Now we can also expand f in an inhomogeneous wavelet series,

$$f = \sum_{k=-\infty}^{\infty} \langle f, \varphi_{l,k} \rangle \varphi_{l,k} + \sum_{j=l}^{\infty} \sum_{k=-\infty}^{\infty} \langle f, \psi_{j,k} \rangle \psi_{j,k}.$$

There exist wavelets with several additional interesting properties. A frequently used tool are wavelets with compact support, which were first developed by Daubechies (1988). Moreover, one can construct wavelets of any given degree of smoothness and with an arbitrary number of vanishing moments. Finally, there exist boundary corrected bases which are appropriate for a wavelet analysis on compact intervals. This turns out to be quite a technical matter and you can consult the literature Daubechies (1992), Kaiser (1995), Härdle et al. (1998), if you are interested in more details.

The principle of estimation of a function in the $L_2(R)$ space by wavelets, the underlying multiresolution analysis (MRA), and the meaning of father and mother wavelets, will be explained also in the context of function approximation in Section 14.3.

14.1 Quantlib twave

You can use XploRe to become more familiar with wavelets. We load the `twave` library by entering the command

```
library("twave")
```

on the XploRe command line. XploRe loads the `twave` quantlib and the interactive menu appears. Quantlib `twave` is an interactive introduction to wavelets consisting of a set of examples with explanations. The quantlib `twave` uses the library `wavelet` which was designed for advanced users who are already familiar with wavelets.

By choosing an item in the interactive menu you can learn more about the corresponding topic.

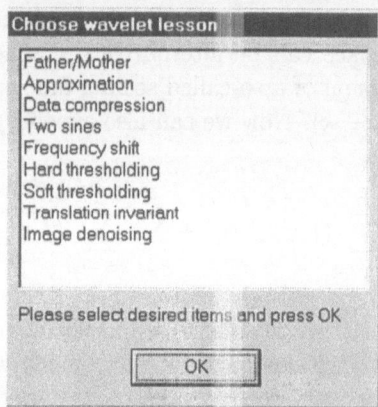

Some choices in the menus are the same in all sections, e.g. the Print menu item. The common menus are described in the following subsections.

14.1.1   Change Basis

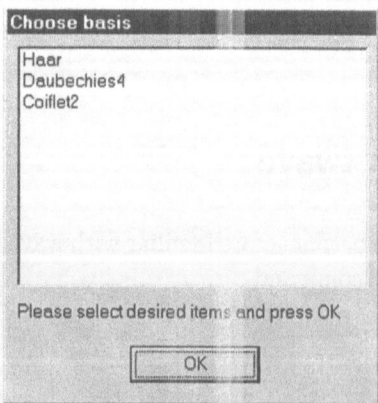

- Haar — which could be used for a discontinuous function or a signal with jumps

- Daubechies 4 — provides a more smooth basis

- Coiflet 2 — provides the "smoothest" basis in our teachware program

If you are familiar with wavelets you will know that much more bases are available. To concentrate on the important problems with wavelet estimation we have restricted ourselves to three typical bases.

14.1.2   Change Function

In some lessons it is useful not just to study the example we provide as default, but also other functions. Try for example the lesson about data compression with the Doppler function. We offer the following functions:

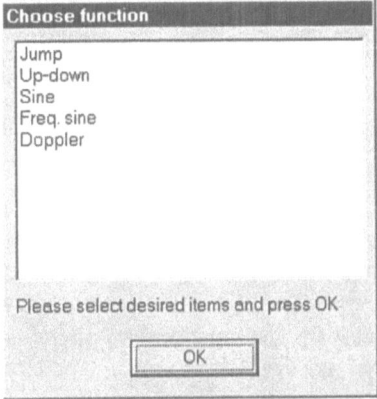

- Jump — function which has a jump between 0.25 and 0.75. Until the jump the function is the identity function and after the jump $1 - x$.

- Up-down — function with a jump between 0.25 and 0.75. The function is 0 till the jump and then 1.

- Sine — sine function

- Freq. sine — sine function which changes its frequency at 0.5.

- Doppler — sine function with a decreasing frequency and an increasing amplitude.

14.1.3   Change View

We provide four different views to the mother wavelet coefficients. Three of the four views are used in appropriate lessons so that they deliver a maximum of information.

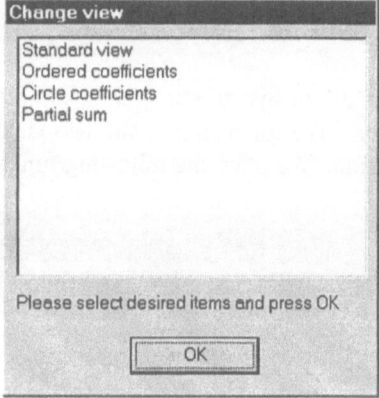

- Standard view — shows the mother wavelet coefficients as vertical bars, where the bar length depends on the magnitude of the coefficient. The coefficients will be shown at a place which corresponds to the position of the basis function in the time-scale plane. See Section 14.3, function approximation.

- Ordered coefficients — shows the mother wavelet coefficients sorted by the absolute magnitude. This allows a fast judgment about the compression properties for chosen basis and function. See Section 14.4, data compression.

- Circle coefficients — circles are used instead of vertical bars. The radius of the circle corresponds to the magnitude of the coefficient. Additionally the circles are drawn in different colors, red if the coefficient is used in the approximation, blue if not; see Subsection 14.7.1, hard thresholding.

- **Partial sum** — shows for each resolution scale how well the function or signal is approximated, if we include all coefficients up to this resolution scale.

14.2   Discrete Wavelet Transform

In this section we describe a periodic version of the discrete wavelet transform.

The display shows a plot of an eigenvector with an index chosen by **Change index** in the interactive menu. Suppose that we have a vector y of the length N, where N is a power of 2. Then the vector w of wavelet coefficients of y is defined by $w = Wy$. The matrix W is orthogonal: all its N eigenvalues are equal to 1. Therefore the kth element of vector w can be represented as the inner product of the data y and the kth eigenfunction e_k: $w_k = \langle y, e_k \rangle$. For a better understanding of the wavelet transform you can look at the plot of the eigenvector e_k, $(k = 1, \ldots, N)$.

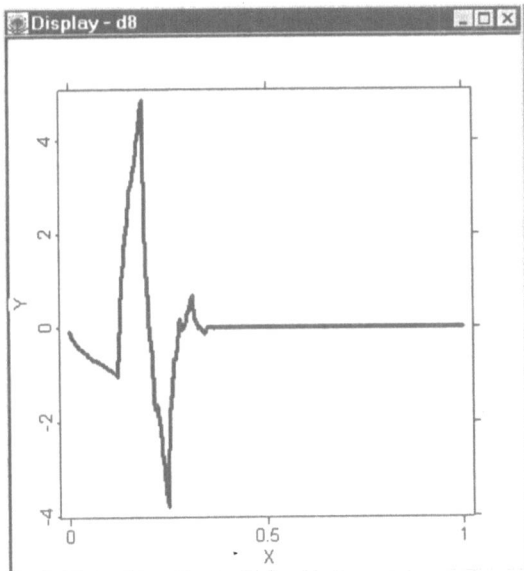

Changing the index k by using the item **Change index** you will see the plots of the eigenvectors on the display. We advise you to change the index k successively starting from 1. If k is less than the number chosen by Change level then you will see the eigenvectors associated to the father wavelet. Otherwise the plot shows the vectors associated to the mother wavelet. Note that the eigenvector approximates a father or mother wavelet of the continuous wavelet transform if the "support" of the eigenvector is strictly embedded in $1, \ldots, N$. It is very important for the speed of the algorithm that the multiplication Wy is implemented not by a matrix multiplication, but by a sequence of special filtering steps which result in $O(N)$ operations.

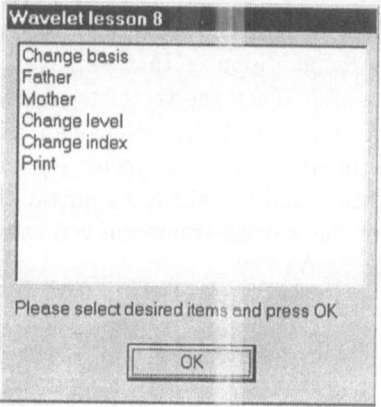

- Change basis — changes the type of the wavelet decomposition. Possible choices are Haar, Daubechies 4 and Coiflet 2. These are particular wavelet bases which are described in more detail e.g. in Daubechies (1992). This menu is also explained in Subsection 14.1.1.

- Father — shows the father wavelet function

- Mother — shows the mother wavelet function

- Change level — changes the number of the father wavelet coefficients on the first resolution level. This number must be a power of 2 in the range $1, \ldots, n/2$

- Change index — changes the index k of the eigenvector e_k to be displayed

- Print — prints the display

14.3 Function Approximation

Imagine a sequence of spaces

$$V_j = \text{span}\{\varphi_{jk}, k \in Z, j \in Z, \varphi_{jk} \in L_2(R)\}.$$

This sequence $\{V_j, j \in Z\}$ is called multiresolution analysis of $L_2(R)$ if $V_j \subset V_{j+1}$ and $\bigcup_{j \geq 0} V_j$ is dense in $L_2(\text{R})$. Further, let us have ψ_{jk} such that for

$$W_j = \text{span}\{\psi_{jk}\},$$

we have

$$V_{j+1} = V_j \oplus W_j,$$

where the circle around plus sign denotes direct sum. Then we can decompose the space $L_2(R)$ in the following way:

$$L_2(R) = V_{j_0} \oplus W_{j_0} \oplus W_{j_0+1} + \cdots$$

and we call φ_{jk} father and ψ_{jk} mother wavelets. This means that any $f \in L_2(R)$ can be represented as a series

$$f(x) = \sum_k \alpha_k \varphi_{j_0 k}(x) + \sum_{j=j_0}^{\infty} \sum_k \beta_{jk} \psi_{jk}(x).$$

According to a given multiresolution analysis, we can approximate a function with arbitrary accuracy. Under smoothness conditions on f, we can derive upper bounds for the approximation error. Smoothness classes which are particularly well suited to the study of approximation properties of wavelet bases are given by the scale of Besov spaces $B_{p,q}^m$. Here m is the degree of smoothness while p and q characterize the norm in which smoothness is measured. These classes contain traditional Hölder and L_2–Sobolev smoothness classes, by setting $p = q = \infty$ and $p = q = 2$, respectively.

For a given Besov class $B_{p,q}^m(C)$ there exists the following upper bound for the approximation error measured in L_2:

$$\sup_{f \in B_{p,q}^m(C)} \left\{ \left\| \sum_{k \in Z} \langle f, \varphi_{J,k} \rangle \varphi_{J,k} - f \right\|_{L_2} \right\} = O\left(2^{-2J(m+1/2-1/\min\{p,2\})} \right).$$

The decay of this quantity as $J \to \infty$ provides a characterization of the quality of approximation of a certain functional class by a given wavelet basis. A fast decay is favorable for the purposes of data compression and statistical estimation.

The following display provides an impression of how a discontinuous function is represented in the domain of coefficients.

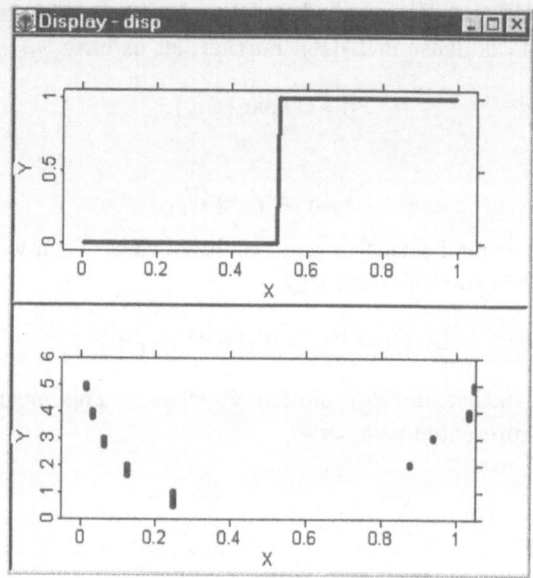

The display contains two plots: the upper shows a jump function, the lower the mother wavelet coefficients corresponding to their position in scale and time. The large coefficients are caused by the discontinuity (center) and by boundary effects since we use a periodic wavelet transform for a nonperiodic function.

You can examine the effects of approximating a wide range of functions using various wavelet bases with the following interactive menu.

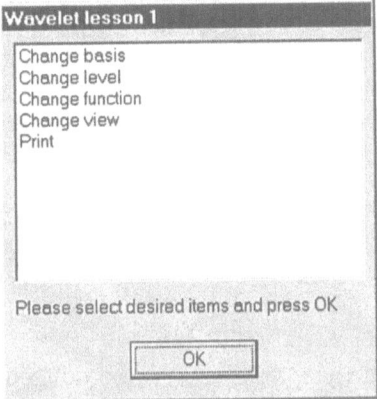

- **Change basis** — changes the type of the wavelet decomposition. Possible choices are Haar, Daubechies 4, Coiflet 2

- **Change level** — changes the number of the father wavelet coefficients on the first resolution level. This number must be a power of 2 in the range $1,\ldots,n/2$

- **Change function** — allows us to examine other functions

- **Change view** — provides other views to the mother wavelet coefficients

- **Print** — prints the display

14.4 Data Compression

The amount of data compression in a certain basis is quite an important feature for several reasons. Obviously it is always preferable to store information with as few as possible bytes. This is even necessary for huge data sets, e.g. for Internet transfers. This can be attained if the information, i.e. a collection of signals, to be stored can be described by a small number of basis functions.

The ability to provide a sparse representation of functions is also important in nonparametric statistics, where an unknown function is observed with a certain noise. Standard problems are nonparametric regression, density estimation and spectral density estimation. In order to improve the rough information about a possibly smooth function given by noisy data one usually applies smoothing methods. A particular class of such methods is given by truncated or otherwise regularized orthonormal series estimators. These estimators are based on empirical versions of the coefficients defining the orthonormal series. Analogous to classical Fourier series, the true wavelet coefficients are given as integrals

$$\alpha_{jk} = \int f(t)\varphi_{jk}(t)dt$$

and their empirical versions based on data $X_1, \ldots, X_n$ are calculated as

$$\tilde{\alpha}_{jk} = \frac{1}{n}\sum \varphi_{jk}(X_i).$$

The smoothing step is then performed in the domain of the coefficients, e.g. by thresholding. Finally, the estimator is synthesized from these modified empirical coefficients by the inverse transform.

Usually, the risk of a truncated orthonormal series estimator is appropriately proportional to the squared noise level times the number of coefficients included in this estimator, plus some approximation error due to the truncation of the expansion. Hence, an efficient data compression will directly provide the possibility of constructing good statistical estimators.

There are two important mathematical characterizations of the ability of data compression.

- First, one can simply observe the L_2-approximation error for a certain functional class in the sequence of approximation spaces given by a multiresolution analysis. The decay of this functional is important for rates of convergence of linear wavelet estimators as the number of observations tends to infinity.

- Second, one can also observe the sum of the minima of the squared wavelet coefficients and the variances of empirical versions of them. This (risk-) functional also describes the amount of data compression in a certain basis. It describes the rates of convergence of appropriate nonlinear wavelet estimators within some logarithmic factor (in the sample size).

The following display provides a comparison of the ability to compress a function with spatially inhomogeneous smoothness properties by a wavelet transform (left) and the Fourier transform (right). The lower windows show the magnitude of the ordered coefficients. The fatter tail on the right reflects that the Fourier transform needs more coefficients to grasp the functional form.

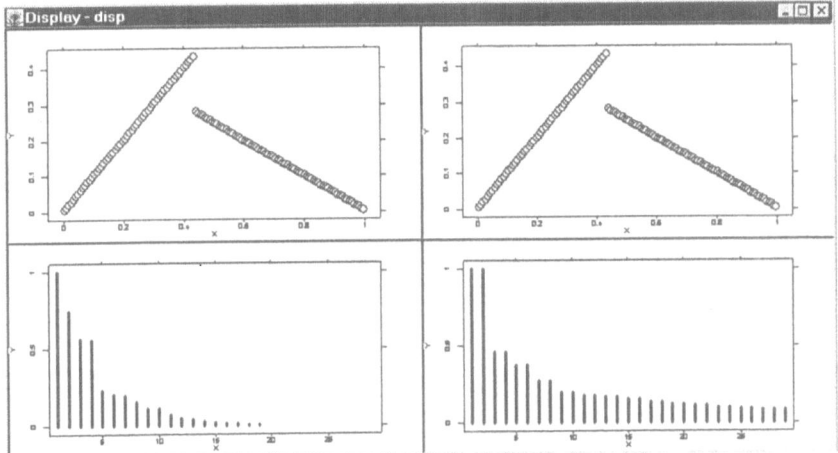

You can eXploRe the possibilities of data compression with the following interactive menu.

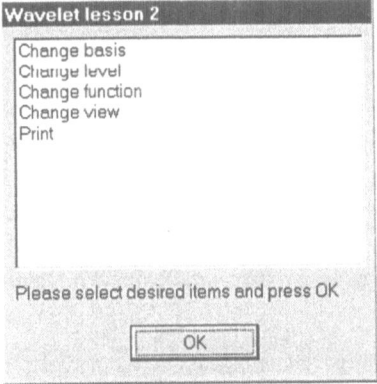

- **Change basis** — changes the type of the wavelet decomposition. Possible choices are Haar, Daubechies 4, Coiflet 2

- **Change level** — changes the number of the father wavelet coefficients on the first resolution level. This number must be a power of 2 in the range $1, \ldots, n/2$

- **Change function** — allows us to examine other functions

- **Change view** — provides other views to the mother wavelet coefficients

- **Print** — prints the display

14.5 Two Sines

In this section we study the distribution of the empirical wavelet coefficients in time and frequency domain.

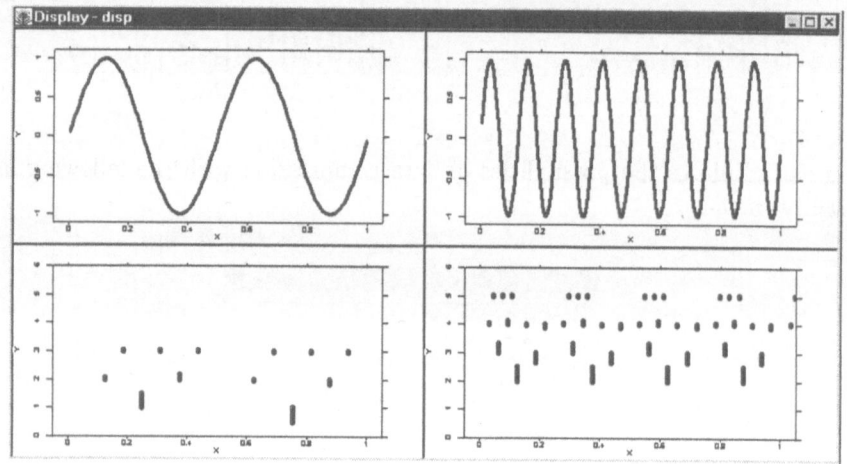

The upper two plots represent two sine wave functions with different frequencies.

The lower two plots show the magnitude of wavelet coefficients with respect to their location in time domain (x-axis) and resolution scale (y-axis). The magnitude of wavelet coefficient is proportional to length of the bar.

We see that the character of the distribution of the wavelet coefficients strongly depends on the frequency of the signal. For the low frequency sine wave we have the largest wavelet coefficients allocated at the lower resolution scale. At the same time the largest wavelet coefficients of the high frequency sine wave are allocated at high resolution scales. Thus the wavelet transform reflects well the properties of the input signal in the frequency domain.

The interactive menu provides you with the possibilities to choose the type of wavelet basis (Haar, Daubechies 4, Coiflet 2) and to change the number of the wavelet coefficients at the first resolution scale. Changing the basis you can change the smoothness of the father and the mother functions. This is synonymous to changing the allocation of these functions in frequency domain. This can help to adjust the wavelet basis to the smoothness properties of the signal. For looking at the energy of the signal at a certain resolution scale use the Change level item and see what happens. The combination of these two possibilities leads to a better compression of the signal.

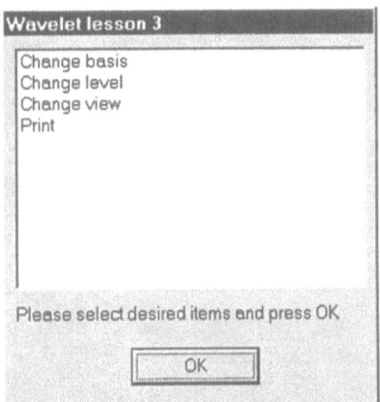

- Change basis — changes the type of the wavelet decomposition. Possible choices are Haar, Daubechies 4, Coiflet 2

- Change level — changes the number of the father wavelet coefficients on the first resolution scale. This number must be a power of 2 in the range $1, \ldots, n/2$

- Change view — provides other views to the mother wavelet coefficients

- Print — prints the display

14.6 Frequency Shift

The goal of this section is to demonstrate how the wavelet transform reflects the properties of the signal simultaneously in frequency and time domains. We

consider the signal composed from two sine waves having different frequencies
on the time intervals [0, 0.5] and (0.5, 1], respectively.

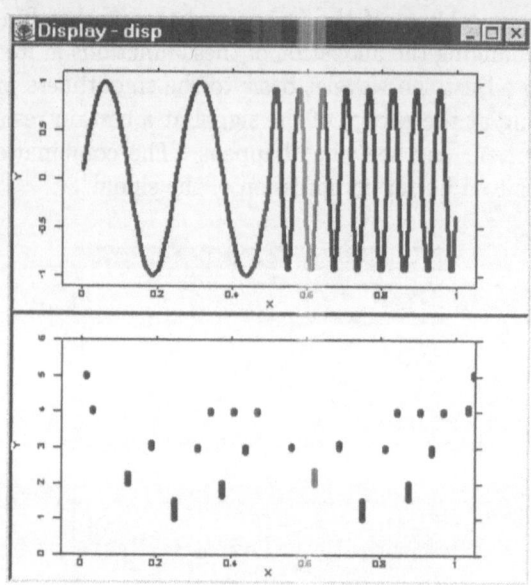

The upper plot in the display above shows the function which consists of two
sine waves with different frequencies. The lower plot shows the magnitude of
the wavelet coefficients with respect to their location in the time domain (x-
axis) and the resolution scale (y-axis). The magnitude of the wavelet coefficient
is proportional to the length of the bar.

The time-frequency structure of the signal is well reflected by the distribution
of the wavelet coefficients. To change the number of father wavelet coefficients
use the Change level item in the interactive menu.

Looking at the first resolution scale we see that the first two coefficients are
greater than the second ones. This is not surprising since the input signal has
low frequency at the first time interval and high frequency at the second one.
On the other hand, this reflects a good allocation of the mother functions on
coarse scales in time domain. The corresponding mother coefficients associated
with the wavelet expansion on the right part of the unit interval are sufficiently
small because they are inner products of highly oscillating functions (signal)
and a sufficiently "smooth" function with finite support (wavelet).

Increasing the resolution scale changes the picture significantly. The wavelet coefficients at the left part of the time domain become small. In this case the mother coefficients are inner products of a smooth function (input signal) and a highly oscillating functions (mother functions). At the same time we observe the increase of wavelet coefficients associated with the time domain on the right, where the input signal has a high frequency.

In this example we clearly see that the wavelet decomposition reflects well the properties of the signal in frequency and time domain. This is not the case for the ordinary Fourier transform which reflects only the frequency properties of the signal.

Using the menu you can change father and mother functions by the Change basis item. The use of higher-order wavelets increases the support of father and mother functions and improves the allocation of these functions in the frequency domain. This leads to a better approximation of the smooth parts of the signal. At the same time the approximation of the signal in a neighborhood of discontinuities becomes worse.

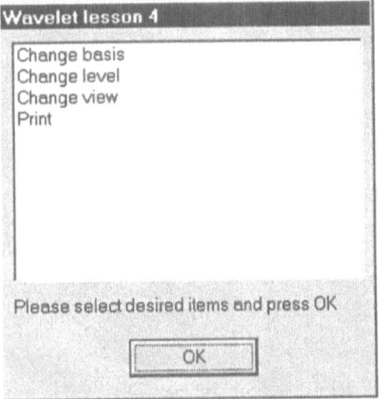

- Change basis — changes the type of the wavelet decomposition. Possible choices are Haar, Daubechies 4, Coiflet 2

- Change level — changes the number of the father wavelet coefficients on the first resolution scale. This number must be a power of 2 in the range $1, \ldots, n/2$

- Change view — provides other views to the mother wavelet coefficients

- Print — prints the display

14.7   Thresholding

In this section we consider the problem of recovering the regression function from noisy data based on a wavelet decomposition.

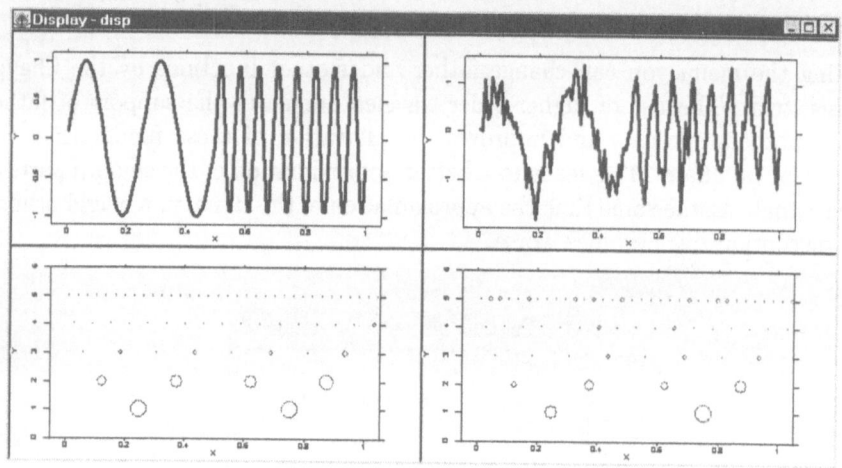

The upper two plots in the display show the underlying regression function (left) and noisy data (right). The lower two plots show the distribution of the wavelet coefficients in the time-scale domain for the regression function and the noisy data.

For simplicity we deal with a regression estimation problem. The noise is assumed to be additive and Gaussian with zero mean and unknown variance. Since the wavelet transform is an orthogonal transform we can consider the filtering problem in the space of wavelet coefficients. One takes the wavelet transform of the noisy data and tries to estimate the wavelet coefficients. Once the estimator is obtained one takes the inverse wavelet transform and recovers the unknown regression function.

To suppress the noise in the data two approaches are normally used. The first one is the so-called linear method. We have seen in Section 14.6 that the wavelet decomposition reflects well the properties of the signal in the frequency domain. Roughly speaking, higher decomposition scales correspond to higher frequency components in the regression function. If we assume that the underlying regression is allocated in the low frequency domain then the filtering procedure becomes evident. All empirical wavelet coefficients beyond some resolution scale are estimated by zero. This procedure works well if the signal is sufficiently smooth and when there is no boundary effect in the data. But for many practical problems such an approach does not seem to be fully appropriate, e.g. images cannot be considered as smooth functions.

To avoid this shortcoming often a nonlinear filtering procedure is used to suppress the noise in the empirical wavelet coefficients. The main idea is based on the fundamental property of the wavelet transform: father and mother functions are well localized in time domain. Therefore one could estimate the empirical wavelet coefficients independently. To do this let us compare the absolute value of the empirical wavelet coefficient and the standard deviation of the noise. It is clear that if the wavelet coefficient is of the same order of the noise level, then we cannot separate the signal and the noise. In this situation a good estimator for the wavelet coefficient is zero. In the case when an empirical wavelet coefficient is greater than the noise level a natural estimator for a wavelet coefficient is the empirical wavelet coefficient itself. This idea is called thresholding.

We divide the different modifications of thresholding in mainly three methods: hard thresholding, soft thresholding and a levelwise thresholding using Stein risk estimator, see Subsection 14.7.3.

14.7.1 Hard Thresholding

To suppress the noise we apply the following nonlinear transform to the empirical wavelet coefficients: $F(x) = x \cdot I(|x| > t)$, where t is a certain threshold. The choice of the threshold is a very delicate and important statistical problem.

On the one hand, a big threshold leads to a large bias of the estimator. But on the other hand, a small threshold increases the variance of the smoother. Theoretical considerations yield the following value of the threshold:

$$t = \sqrt{2\sigma^2 \log(n)/n},$$

where n is the length of the input vector and σ^2 is the variance of the noise. The variance of the noise is estimated based on the data. We do this by averaging the squares of the empirical wavelet coefficients at the highest resolution scale.

We provide two possibilities for choosing a threshold. First of all you can do it by "eye" using the Hard threshold item and entering the desired value of the threshold. The threshold offered is the one described in the paragraph before. Note that this threshold value is in most cases conservative and therefore we should choose a threshold value below the offered one. The item Automatic means that the threshold will be chosen as $\sqrt{2\sigma^2 \log(n)/n}$ with a suitably estimated variance.

The plots of the underlying regression function and the data corrupted by the noise are shown in top of the display. You can change the variance of the noise and the regression function by using the Noise level and the Change function items.

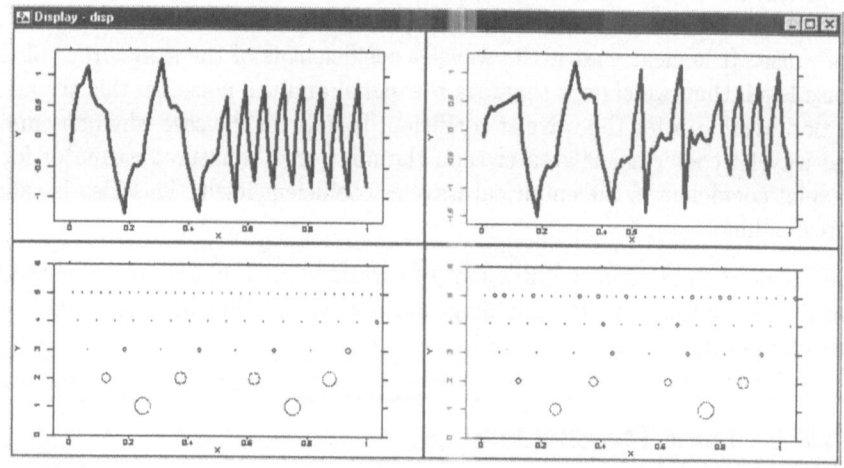

On the display above you can see the estimated regression function after the thresholding procedure. The left upper window shows how the threshold procedure distorts the underlying regression function when there is no noise in the data. The final estimator is shown in the right upper window.

Two plots in the bottom of the display show the thresholded and the nonthresholded wavelet coefficients. The thresholded coefficients are shown as blue circles and the nonthresholded as red ones. The radii of the circles correspond to the magnitudes of the empirical wavelet coefficients.

You can try to improve the estimator using the items Change basis and Change level. Under an appropriate choice of the parameters you can decrease the bias of your smoother.

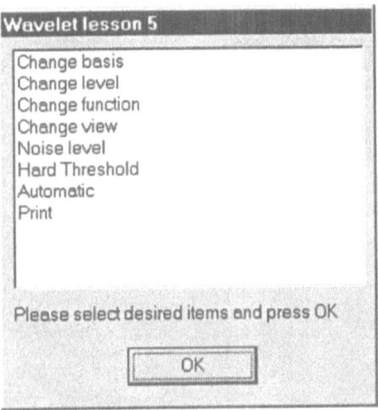

- Change basis — changes type of wavelet decomposition. Possible choices are Haar, Daubechies 4, Coiflet 2

- Change level — changes the number of father wavelet coefficients on the first resolution scale. This number must be a power of 2 in the range $1, \ldots, n/2$

- Change view — provides other views to the mother wavelet coefficients

- Noise level — changes variance of white Gaussian noise

- Hard Threshold — changes magnitude of the threshold

- Automatic — threshold is chosen as $\sqrt{2\mathrm{Var}\,\log(n)/n}$

- Print — prints the display

14.7.2   Soft Thresholding

Along with hard thresholding in many statistical applications soft thresholding procedures are often used. In this section, we study the so-called wavelet shrinkage procedure for recovering the regression function from noisy data.

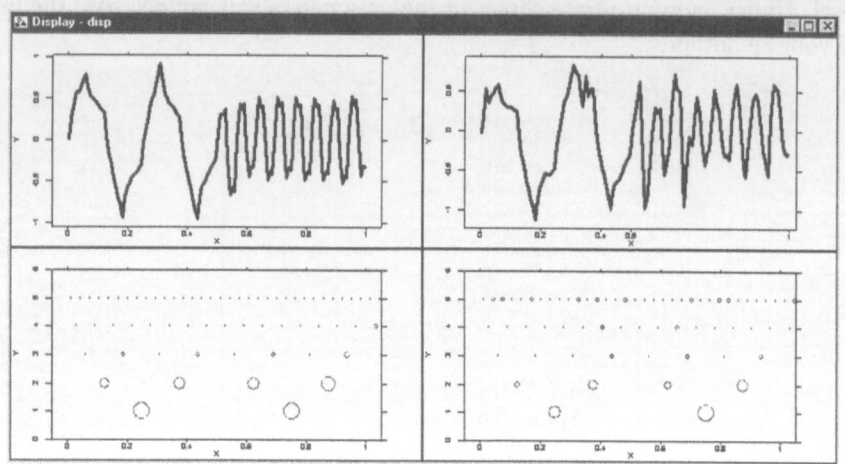

The display shows the same regression function as in the lesson before, without and with noise, after soft thresholding.

The only difference between the hard and the soft thresholding procedures is in the choice of the nonlinear transform on the empirical wavelet coefficients. For soft thresholding the following nonlinear transform is used:

$$S(x) = \text{sign}(x)(|x| - t)I(|x| > t),$$

where t is a threshold. The menu provides you with all possibilities for choosing the threshold and exploring the data.

In addition you can use a data driven procedure for choosing the threshold. This procedure is based on Stein's principle of unbiased risk estimation (Donoho and Johnstone, 1995). The main idea of this procedure is the following. Since $S(x)$ is a continuous function, given t, one can obtain an unbiased risk estimate for the soft thresholding procedure. Then the optimal threshold is obtained by minimization of the estimated risks.

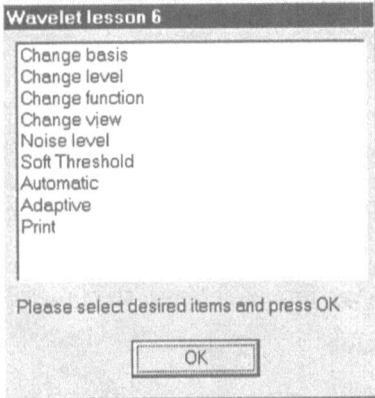

- **Change basis** — changes the type of the wavelet decomposition. Possible choices are Haar, Daubechies 4, Coiflet 2

- **Change level** — changes the number of the father wavelet coefficients on the first resolution scale. This number must be a power of 2 in the range $1, \ldots, n/2$

- **Change view** — provides other views to the mother wavelet coefficients

- **Noise level** — changes the variance of the white Gaussian noise

- **Soft Threshold** — changes the magnitude of the threshold

- **Automatic** — threshold is chosen as $\sqrt{2\mathrm{Var}\,\log(n)/n}$

- **Adaptive** — adaptive thresholding using Stein's principle

- **Print** — prints the display

14.7.3 Adaptive Thresholding

We first give an intuitive idea about adaptive thresholding using Stein's principle. Later on we give a detailed mathematical description of the procedure.

We define a risk function $R(s_k, Z_k, t)$ with $Z_k = c_k + s_k e_k, k = 1, \ldots, M$, where c_k is the unknown coefficient, s_k known (or to be estimated) scale parameters, e_k i.i.d. $N(0,1)$ random variables and t the threshold.

Since Stein enables us to estimate the risk

$$\sum_{k=1}^{M} R(s_k, Z_k, t),$$

the risk minimizing argument t^* can also be estimated and will be taken then as the "optimal" adaptive threshold. But still we have to decide for the definition of the risk function (and therefore for the thresholding) what kind of threshold function (hard or soft) we want to choose.

This procedure can be performed for each resolution level.

Mathematical derivation

Now we discuss the data driven choice of threshold, initial level j_0 and the wavelet basis by the Stein (1981) method of unbiased risk estimation. The argument below follows Donoho and Johnstone (1995).

We first explain the Stein method for the idealized one-level observation model discussed in the previous section: $Z_k = c_k + s_k e_k$, for $k = 1, \ldots, M$, where $c = (c_1, \ldots, c_M)$ is the vector of unknown parameters, $s_k > 0$ are known scale parameters and e_k are i.i.d. $N(0, 1)$ random variables.

Let $c' = (c'_1, ..., c'_M)$ be an estimator of c. Introduce the mean squared risk of the estimate of c: $R = \sum_{k=1}^{M} E(c'_k - c_k)^2$.

Assume that the estimators c'_k have the form

$$c' = Z_k + H_t(Z_k), \tag{14.1}$$

where t is a parameter and H_t is a weakly differentiable real valued function for any fixed t. One may think initially of t to be a threshold, but Stein's argument works in the general case as well. The parameter t can be chosen by the statistician. In other words, (14.1) defines a family of estimators, indexed by t, and the question is how to choose the optimal $t = t^*$. Define the optimal t^* as a minimizer of the risk function with respect to t.

If the true parameters c_k were known, one could compute t^* explicitly. In practice this is not possible, and one chooses a certain approximation t' of t^* as a minimizer of an unbiased estimator R' of the risk R. To construct R', note that

$$E(c'_k - c_k)^2 = E\{R(s_k, Z_k, t)\}, \tag{14.2}$$

where

$$R(s, x, t) = s^2 + 2s^2(d/dx)H_t(x) + H_t^2(x).$$

In fact,

$$E(c'_k - c_k)^2 = s_k^2 + 2s_k E(c_k H_t(Z_k)) + E(H_t^2(Z_k)), \qquad (14.3)$$

and, by partial integration, (14.2) follows. The relation (14.2) yields $R = E(R')$, where the value $R' = \sum_{k=1}^{M} R(s_k, Z_k, t)$ is an unbiased risk estimator, or risk predictor. It is called Stein's unbiased risk estimator (SURE).

The Stein principle is to minimize R' with respect to t and take the minimizer

$$\hat{t} = \arg\min_{t \geq 0} \sum_{k=1}^{M} R(s_k, Z_k, t) \qquad (14.4)$$

as a data driven estimator of the optimal t^*. The unbiasedness relation $E(R') = R$ (for every t) alone does not guarantee that the estimate is close to t^*. Some more developed argument is used to prove this, see Donoho and Johnstone (1995).

In the rest of this paragraph we formulate the Stein principle for the example of soft thresholding wavelet estimators. For soft thresholding (see above) we have

$$H_t(x) = -xI\{|x| < t\} - tI\{|x| \geq t\}\operatorname{sign}(x)$$

and

$$R(s, x, t) = (x^2 - s^2)I\{|x| < t\} + (s^2 + t^2)I\{|x| \geq t\}$$

$$= [x^2 - s^2] + (2s^2 - x^2 + t^2)I\{|x| \geq t\}. \qquad (14.5)$$

An equivalent expression is

$$R(s, x, t) = \min(x^2, t^2) - 2s^2 I\{x^2 \leq t^2\} + s^2.$$

The expression in square brackets in (14.5) does not depend on t. Thus, the definition (14.2) is equivalent to

$$\hat{t} = \arg\min_{t \geq 0} \sum_{k=1}^{M} (2s_k^2 + t^2 - Z_k^2)I\{|Z_k| \geq t\}. \qquad (14.6)$$

Let $(p_1, \ldots, p_M)$ be the permutation ordering the array such that $|Z_k|$, $k = 1, \ldots, M$: $|Z_{p_1}| \leq |Z_{p_2}| \leq \cdots \leq |Z_{p_M}|$ and $|Z_{p_0}| = 0$. According to (14.6) one obtains $t' = |Z_{p_l}|$ where

$$l = \arg \min_{0 \leq k \leq M} \sum_{s=k+1}^{M} (2s_{p_s}^2 + Z_{p_k}^2 - Z_{p_s}^2).$$

In particular for $M = 1$ the above equation yields the following estimator

$$\hat{c}_1 = \begin{cases} Z_1; & Z_1^2 \geq 2s_1^2, \\ 0; & Z_1^2 < 2s_1^2. \end{cases}$$

It is easy to see that the computation of the estimate of t^* defined above requires approximately $M \ln(M)$ operations provided that a quick sort algorithm is used to order the array $|Z_k|$, $k = 1, \ldots, M$.

We proceed from the idealized model $Z_k = c_k + s_k e_k$ to a more realistic density estimation model. In the context of wavelet smoothing the principle of unbiased risk estimation gives the following possibilities for adaptation:

1. adaptive threshold choice at any resolution level $j \geq j_0$.

2. adaptive choice of j_0 plus 1.

3. adaptive choice of father wavelet and mother wavelet plus 2.

Please notice that in XploRe we have implemented the first version. To demonstrate these possibilities consider the family of wavelet estimators

$$f^*(x, t, j_0, \psi) = \sum_k \alpha_{j_0 k}^*[\varphi, t]\varphi_{j_0 k}(x) + \sum_{j=j_0}^{j_1} \sum_k \beta_{jk}^*[\psi, t_j]\psi_{jk}(x), \qquad (14.7)$$

where $\alpha_{j_0 k}^*[\varphi, t] = \hat{\alpha}_{j_0 k} + H_t(\hat{\alpha}_{j_0 k})$ and $\beta_{jk}^*[\psi, t_j] = \hat{\beta}_{jk} + H_t(\hat{\beta}_{jk})$ are soft thresholded empirical wavelet coefficients with H_t. Here $t = (t, t_{j_0}, \ldots, t_{j_1})$ is a vector of thresholds. The dependence of f^* on the mother wavelet is skipped in the notation since the mother wavelet is supposed to be canonically associated with the father wavelet. As in (14.2) it can be shown that, under certain general conditions, $E\|f^* - f\|_2^2 = E\left(\hat{R}(t, j_0, \varphi)\right)$. Here Stein's unbiased risk estimator is given by

$$\hat{R}(t, j_0, \varphi) = \sum_k R(s_{j_0 k}[\varphi], \hat{\alpha}_{j_0 k}, t) + \sum_{j=j_0}^{j_1} \sum_k R(s_{jk}[\psi], \hat{\beta}_{jk}, t_j),$$

where $R(s, x, t)$ is defined above, and $s_{jk}^2[\varphi]$ and $s_{jk}^2[\psi]$ are variances of the corresponding empirical wavelets coefficients. To obtain the best estimator from the family (14.7), one can choose the unknown parameters of the estimator minimizing $'R(t, j_0, \varphi)$. For the cases 1, 2 and 3 these parameters can be chosen respectively as follows:

1. adaptive choice of thresholds: $\hat{t} = \arg \min_t \hat{R}(t, j_0, \varphi)$.

2. adaptive choice of thresholds and j_0: $(\hat{t}, \hat{j}_0) = \arg \min_{t, j_0} \hat{R}(t, j_0, \varphi)$.

3. adaptive choice of thresholds, j_0 and wavelet basis:
   $(\hat{t}, \hat{j}_0, \hat{\varphi}) = \arg \min_{t, j_0, \phi} \hat{R}(t, j_0, \varphi)$.

In the third case we assume that the minimum is taken over a finite number of given wavelet bases.

Remark 1: Since in practice the values of the different variances are not available, one can use their empirical versions instead. For example if (14.7) is the wavelet density estimator based on the sample $X_1, \ldots, X_n$, one can replace $s_{jk}^2[\psi]$ by its estimator.

It is clear that this yields a consistent estimator for the variances under rather general assumptions on the basis functions and on the underlying density of the X_i's.

Remark 2: If one wants to threshold only the coefficients β_{jk}, which is usually the case, the function H_t for α_{jk} should be identically zero. Therefore, $R(s_{j_0k}[\varphi], \hat{\alpha}_{jk}, t)$ should be replaced by $s_{j_0k}[\varphi]$ and Steins unbiased risk estimator takes the form

$$\hat{R}((t_{j_0}, \ldots, t_{j_1}), j_0, \varphi) = \sum_k s_{jk}^2[\varphi] + \sum_{j=j_0}^{j_1} \sum_k R\left(s_{jk}[\psi], \hat{\beta}_{jk}, t_j\right).$$

Let us now apply the Stein principle to a regression estimation example. We choose the following step function for $x \in [0, 1]$:

$$f(x) = 0.1I(x < 0.4) + 2I(x \in [0.4, 0.6] + 0.5I(x \in [0.6, 0.8]s_{jk}^2[\varphi]).$$

The function was observed at 128 equispaced points and disturbed with Gaussian noise with variance 1/128. We use the Stein rule only for threshold choice 1 (level by level) and not for the cases 2 and 3 where the adaptive choice of j_0

and of the basis is considered. We thus choose the threshold t' as the minimizer with respect to $t = (t_{j_0}, \ldots, t_{j_1})$ of

$$\hat{R}(t) = \sum_{j=j_0}^{j_1} \sum_k R\left(\hat{s}_{jk}[\psi], \hat{\beta}_{jk}, t_j\right),$$

where $R(s, x, t)$ is defined as above and $\hat{s}_{jk}[\psi]$ is an empirical estimator of the variance of the wavelet regression coefficients. For computation we use the discrete wavelet transform.

14.8   Translation Invariance

Commonly used wavelet estimators are, in contrast to kernel estimators, not translation-invariant: if we shift the underlying data set by a small amount s, apply nonlinear thresholding and shift the estimator back by s, then this new estimator $f(s)$ is usually different from the estimator without the shifting and backshifting operation. Coifman and Donoho (1995) and Nason and Silverman (1994) proposed to make the wavelet estimators translation-invariant and defined, with shifts s_i, $i = 1, \ldots, I$, the following new estimator $f^*(x) = \sum_i f^{(s_i)}(x)/I$.

This estimator possesses some advantages over the usual estimation scheme. First, it follows immediately by Jensen's inequality that the L_2-loss of f^* is not greater than the average loss of the $f^{(s_i)}$'s. Second, wavelet estimators sometimes have a quite irregular visual appearance. Often there are some spurious features caused by random fluctuations. This effect is weakened by averaging over different shifts as described above. In a small simulation, Neumann (1996) observed a considerable improvement over the standard estimation scheme, even by taking only a small number of shifts.

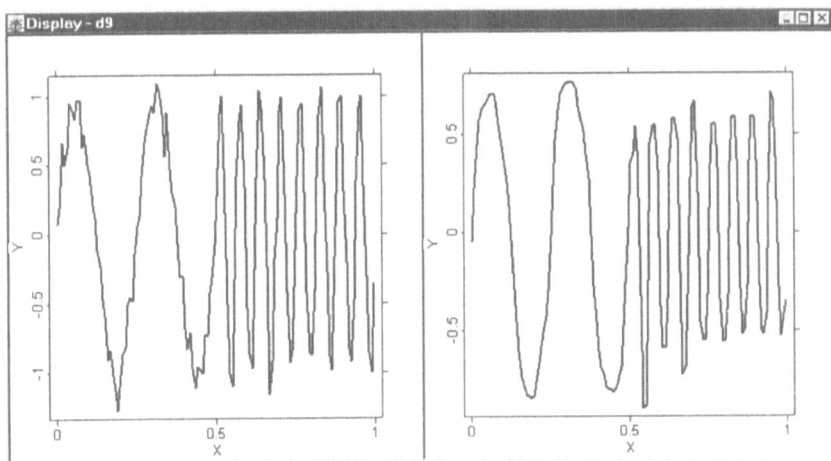

In our example the number of shifts is always $\log_2(n)$ with n the number of observations.

The interactive menu provides you the opportunity to further improve the estimate.

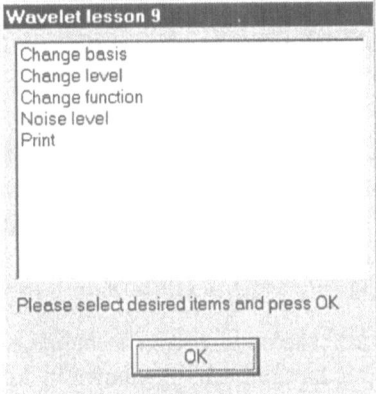

- **Change basis** — changes the type of the wavelet decomposition. Possible choices are Haar, Daubechies 4, Coiflet 2

- **Change level** — changes the number of the father wavelet coefficients on the first resolution level. This number must be a power of 2 in the range $1, \ldots, n/2$

- Change function — allows us to examine other functions

- Noise level — changes variance of white Gaussian noise

- Print — prints the display

14.9   Image Denoising

Once one has the wavelet tools for the univariate case available, the generalization of these methods to the multivariate case is not very complicated. Besides specific higher-dimensional wavelet bases which we do not consider here, one can easily construct multidimensional bases by taking tensor products of one-dimensional basis functions. There exist two important schemes of constructing such bases.

First, as it is usually done, one can devise a multivariate multiresolution analysis in the following way: At a given resolution scale one takes tensor products of dyadically translated father and mother wavelets from a one-dimensional multiresolution analysis. This gives a d-dimensional isotropic wavelet basis with a one-dimensional scale index and a d-dimensional location index.

Second, one can also take all possible tensor products of univariate basis functions. This yields a d-dimensional anisotropic wavelet basis with a d-dimensional scale index and a d-dimensional location index.

The isotropic basis provides a d-dimensional multiresolution analysis in the usual sense. On first sight it seems to be more appealing than the anisotropic basis and it is almost exclusively used in statistics. Appropriate wavelet estimators based on the isotropic basis can attain minimax rates of convergence in isotropic smoothness classes, which justifies its use in statistics.

Inbetween there exists also the possibility of a smooth transition from an iso- to an anisotropic basis. This method is implemented in XploRe (see the command fwt2).

However, it was shown by Neumann and Sachs (1997) in the two-dimensional case that the isotropic basis is not really able to adapt to different degrees of smoothness in different directions. Expressed in terms of the kernel-estimator language, a projection estimator using basis functions from this basis cannot mimic a multivariate kernel estimator based on a product kernel with different (directional) bandwidths. It is shown by Neumann and Sachs (1997) and Neumann (1996a) that estimators based on the anisotropic wavelet basis can attain minimax rates of convergence in anisotropic smoothness classes. Moreover, thresholded estimators in this basis are able to recognize a lower-dimensional structure in the data. For example, if the underlying function is of additive or multiplicative structure in certain univariate components, then appropriately thresholded estimators attain one-dimensional rates of convergence. These results obtained under structural restrictions can be extended to really nonparametric functional classes. Smoothness classes with dominating mixed derivatives also allow nearly one-dimensional rates of convergence.

A quite natural application of this methodology (to the particular problem of estimating the time-varying spectral density of a locally stationary process) can be found in Neumann and Sachs (1997). In this case the two axes on the plane, time and frequency, have a specific meaning. Accordingly, one cannot expect the same degrees of smoothness in both directions. Hence, the use of the anisotropic basis seems to be more natural than the use of the isotropic one.

In principle, this methodology derived in quite a general setting can also be applied to image denoising. Assume we have a certain grey-scale image given on a grid of pixels. Assume further that this image has been blurred by some additive noise. Now one can use wavelet thresholding to perform some (partial) denoising. Since the anisotropic basis is never essentially worse but sometimes better than the isotropic basis, it seems to be advisable to employ the former construction.

Note, however, that neither the anisotropic nor the isotropic basis is devised for the specific purpose of image reconstruction. Denoising in one of these bases is an all-purpose method which can be applied to a wide variety of problem settings. We do not make the attempt to compare this method with the many denoising techniques developed by engineers for the specific purpose of image reconstruction. We think that some of these special-purpose methods will certainly outperform our simple all-purpose approach.

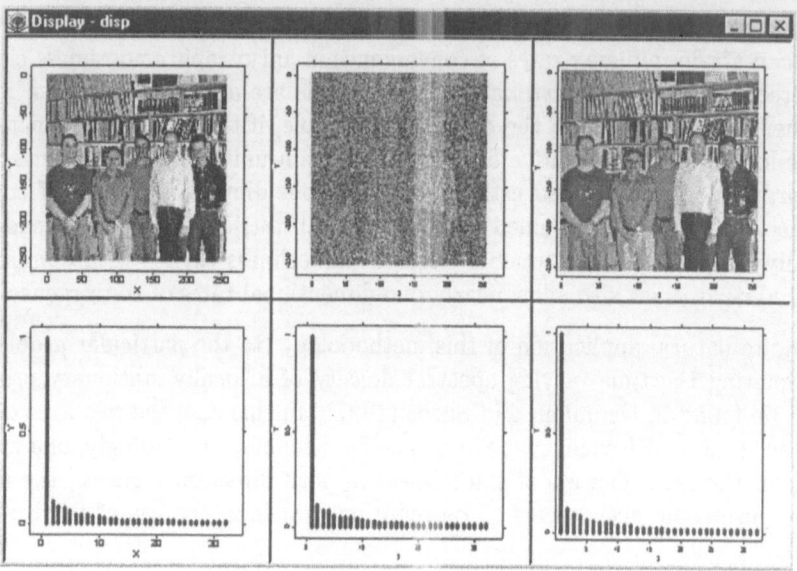

The display above shows in the top row three versions of an image. The left image is the original, the middle is the original one plus noise and the right one is the denoised image. The threshold value is chosen by an experienced data analyst. The lower row shows the magnitude of the ordered wavelet coeffcients. You can try to change the noise level and threshold of the wavelet basis with the following interactive menu.

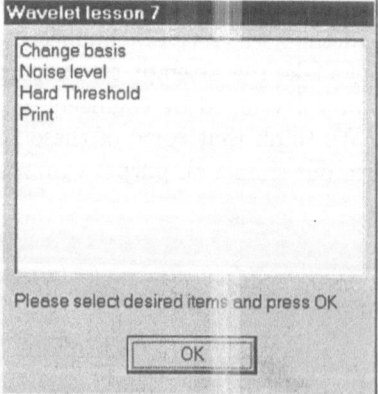

• Change basis — changes the type of the wavelet decomposition. Possible choices are Haar, Daubechies 4, Coiflet 2

- Hard Threshold — changes the magnitude of the threshold
- Print — prints the display

Note that this lesson does not work in the Academic Edition of XploRe.

Bibliography

Coifman, R. R. and Donoho, D. (1995). Translation-invariant denoising, in Antoniadis and Oppenheim: *Wavelets and Statistics, Lecture Notes in Statistics* **103**: 125–150.

Daubechies, I. (1988). Orthonormal bases of compactly supported wavelets, *Comm. Pure and Appl. Math.* **41**: 909–996.

Daubechies, I. (1992). *Ten Lectures on Wavelets*, SIAM, Philadelphia.

Donoho, D. and Johnstone, I. (1995). Adapting to unknown smoothness via wavelet shrinkage, *Journal of the American Statistical Association* **90**: 1200–1224.

Donoho, D. and Johnstone, I. (1998). Minimax estimation via wavelet shrinkage, *Annals of Statistics* **26**: 879–921.

Härdle, W., Kerkyacharian, G., Picard, D. and Tsybakov, A. (1998). *Wavelets, Approximation, and Statistical Applications, Lecture Notes in Statistics* **129**, Springer Verlag, New York.

Kaiser, G. (1995). *A Friendly Guide to Wavelets*, Birkhäuser, Basel

Nason, G. and Silverman, B. W. (1994). The discrete wavelet transform in S, *Journal of Computer and System Science* **3**: 163–191.

Neumann, M. (1996). Spectral density estimation via nonlinear wavelet methods for stationary non-gaussian time series, *Journal of Time Series Analysis* **17**: 601–633.

Neumann, M. (1996). *Multivariate wavelet thresholding: a remedy against the curse of dimensionality?* Discussion Paper 42, SFB 373, Humboldt-Universität zu Berlin, Germany.

Neumann, M. and Sachs, R. (1997). Wavelet thresholding in anisotropic function classes and application to adaptive estimation of evolutionary spectra, *Annals of Statistics* **25**: 38–76.

Stein, C. M. (1981). Estimation of the mean of a multivariate normal distribution, *Annals of Statistics* **9**: 1135–1151.

Part III: Programming

15 Reading and Writing Data

Sigbert Klinke, Jürgen Symanzik and Marlene Müller

This chapter describes the use of XploRe's commands for reading and writing data files as well as how to set the output format for the output window. Section 15.1 introduces reading and writing simple ASCII data sets. Section 15.2 explains the input format strings which can be used to read arbitrary ASCII data files. Section 15.3 explains the format strings which can be used to write formatted numeric data. Finally, the last Section 15.4 explains how to customize the output window of XploRe.

15.1 Reading and Writing Data Files

```
x = read ("file")
        reads numeric data from a file file.dat into a matrix x, each
        column of the file will be interpreted as column vector of x

x = readm ("file")
        reads mixed text and numeric data from file.dat into a list x,
        which contains text and numeric matrices separately

write (x, "file.dat" {,mode {,format}})
        writes text or numeric arrays to the file file.dat
```

This section introduces reading and writing simple ASCII data sets. All XploRe code for this section can be found in the quantlet ❑ iofmt01.xpl.

The function **read** is the command to read numeric data from a file. Each column of the file will be interpreted as a vector of the resulting matrix. Let us consider the data file **geo.dat** which has the following contents:

```
86     5
87    14
88    24
89    64
90   100
91   129
92   200
93   176
94   177
```

We can read this data file and store it into the matrix x by the following instruction:

```
x=read("geo")
```

Typing just x at the command line will show the contents of x as follows:

```
Contents of x
[1,]     86        5
[2,]     87       14
[3,]     88       24
[4,]     89       64
[5,]     90      100
[6,]     91      129
[7,]     92      200
[8,]     93      176
[9,]     94      177
```

Data files often contain mixed text and numeric columns, as in popul.dat:

```
778 New_York
355 Chicago
248 Los_Angeles
200 Philadelphia
167 Detroit
 94 Baltimore
 94 Houston
 88 Cleveland
 76 Washington_DC
```

```
75 Saint_Louis
74 Milwaukee
74 San_Francisco
70 Boston
68 Dallas
63 New_Orleans
```

which stores population data about US cities. Loading this data file with read would give missing values (NaN) in the second column, since read is not able to decode the text strings. This file should be read by readm. Since readm is part of the xplore library, we load this library first:

```
library("xplore")
x=readm("popul")
```

creates a list x which consists of three components. The instruction names(x) shows

```
Contents of names
[1,] "type"
[2,] "double"
[3,] "text"
```

in the output window. The first component, the vector x.type indicates the type of the column in the original data file:

```
Contents of type
[1,]       0
[2,]       1
```

Here, 0 stands for numeric and 1 for text. The second and third components contain the data: x.double is the numeric column, x.text is the text column.

In addition to read and readm, XploRe provides the command readascii to read any type of ASCII data. readm is actually based on readascii. We will have a closer look at readascii in the next section when we discuss input format strings.

To write data to a data file, the function write can be used. write can save both, numeric or text data. Let us consider the following example:

```
x=(1:4)~(2:5)
write(x,"mydata.dat")
```

writes the matrix to the data file `mydata.dat`. This file has then the following contents:

```
1.000000 2.000000
2.000000 3.000000
3.000000 4.000000
4.000000 5.000000
```

Note that `write` optionally accepts a writing mode and a format string to produce formatted output. The possible modes and format strings will be explained in detail when we come to string output, see Section 15.3.

15.2   Input Format Strings

```
x = readascii ("file")
    reads any type of ASCII data from file.dat
```

The function `readascii` is a powerful tool to read arbitrary ASCII data files. The use of format strings allows us to handle different delimiters when reading sequences of ASCII strings.

You find the XploRe code for the following examples in the ◘ iofmt02.xpl. As input, the data file `testascii.dat` is used:

```
1;2;3;4
5;6;7;8
ab;cd;de;fg
```

We can read this file and print the resulting list by

```
dat = readascii ("testascii.dat", "\t\n;")
dat
```

The list `dat` is a list that consists of two components, `dat.data` and `dat.type` which print as follows:

```
Contents of dat.data
[ 1,]  "1"
[ 2,]  "2"
[ 3,]  "3"
[ 4,]  "4"
[ 5,]  "
"
[ 6,]  "5"
[ 7,]  "6"
[ 8,]  "7"
[ 9,]  "8"
[10,]  "
"
[11,]  "ab"
[12,]  "cd"
[13,]  "de"
[14,]  "fg"
[15,]  "
"
Contents of dat.type
[ 1,]        0
[ 2,]        0
[ 3,]        0
[ 4,]        0
[ 5,]       20
[ 6,]        0
[ 7,]        0
[ 8,]        0
[ 9,]        0
[10,]       20
[11,]       10
[12,]       10
[13,]       10
[14,]       10
[15,]       20
```

The data component is a string vector, based on the contents of the input file.
In this example, the format string used consists of \t\n;. Here, \t stands for
tabulator and \n for newline. Important is the ";" which tells XploRe that
a semicolon is used as the delimiter. This way, XploRe is able to read ASCII

strings from files that contain delimiters other than the default blank delimiter.

The `type` component indicates which data type has been read for each cell of the vector. Possible types are 0 (number), 1 (missing), 10 (text), and 20 (newline). Thus, cells 5, 10, and 15 in the example above are newlines. Cells 1 through 4 and 6 through 9 are numbers, and cells 11 through 14 are text lines.

To see the difference, we study what happens when using another format string for the same data file:

```
dat = readascii ("testascii.dat", "\t\n")
dat
```

produces

```
Contents of dat.data
[1,] "1;2;3;4"
[2,] "
"
[3,] "5;6;7;8"
[4,] "
"
[5,] "ab;cd;de;fg"
[6,] "
"
Contents of dat.type
[1,]        10
[2,]        20
[3,]        10
[4,]        20
[5,]        10
[6,]        20
```

Here, each line of the data file is read as a string. The ";" is considered as a regular character but not as a delimiter as in the previous example. Obviously, only three text lines (cells 1, 3, and 5) are read, each followed by a newline (cells 2, 4, and 6).

15.3   Output Format Strings

```
x = string (format, v1 {, v2, ...})
      converts the vector v1 (or all input vectors) to a string using the
      format string format

write (x, "file.dat" {,mode {,format}})
      writes text or numeric arrays to the file file.dat; optionally a
      mode mode and a format string format can be given
```

The use of format strings in XploRe's **string** command comes close to the usage of C format strings. **string** is mainly used to convert numeric vectors to strings in order to append them to string objects.

Let us consider some examples which can also be found in ⊕ iofmt03.xpl. The following code line shows how to format all elements of a vector in the same way:

```
fmt = "Number   : %10.3f"
i   = 1:3
str = string (fmt, i)
str
```

gives

```
Contents of str
[ 1,] "Number  :       1.000"
[ 2,] "Number  :       2.000"
[ 3,] "Number  :       3.000"
```

Here we assigned a total of 10 characters to the float number. Three of these characters are reserved for digits after the ".".

The following example shows how different format strings effect the output format of **pi** (the constant π):

```
string ("%8.5g", pi)
string ("%1.0f", pi)
string ("%22.20f", pi)
```

shows

```
Contents of string
[1,] "3.1416
Contents of string
[1,] "3"
Contents of string
[1,] "3.14159265358979311600"
```

The first string is the default representation when a float number is converted into a string. (One can check this default by issuing `getenv("outputformat")`, see Section 15.4.) The second string shows `pi` rounded to an integer, since 0 decimals are required. In the third string, 22 digits are converted into a string with 20 of them reserved for decimals.

Finally, let us consider the differences between the f-, e- and g-formats, which produce the usual decimal notation, exponential notation or a combination of both, respectively. The following example makes use of the possibility to use more than one format in the format string.

```
string ("%10.5f and %10.5g and %10.5e", pi, pi, pi)
```

will print

```
Contents of string
[1,] "   3.14159 and      3.1416 and 3.14159e+00"
```

in the output window. The first format requires a string of length 10, with exactly 5 decimals. The second format is used to print pi with a precision of 5 digits, as before the string length is set to 10. In comparison with the f-format, the g-format has the advantage that it would automatically use the exponential representation if the string length — here 10 — is exceeded. The third format forces the exponential representation, with a precision of 5 decimals. The string length 10 is not useful here and hence ignored.

As we have seen in the previous section, the function `write` stores data to a data file. `write` can save both numeric or text data. The `write` function has two optional parameters, the writing mode and the format string.

The writing mode can be set to `"a"` if the data to be written should be appended to an existing file. If the writing mode is different from `"a"`, the file will be reset.

For numeric data, a format string can be given as an optional parameter, which will save the data in the specified format. For example,

```
x=(1:4)~(2:5)
write(x,"mydata.dat","r","%2.0f")
```

writes the matrix to the data file `mydata.dat` using the format "`%2.0f`". This file has then the following contents:

```
1  2
2  3
3  4
4  5
```

15.4 Customizing the Output Window

```
getenv ({variable})
```
> gets the value of the environment variable `variable`, if no input is given the values of all variables are listed

```
setenv (variable, value)
```
> sets the environment variable `variable` to the value `value`

```
output ("file", option)
```
> saves contents of the output window to the file `file`

XploRe provides a number of variables that customize the environment. Among others, a set of variables for modifying the output style in the output window is available:

Variable	Default
outheadline headline for arrays	"\r\nContents of %s\r\n\r\n"
outlayerline layerline for arrays	"[,,%li,%li,%li,%li,%li,%li]\r\n"
outlineno line number style	"[%*li,]"
outmaxdata maximal output elements	"2048"
outputformat numeric output format string	"% 8.5g"
outputstringformat string output style	""%s""

The defaults can be checked by using **getenv** to print out the value of a variable. For example,

```
getenv("outmaxdata")
```

prints the string

```
Contents of getenv
[1,] "2048"
```

which implies that not more than 2048 data items are printed.

The variable **outmaxdata** can now be modified by using **setenv**:

```
setenv("outmaxdata","100")
```

changes the number of data items to 100. Of course, several of the above variables can be modified simultaneously.

Let us now consider the format strings which control the style of the headline, the line number and the array dimension. Consider a $3 \times 3 \times 3$ array of normal random numbers as an example:

```
randomize(0)
```

```
x = normal(3,3,2)
x
```

The default settings used in this extract from the output window:

```
outheadline  ⟶   Contents of x
outlayerline ⟶   [,,1,1,1,1,1,1]
                 [1,] -0.21293 -1.3052 -0.38946
   outlineno  ⟶   [2,] -1.0078 -1.4341 -1.5746
                 [3,] 1.9502 0.69812 -0.26595
outlayerline ⟶   [,,2,1,1,1,1,1]
                 [1,] 0.19947 -1.2245 0.20864
   outlineno  ⟶   [2,] -2.5033 -0.42123 -0.14358
                 [3,] -0.49264 -0.72973 0.6273
```

15.4.1 Headline Style

The variable `outheadline` is used to influence the appearance of the headline in the output window. The default value is `"\r\nContents of %s\r\n\r\n"` where `%s` is a placeholder for the name of the variable. This results in the output of the words `Contents of x`, assuming x is the variable to be printed, followed by a blank line. Obviously, the combination of `\r\n` causes a line break.

It is possible to produce different outputs through the use of different format strings. No headline will be printed, when `setenv ("outheadline", "")` is used. Alternatively, an arbitrary headline, in this example a German title, can be produced.

```
randomize(0)
x = normal(2,2,2)
x
setenv ("outheadline", "Inhalt von %s\r\n\r\n")
x
```

Q iofmt04.xpl

prints now x in the following way:

```
Contents of x
[,,1,1,1,1,1,1]
[1,] -0.21293    1.9502
[2,]  -1.0078  -1.3052
[,,2,1,1,1,1,1]
[1,]   -1.4341 -0.38946
[2,]   0.69812  -1.5746
Inhalt von x
[,,1,1,1,1,1,1]
[1,] -0.21293    1.9502
[2,]  -1.0078  -1.3052
[,,2,1,1,1,1,1]
[1,]   -1.4341 -0.38946
[2,]   0.69812  -1.5746
```

Note that we use the special characters \r\n to force a new line in the output window. \n is the new line character used in C. \r is the special C character which forces the cursor to jump to the first character in the current line. Throughout this book, we have mostly used

```
setenv ("outheadline", "Contents of %s\r\n")
```

to suppress the blank line after the headline.

15.4.2 Layer Style

With outlayerline it is possible to influence the appearance of the layer style in the output window. It should be noted that the layer line only appears in the output window if the array consists of more than one layer. In the following example, the layer line does not appear, since x is a matrix:

```
randomize(0)
x = normal(2,2)
x
```

shows

```
Contents of x
[1,] -0.21293    1.9502
[2,]  -1.0078   -1.3052
```

The default value for `outlayerline` is `"[,,%li,%li,%li,%li,%li,%li]\r\n"` where the %li are placeholders for the layer number. XploRe supports arrays up to eight dimensions. The first two dimensions are regular rows and columns of a matrix. These first two dimensions cannot be modified. Thus, we need at most six placeholders of the form %li (which represents a long integer, i.e. 64 bit on most machines) for the remaining dimensions.

It is possible to produce different outputs through the use of format strings. The following example shows the default output and the output after setting the layerline to the string `"Layer:   %s\r\n"`:

```
randomize(0)
x = normal(2,2,2)
x
setenv ("outlayerline", "Layer: %li\r\n")
x
```

iofmt05.xpl

The result is

```
Contents of x
[,,1,1,1,1,1,1]
[1,] -0.21293    1.9502
[2,]  -1.0078   -1.3052
[,,2,1,1,1,1,1]
[1,]  -1.4341 -0.38946
[2,]   0.69812 -1.5746
Contents of x
Layer: 1
[1,] -0.21293    1.9502
[2,]  -1.0078   -1.3052
Layer: 2
[1,]  -1.4341 -0.38946
[2,]   0.69812 -1.5746
```

Note that in this last example we use only one placeholder %li instead of the potential six. Since this example contains only one additional layer beyond

the regular rows and columns of a matrix — normal(2,2,2) means 2 rows,
2 columns and 2 layers — one placeholder is sufficient.

15.4.3   Line Number Style

With outlineno it is possible to influence the appearance of the line numbers
of an array in the output window. The default value for this parameter is
"[%*li,]", where %*li is a placeholder for the line number.

The following example shows the default output and the output after setting
the layerline to the string "Line: %*li:

```
randomize(0)
x = normal(10,2)
x
setenv ("outlineno", "Line %*li:      ")
x
```

                                                  🔍 iofmt06.xpl

This yields

```
Contents of x
[ 1,] -0.21293  -2.5033
[ 2,]  -1.0078 -0.49264
[ 3,]   1.9502  -1.2245
[ 4,]  -1.3052 -0.42123
[ 5,]  -1.4341 -0.72973
[ 6,]  0.69812  0.20864
[ 7,] -0.38946 -0.14358
[ 8,]  -1.5746   0.6273
[ 9,] -0.26595 0.059032
[10,]  0.19947  0.44835
Contents of x
Line  1:     -0.21293  -2.5033
Line  2:      -1.0078 -0.49264
Line  3:       1.9502  -1.2245
Line  4:      -1.3052 -0.42123
Line  5:      -1.4341 -0.72973
Line  6:      0.69812  0.20864
Line  7:     -0.38946 -0.14358
```

```
Line  8:     -1.5746   0.6273
Line  9:     -0.26595  0.059032
Line 10:      0.19947  0.44835
```

For matrices with more than 9 rows, a right-justified line number should be used. This is accomplished by including a * into the placeholder: The default setting [%*li,] uses this.

15.4.4 Value Formats and Lengths

This subsection considers the variables outputformat, outputstringformat and outmaxdata, which influence the format and the length of objects, that can be printed to the output window. All code examples for this section can be found in the quantlet Q iofmt07.xpl.

outputformat sets the default output format for numeric arrays to % 8.5g. This format string means that the g-format is used with 5 decimals a total string length of 8. The variable outputformat can be changed now by using setenv:

```
setenv("outputformat","% 8.6g")
```

changes the number of decimals to 6. For outputformat one can use the same string format as for the function string. See Section 15.3 for examples on formatting numeric output.

The variable outputstringformat sets the default format for text arrays. The default is ""%s"". For example,

```
x="a"~"bb"
x
```

gives

```
Contents of x
[1,] "a" "bb"
```

in the output window. It is sometimes convenient to suppress the quotation marks. Using

```
setenv("outputstringformat","%s")
x
```

we obtain

```
Contents of x
[1,] a bb
```

Finally, the variable `outmaxdata` can change the number of maximal data items in the output window. The default is 2048. Note that the number of data items is also restricted by a size restriction of 32 Kilobytes for each output item.

15.4.5 Saving Output to a File

By default, all output is printed in the output window. The function `output` can save output additionally to a file. The function `output` must be called with one of the following options:

- `"reset"` to open a new file or to reset an existing one,

- `"append"` to open a new file or to append to an existing one,

- `"close"` to close the output file.

For example,

```
x=uniform(2,2)
x                                 ; usual output
output("output.txt","reset")
x+2                               ; output goes also to output.txt
output("output.txt","close")
x+4                               ; usual output
output("output.txt","append")
x+8                               ; output goes also to output.txt
output("output.txt","close")
```

iofmt08.xpl

This quantlet shows the first output of x as usual in the output window. The second output of x+2 goes to the output window and the file output.txt. Then the file is closed, such that the third output of x+4 is only printed in the output window. Now we open output.txt for appending. The output of x+8 goes to the output window and the file output.txt again. Finally, we close the file.

16 Matrix Handling

Yasemin Boztug and Marlene Müller

XploRe offers a large variety of commands and tools for creating and manipulating **multidimensional objects** called **matrices** and **lists**. The first part of this chapter presents the basic instructions for matrix handling. The second part illustrates some topics in matrix algebra with XploRe. The last part of this chapter presents the list object, which is a useful tool for handling data sets of heterogeneous formats (e.g. text and numeric).

16.1 Basic Operations

This section shows us how to create matrices and arrays and how to do simple matric calculations. You find all the following examples in ⬛ matrix01.xpl and ⬛ matrix02.xpl.

16.1.1   Creating Matrices and Arrays

z = matrix (d1 {, ..., dn})
      generates an array z of ones up to eight dimensions

randomize (seed)
      sets the seed for the random number generator

z = uniform (d1 {, ..., dn})
      generates an array z of uniform pseudorandom numbers up to
      eight dimensions

z = normal (d1 {, ..., dn})
      generates an array z of standard pseudorandom numbers up to
      eight dimensions

z = diag (x)
      creates a diagonal matrix z from a vector x

x = unit (d)
      generates a d-dimensional identity matrix x

d = dim (x)
      shows the dimension of an array x

n = rows (x)
      shows the number of rows of an array x

p = cols (x)
      shows the number of columns of an array x

z = x |y
      concatenate two arrays x and y rowwise

z = x ~ y
      concatenate two arrays x and y columnwise

We create a matrix or an array of dimensions $d_1 \times d_2 \times \ldots \times d_8$, i.e. up to eight dimensions with the function matrix. All the elements of the created array are set to 1. The instruction

```
mat=matrix(2,3)
mat
```

generates a 2 × 3 matrix, the elements of which are all equal to one:

```
Contents of mat
[1,]        1          1          1
[2,]        1          1          1
```

The definition of higher-dimensional matrices is straightforward. However, as we cannot represent three-dimensional (or higher-dimensional) objects in the two-dimensional space of this page, or on your screen, the first two dimensions are displayed in the standard format, the projections of the higher ones are then displayed. Consider the following matrix of dimension 2 × 3 × 2:

```
array = matrix(2,3,2)
array
```

This matrix is displayed in the form of two submatrices, the **layers** of the matrix: The first layer contains the elements of $\text{array}_{i,j,1}$, the second layer contains the elements of $\text{array}_{i,j,2}$:

```
Contents of array
[,,1,1,1,1,1,1]
[1,]        1          1          1
[2,]        1          1          1
[,,2,1,1,1,1,1]
[1,]        1          1          1
[2,]        1          1          1
```

The functions `uniform` and `normal` have the same syntax as `matrix`. They are used to generate **pseudorandom** uniform or normal (Gaussian) numbers, respectively. The seed for the the random number generator can be set by `randomize`. For example,

```
randomize(111222)
normal(2,2,2)
```

produces

```
Contents of rnorm
[,,1,1,1,1,1,1]
[1,]   0.23144   1.1671
[2,] -0.38991 -0.0089669
[,,2,1,1,1,1,1]
[1,]   0.60053 -0.54146
[2,] -0.28422   1.0531
```

A matrix A, with typical element $a_{i,j}$, is said to be diagonal if all elements outside its main diagonal are equal to zero, i.e. $a_{i,j} = 0, \forall i \neq j$. We create a diagonal matrix with the function diag, whose argument is the ordered sequence of elements in the main diagonal:

```
x = #(1, 2, 3)
diag (x)
```

which yields

```
[1,]        1         0         0
[2,]        0         2         0
[3,]        0         0         3
```

A particular diagonal matrix is the identity matrix denoted by I: The elements of its main diagonal are all equal to one. We create a unit matrix with the function unit, which has as argument the dimension of the matrix. Since this function belongs to the xplore library, we need to load that library before using it. The commands

```
library("xplore")
id=unit(3)
```

create the three-dimensional identity matrix:

```
Contents of id
[1,]        1         0         0
[2,]        0         1         0
[3,]        0         0         1
```

We display the dimension of the matrix, i.e. the number of elements in each direction, with the function dim. This function has the matrix as its argument and returns its dimensions. Thus, the instruction

```
dim(mat)
```

returns

```
Contents of dim
[1,]      2
[2,]      3
```

which means that the first dimension of mat is equal to 2, and the second dimension is equal to 3.

The function dim nests the functions rows and cols, which respectively return the number of rows and columns of a matrix. Thus,

```
rows(mat)
```

yields

```
Contents of rows
[1,]      2
```

You can check that cols(mat) returns 3.

The operators | and ~ can be used to concatenate matrices or arrays in vertical or horizontal direction. We have used this already for creating simple data matrices in Chapter 2. Both operators also work for matrices:

```
unit(3)~matrix(3)
```

returns

```
Contents of _tmp
[1,]      1        0        0        1
[2,]      0        1        0        1
[3,]      0        0        1        1
```

whereas

```
diag(1:3)|matrix(1,3)
```

gives

```
Contents of _tmp
[1,]        1         0         0
[2,]        0         2         0
[3,]        0         0         3
[4,]        1         1         1
```

The functions dim, rows, cols as well as the operators | and ~ can also be applied to alphanumeric matrices, i.e. multidimensional objects containing text. For example,

```
textmat = #("aa","c") ~ #("b","d2")
dim(textmat)
```

creates an alphanumeric matrix and prints its dimension

```
Contents of dim
[1,]        2
[2,]        2
```

However, matrices cannot mix alphanumeric and numeric elements. We will see in Section 16.7 that numeric and alphanumeric elements can only be mixed in lists.

16.1.2   Operators for Numeric Matrices

```
x + y
    elementwise addition of x and y

x - y
    elementwise subtraction of x and y

x .* y
    elementwise multiplication of x by y

x / y
    elementwise division of x by y

x ∧ y
    elementwise exponentiation of x by y

x % y
    modulo division of x by y

x * y
    matrix-wise multiplication of x and y

y = inv (x)
    computes y as the inverse matrix of x

y = trans (x) or y = x'
    computes the transpose of an array x
```

Let A and B be two matrices of the same dimension $k \times n$, with typical elements a_{ij} and b_{ij}. Given that these two matrices are of the same dimension, they are said to be **conformable** for addition and subtraction. The matrix C, of dimension $k \times n$, defined as $C = A + B$, obtained by the addition of matrices A and B, is the matrix with typical element c_{ij} such that $c_{ij} = a_{ij} + b_{ij}$. The matrix D, of dimension $k \times n$, defined as $D = A - B$, is the matrix with typical element d_{ij} such that $d_{ij} = a_{ij} - b_{ij}$. For example

```
A = (2|4)~(8|6)
B = (5|6)~(7|8)
A + B
```

returns

```
Contents of _tmp
[1,]          7          15
[2,]         10          14
```

You can use the elementwise operators also for adding and subtracting scalars and conformable vectors. For example

```
A + 3
A - (1|1)
```

gives

```
Contents of _tmp
[1,]          5          11
[2,]          7           9
Contents of _tmp
[1,]          1           7
[2,]          3           5
```

Two matrices E and F are said to be conformable for the matrix multiplication EF, if the number of columns of E is equal to the number of rows F. The matrix G defined as $G = EF$ has as typical element g_{ij}, with $g_{ij} = \sum_k e_{ik} f_{kj}$.

Let's define the following matrices:

```
E = #(2, 4)~#(8, 6)
F = #(1, 2)~#(3, 4)
```

Since both matrices are conformable for the two multiplications EF and FE, we can verify that matrix multiplication is not commutative:

```
mult1 =  F * F
mult2 =  E * E
mult1
mult2
```

shows

```
Contents of mult1
[1,]      18        38
[2,]      16        36
Contents of mult2
[1,]      14        26
[2,]      20        40
```

The two matrices `mult1` and `mult2` are indeed different.

Matrix multiplication is different from matrix elementwise multiplication, the operator of which is the compound operator `.*` obtained by concatenating the dot and star symbols. Two matrices are said conformable for elementwise multiplication if they are of the same dimension. The matrix M, defined by $M = A. * B$, has typical element m_{ij}, where $m_{ij} = a_{ij}b_{ij}$. Matrix elementwise multiplication is obviously commutative:

```
mult3 = E.*F
mult4 = F.*E
mult3
mult4
```

shows

```
Contents of mult3
[1,]       2        24
[2,]       8        24
Contents of mult4
[1,]       2        24
[2,]       8        24
```

The matrices `mult3` and `mult4` are the same, and differ from both `mult1` and `mult2`.

Note that all elementwise operations extend to arrays in a natural way. Typical matrix operations, such as matrix multiplication, are performed for each layer of an array. This holds also for the determinant, the inverse and the transpose which we will consider now.

A square matrix is such that its number of rows is equal to its number of columns. A square matrix A is said to be invertible if its **determinant**, denoted by $\det(A)$ or $|A|$, is different from zero. The determinant of a matrix measures

the "volume" of the space spanned by its column vectors. The determinant of
a matrix is thus equal to zero if its rows, or column vectors, are collinear.

We calculate the determinant of a matrix with the function det, which has the
matrix as its argument. Thus,

    det(A)

returns

    Contents of det
    [1,]   -20

The **inverse** of an invertible matrix A is the unique matrix, denoted by A^{-1}
of the same dimension of A, such that $A^{-1}A = AA^{-1} = I$, the identity matrix.
We evaluate the inverse of a matrix with the function inv. The matrix A is
invertible since its determinant is different from zero. We calculate this inverse
as follows:

    inv(A)

yields

    Contents of cinv
    [1,]   -0.3   0.4
    [2,]   0.2   -0.1

We verify that the product of A and its inverse is the two-dimensional identity
matrix:

    A*inv(A)

gives

    Contents of _tmp
    [1,]   1   0
    [2,] -1.1102e-16   1

which is numerically close to the two-dimensional identity matrix. The difference in the element in the second row and the first column is caused by
rounding errors.

The **transpose** of a matrix A, with typical element a_{ij}, is the matrix denoted by A^T: the element in the ith row and jth column of A^T is a_{ji}, the element in the jth row and ith column of A. The matrix A is said to be symmetric if and only if $A^T = A$.

We calculate the transpose of a matrix with either the function **trans** or the operator '. The following two instructions are equivalent:

```
trans(A)
A'
```

and return the transpose of the matrix A.

16.2   Comparison Operators

---

x == y
>   elementwise equality of x and y

x !=y or x <> y
>   elementwise inequality of x and y

x > y
>   checks if x is elementwise greater than y

x >= y
>   checks if x is elementwise greater than or equal to y

x < y
>   checks if x is elementwise less than y

x <= y
>   checks if x is elementwise less than or equal to y

x && y
>   elementwise logical AND operator for x and y

x ||y
>   elementwise logical OR operator for x and y

!x
>   elementwise logical NOT operator for x

XploRe has no specific data type for **boolean** values. The boolean value **true** is encoded by 1, whereas the value **false** is encoded by 0. Hence, all matrix operations for numeric matrices can be used for boolean expressions as well. Note that many functions which require boolean input only check if the input is false or not.

Let us consider which can be found in ◘matrix03.xpl.

```
A = (2|4)~(8|8)
B = (5|6)~(4|8)
A > B
```

The last instruction, i.e. the comparison yields

```
Contents of _tmp
[1,]        0        1
[2,]        0        0
```

which means that only the element in the first row and second column of A is greater than the corresponding element of B. Similarly,

```
A == B
```

gives

```
Contents of _tmp
[1,]        0        0
[2,]        0        1
```

Note that A == B is different from the assignment A = B, which would create A as a copy of the matrix B.

You may check now that the instruction

```
(A > B) || (A == B)
```

which checks if "A > B or A == B" is identical to

```
A >= B
```

Both instructions return

```
Contents of _tmp
[1,]        0        1
[2,]        0        1
```

which means that both values in the second column fulfill the >= condition.

Let us point out again that all introduced comparison operators work elementwise. They extend to arrays as usual. To verify a condition for all elements of a matrix or an array, the prod function can be used:

```
prod(prod(A==B),2)
```

would return 1 in the case that all elements of the matrices A and B are identical.
We will learn more about prod in Section 16.4.

16.3 Matrix Manipulation

Once a matrix or an array has been defined it might be necessary to either
extract some submatrices from the matrix, to **reshape** it, i.e. change its dimensions, or to modify the order of its elements. You find all the following
examples in ⌑matrix04.xpl and ⌑matrix05.xpl.

16.3.1 Extraction of Elements

```
y = x[i,j] or y = x[i,] or y = x[,j]
      extracts element i,j or row i or column j from x

y = index (x, ind)
      generates a new matrix y from an old matrix x by extracting the
      rows of x indicated in the index vector ind

y = paf (x, ind)
      selects rows of an object x depending on a given vector ind re-
      sulting in the matrix y

y = sort (x, c)
      sorts the rows of a matrix x by the vector c
```

We extract elements or submatrices of a matrix with the operator []. Consider
the 4×3 matrix

```
mat=(1|2|3|4)~(5|6|7|8)~(9|9|9|9)
```

The following three lines extract respectively the first row, the second column,
and the element in the fourth row and third column:

```
mat[1,]                        ; extract first row
```

```
mat[,2]                        ; extract second column
mat[4,3]                       ; extract element (4,3)
```

The [] operator can be combined with the **range operator**, denoted by the colon symbol :, and the | operator. The command

```
mat[1:3,1|3]
```

extracts the elements which are between the first and the third rows of mat and in the 1st and 3rd columns.

In a similar way, we extract certain rows from a matrix with the function index. Its first argument is the matrix, its second argument is the index of the considered rows. This second argument can be a vector of indices. In that case, all rows which are in the index vector are extracted. The instructions

```
x = #(1,3,5)~#(2,4,6)
index (x,#(3,2))
```

create a 3 × 2 matrix, and respectively extract its third and second rows:

```
Contents of index
[1,]        5          6
[2,]        3          4
```

Note that index(x,#(3,2)) is identical to x[#(3,2)] as long as the values in the index vector — here #(3,2) — do not exceed the dimension of x. If this is the case, x[#(3,2)] would return an error message while the index function returns NaN values.

We can select rows of a matrix on the basis of a logical condition with the function paf. The first argument of the function is the matrix, the second argument is an indicator vector with the same row dimension as the matrix. The function paf selects the rows of the matrix whose corresponding element in the indicator vector is different from zero. Note that paf returns an error, if no element of the indicator vector is different from zero!

The following example illustrates the use of this function:

```
x = #(4, 2, 5)
ind = x > 3
paf(x, ind)
```

The three-dimensional vector x and the three-dimensional indicator vector ind are created. Since the vector ind is created with the logical condition x > 3, its elements are either equal to 1 if the corresponding element of the vector x is strictly greater than 3, or to zero otherwise. As only the second element of x is less than 3, the second element of ind is set to zero, the function paf selects the remaining first and third rows of x and returns them:

```
Contents of paf
[1,]          4
[2,]          5
```

The function sort sorts the rows of a matrix. Its first argument is the matrix, its second argument is the column with respect to which the sorting is done. If this column order is positive, the matrix will be sorted by that column in ascending order, while if the column order is negative, the matrix will be sorted in descending order. The column order can be a vector. In that case, the matrix is first sorted by the first element of the sorting vector, next the matrix is sorted by the second element of the sorting vector provided that this second sorting does not modify the previous one. The instructions

```
x = #(4,2,5)~#(1,7,4)
x
sort(x,-2)
```

create a 3×2 matrix. Since the second argument of sort is equal to -2, this matrix is sorted in descending order by the second column:

```
Contents of x
[1,]          4          1
[2,]          2          7
[3,]          5          4
Contents of sort
[1,]          2          7
[2,]          5          4
[3,]          4          1
```

If the second argument of the function sort is set to 0, all columns of the matrix are independently sorted in ascending order. This means,

```
sort(x,0)
```

gives

```
Contents of sort
[1,]        2        1
[2,]        4        4
[3,]        5        7
```

Let us remark that all functions from this section can also be applied to alphanumeric matrices. Also, the extension to higher-dimensional arrays is straightforward.

16.3.2 Matrix Transformation

```
y = reduce (x)
      deletes all dimensions of x with only a single component

y = reshape (x, d)
      transforms an array x into a new one y with given dimensions d

y = vec (x1 {,x2 ... })
      vectorizes the arrays x1, x2, ...
```

If some of the dimensions of a multidimensional matrix are equal to one, we can discard them with the function **reduce**. Since this function is part of the library **xplore**, this library should be loaded before the use of the function. The commands

```
library("xplore")
x = matrix(3,1,2,1)
reduce(x)
```

create a $3 \times 1 \times 2 \times 1$ matrix of ones, discard the two dimensions equal to one, and leave a 3×2 matrix which is displayed as follows:

```
Contents of y
[1,]        1        1
[2,]        1        1
[3,]        1        1
```

The dimensions of an array can be modified with the function reshape although the modified array does contain the same number of elements as the original array. The first argument of the function is the array to be modified, i.e. the source array, the second argument is the dimension of the target array. The function reshape reads rowwise the elements of the source array and stores them rowwise into the target array. The commands

```
x = #(1, 1, 1, 1, 1)~#(2, 2, 2, 2, 2)
x
d = #(2, 5)
reshape(x, d)
```

create a 5 × 2 matrix, display it, reshape it into a 2 × 5 matrix, and display it:

```
Contents of x
[1,]        1        2
[2,]        1        2
[3,]        1        2
[4,]        1        2
[5,]        1        2
Contents of reshape
[1,]        1        1        1        2        2
[2,]        1        1        2        2        2
```

The function vec reshapes a matrix into a vector:

```
vec(x)
```

yields

```
Contents of #
[ 1,]        1
[ 2,]        1
[ 3,]        1
[ 4,]        1
[ 5,]        1
[ 6,]        2
[ 7,]        2
[ 8,]        2
```

```
[ 9,]        2
[10,]        2
```

The vec function can optionally add some elements to the reshaped vector:

```
vec(x,5)~vec(3,x)
```

returns

```
Contents of _tmp
[ 1,]        1        3
[ 2,]        1        1
[ 3,]        1        1
[ 4,]        1        1
[ 5,]        1        1
[ 6,]        2        1
[ 7,]        2        2
[ 8,]        2        2
[ 9,]        2        2
[10,]        2        2
[11,]        5        2
```

The instruction vec(x,5) reshapes the matrix x into a vector and adds the number 5 as the last element of the created vector. However, since the number 3 is the first argument of the function vec in the instruction vec(3,x), the first element of the created vector is this number to which the "vectorized" matrix is added.

As a consequence, both instructions

```
z1 = #(1,2,3)
z2 = vec(1,2,3)
```

are equivalent.

16.4   Sums and Products

```
y = sum (x {,d})
      computes the sum of the elements of x, optionally with respect
      to dimension d

y = cumsum(x {,d})
      computes the cumulative sum of the elements of x, optionally
      with respect to dimension d

y = prod(x {,d})
      computes the product of the elements of x, optionally with respect
      to dimension d

y = cumprod(x {,d})
      computes the cumulative product of the elements of x, optionally
      with respect to dimension d
```

The function sum computes the sum of the elements of an array with respect
to a given dimension. The default dimension is the first one, i.e. the elements
of the matrix are summed columnwise. The following and all other examples
of this section can be found in ◻ matrix06.xpl.

```
x=#(1,3)~#(2,4)
sum(x)
```

displays

```
Contents of sum
[1,]        4        6
```

while

```
sum(x,2)
```

gives

```
[1,]        3
[2,]        7
```

Similarly, the function `cumsum` computes the cumulative sum of the elements of an array with respect to a given dimension:

```
cumsum(#(5,4,3)~#(1,2,3))
```

yields

```
Contents of cumsum
[1,]       5        1
[2,]       9        3
[3,]      12        6
```

The functions `prod` and `cumprod` evaluate respectively the product and cumulative product of the elements of a matrix with respect to a given dimension. The syntax of these functions is the same as the functions `sum` and `cumsum`. Thus,

```
prod(#(5,4,3)~#(1,2,3))
```

returns

```
Contents of mul
[1,]      60        6
```

while

```
cumprod(#(5,4,3)~#(1,2,3))
```

gives

```
Contents of cumprod
[1,]       5        1
[2,]      20        2
[3,]      60        6
```

16.5   Distance Function

```
y = distance (x, metric)
      computes the distance between p-dimensional data points in x
      depending on a specified metric
```

The function **distance** evaluates the distance between two p-dimensional vectors with respect to a specified metric. The possible metrics are given in the following table:

Distance	XploRe Name
L_1	"l1"
L_2	"l2", "euclid"
Maximum	"maximum"
Cosine	"cosine"
χ^2	"chisquare"
Centroid	"centroid"
Tanimoto	"tanimoto"
Matching	"matching"

Several of these distances (such as Tanimoto and Matching) are designed for binary data.

The first argument of **distance** is the matrix, the second argument the XploRe name. By default the second argument is "l2". The commands

```
x = #(1, 4)~#(1, 5)
distance(x,"l1")
```

define the matrix x and evaluate the L_1 distance between the rows of that matrix:

```
Contents of distance
[1,]   0.000000   7.000000
[2,]   7.000000   0.000000
```

while the command

```
distance(x)
```

yields the L_2 distance, since the second argument of the function `distance` is missing:

```
Contents of distance
[1,]  0.000000  5.000000
[2,]  5.000000  0.000000
```

The examples for this section are collected in ◻ `matrix07.xpl`.

16.6   Decompositions

Matrix decompositions are useful in numerical problems, in particular for solving systems of linear equations. All XploRe examples for this section can be found in ◻ `matrix08.xpl`.

16.6.1   Spectral Decomposition

> {values,vectors} = eigsm (x)
>   computes the eigenvalues and vectors of a symmetric matrix x
>
> {values,vectors} = eiggn (x)
>   computes the eigenvalues and vectors of a quadratic matrix x

Let A be a square matrix of dimension $n \times n$. A scalar λ is an **eigenvalue** and a nonzero vector v is an **eigenvector** of A if $Av = \lambda v$.

The eigenvalues are the roots of the characteristic polynomial of order n defined as $(|A| - \lambda I_n) = 0$, where I_n denotes the n-dimensional identity matrix. The determinant of the matrix A is equal to the product of its n eigenvalues: $|A| = \Pi_{i=1}^{n} \lambda_i$.

The function `eigsm` calculates the eigenvectors and eigenvalues of a given symmetric matrix. We evaluate the eigenvalues and eigenvectors of nonsymmetric matrices with the function `eiggn`.

The function **eigsm** has as its unique argument the matrix and returns the eigenvalues and vectors. The returned arguments are unsorted with respect to the eigenvalues. Consider the following example:

```
x = #(1, 2)~#(2, 3)
y = eigsm(x)
y
```

in which we define a matrix x, and calculate its eigenvalues and eigenvectors. XploRe stores them in a variable of list type, y: The variable y.values contains the eigenvalues, while the variable y.vectors contains the corresponding eigenvectors:

```
Contents of y.values
[1,] -0.23607
[2,]   4.2361
Contents of y.vectors
[1,]  0.85065   0.52573
[2,] -0.52573   0.85065
```

We verify that the determinant of the matrix x is equal to the product of the eigenvalues of x:

```
det(x) - y.values[1] * y.values[2]
```

gives

```
Contents of _tmp
[1,]   4.4409e-16
```

i.e. something numerically close to zero.

If the n eigenvalues of the matrix A are different, this matrix can be decomposed as $A = P\Lambda P^T$, where Λ is the diagonal matrix the diagonal elements of which are the eigenvalues, and P is the matrix obtained by the concatenation of the eigenvectors. The transformation matrix P is **orthonormal**, i.e. $P^T P = PP^T = I$. This decomposition of the matrix A is called the spectral decomposition.

We check that the matrix of concatenated eigenvectors is orthonormal:

```
y.vectors'*y.vectors
```

yields a matrix numerically close to the identity matrix:

```
Contents of _tmp
[1,]           1 -1.0219e-17
[2,] -1.0219e-17           1
```

We verify the spectral decomposition of our example:

```
z = y.vectors *diag(y.values) *y.vectors'
z
```

which gives the original matrix x:

```
Contents of z
[1,]       1       2
[2,]       2       3
```

If the matrix A can be decomposed as $A = P\Lambda P^T$, then $A^n = P\Lambda^n P^T$. In particular, $A^{-1} = P\Lambda^{-1}P^T$. Therefore, the inverse of x could be calculated as

```
xinv = y.vectors*inv(diag(y.values))*y.vectors'
xinv
```

which gives

```
Contents of xinv
[1,]       -3       2
[2,]        2      -1
```

which is equal to the inverse of x

```
inv(x)
```

```
Contents of cinv
[1,]       -3       2
[2,]        2      -1
```

16.6.2   Singular Value Decomposition

{u, l, v} = svd (x)
  computes the singular value decomposition of a matrix x

Let B be a $n \times p$ matrix, with $n \geq p$, and rank r, with $r \leq p$. The singular value decomposition of the matrix B decomposes this matrix as $B = ULV$, where U is the $n \times r$ orthonormal matrix of eigenvectors of $B B^T$, V is the $p \times r$ orthonormal matrix of eigenvectors of $B^T B$ associated with the nonzero eigenvalues, and L is a $r \times r$ diagonal matrix.

The function svd computes the singular value decomposition of a $n \times p$ matrix x. This function returns the matrices u, v and Λ in the form of a list.

```
x = #(1, 2, 3)~#(2, 3, 4)
y = svd(x)
y
```

XploRe returns the matrix U in the variable y.u, the diagonal elements of L in the variable y.l, and the matrix V in the variable y.v:

```
Contents of y.u
[1,]   0.84795    0.3381
[2,]   0.17355   0.55065
[3,] -0.50086    0.7632
Contents of y.l
[1,]   0.37415
[2,]    6.5468
Contents of y.v
[1,] -0.82193  0.56959
[2,]   0.56959  0.82193
```

We test that y.u *diag(y.l) *y.v' equals x with the commands

```
xx = y.u *diag(y.l) *y.v'
xx
```

This displays the matrix x:

```
Contents of xx
[1,]        1        2
[2,]        2        3
[3,]        3        4
```

16.6.3   LU Decomposition

{1, u, index} = ludecomp (x)
        computes the LU decomposition of a matrix x

The LU decomposition of an n-dimensional square matrix A is defined as

$$A = LU,$$

where U is an n-dimensional upper triangular matrix, with typical element u_{ij} such that $u_{ij} = 0 \; \forall i > j$, and L is an n-dimensional lower triangular matrix, with typical element l_{ij} such that $l_{ij} = 0 \; \forall j > i$. The LU decomposition is used for solving linear equations and inverting matrices. The function ludecomp performs the LU decomposition of a square matrix. It takes the matrix as its argument, returns in the form of a list the upper and lower triangular matrices, and an index vector which records the row permutations in the LU decomposition:

```
x = #(1, 2)~#(2, 3)
lu = ludecomp(x)
lu
```

gives

```
Contents of lu.l
[1,]        1        0
[2,]      0.5        1
Contents of lu.u
[1,]        2        3
[2,]        0      0.5
Contents of lu.index
[1,]        2
[2,]        1
```

We re-obtain the original matrix x by using the function index, which takes as its argument the row-permutations in the LU decomposition. The instruction

```
index(lu.l*lu.u,lu.index)
```

returns the matrix x

```
Contents of index
[1,]          1          2
[2,]          2          3
```

16.6.4   Cholesky Decomposition

```
bd = chold (x, 1)
     calculates the triangularization of a matrix x, such that b'*d*b=x;
     1 is the number of lower nonzero subdiagonals including the di-
     agonal
```

Let A be an n-dimensional square matrix, symmetric, i.e. $A_{ij} = A_{ji}$, and positive definite, i.e. $x^T A x > 0$, $\forall$ n-dimensional vectors v. The Cholesky decomposition decomposes the matrix A as $A = L L^T$, where L is a lower triangular matrix.

The function chold computes a triangularization of the input matrix. The following steps are necessary to find the lower triangular L:

```
library("xplore")          ; unit is in library xplore!
x = #(2,2)~#(2,3)
tmp = chold(x, rows(x))
d = diag(xdiag(tmp))       ; diagonal matrix
b = tmp-d+unit(rows(tmp))  ; lower triangular
L = b'*sqrt(d)             ; Cholesky triangular
```

We can check that the decomposition works by comparing x and L*L'. The instruction

```
x - L*L'
```

displays

```
Contents of _tmp
[1,] -4.4409e-16 -4.4409e-16
[2,] -4.4409e-16 -4.4409e-16
```

which means that the difference between both matrices is practically zero.

The Cholesky decomposition is used in computational statistics for inverting the Hessian matrices and the matrices $X^T X$ in regression analysis.

16.7 Lists

A list is an ordered collection of objects of different type, i.e. scalars, matrices, string variables. You can find all the following examples in ⌂matrix09.xpl.

16.7.1 Creating Lists

```
l = list (x1 {,x2{,...}})
        generates lists l from given objects (x1 {,x2{,...}})

y = x.elem
        gives the element elem of a list x
```

We create a list with the function list, which takes as arguments the ordered elements of the list. The commands

```
seller1 = list("Ludwig","Beethoven",20)
seller1
```

create a list, with the string "Ludwig" as its first element, "Beethoven" as its second element, and the number 20 as its third element. This list is stored in the variable seller1, the content of seller1 is displayed as follows:

```
Contents of seller1.el1
[1,] "Ludwig"
```

```
Contents of seller1.el2
[1,] "Beethoven"
Contents of seller1.el3
[1,]        20
```

The three components of the list are displayed. The default label of the ith element of the list seller1 is seller1.el$i$. Thus, the second element of the list seller is labeled as seller1.el2. The dot operator allows us to access any component of the list. We access the second component as

```
seller1.el2
```

```
Contents of el2
[1,] "Beethoven"
```

However, we could give a label to the components of a list. Consider the following list:

```
age = 35
seller2 = list("Claudio","Arrau",age)
seller2
```

Since the third component of the list seller2 is the assigned variable age, this element is labeled age:

```
Contents of seller2.el1
[1,] "Claudio"
Contents of seller2.el2
[1,] "Arrau"
Contents of seller2.age
[1,]        35
```

A list can be a component of another list: We define the list sellers, which contains the lists seller1 and seller2:

```
sellers = list(seller1,seller2)
sellers
```

returns

```
Contents of sellers.seller1.el1
[1,] "Ludwig"
Contents of sellers.seller1.el2
[1,] "Beethoven"
Contents of sellers.seller1.el3
[1,]        20
Contents of sellers.seller2.el1
[1,] "Claudio"
Contents of sellers.seller2.el2
[1,] "Arrau"
Contents of sellers.seller2.age
[1,]        35
```

In that case, we access the element **age** of the variable `seller2` of the list `sellers` by using twice the dot operator:

```
sellers.seller2.age = 36
sellers.seller2.age
```

which returns

```
Contents of sellers.seller2.age
[1,]        36
```

16.7.2 Handling Lists

```
insert (l, pos, x)
       inserts an object x at the specified position pos within a list l

delete (l, pos)
       deletes the object at the given position pos from the list l

append (l, x)
       appends an object x to the specified list l

y = x{n}
       gives an element of a list x with number n
```

You may use any kind of object as an element of a list, e. g. strings, matrices or even lists. To insert an object within a list one can use the **insert** function.

```
name = "Andrew"
insert(sellers, 1, name)
sellers
```

inserts a new element before el1 .This element contains the name **"Andrew"**. Unfortunately, the element is not assigned the object name but the name according to the position notation (**el1** in the example). Because the former first element retains its old name **el1** we now have two elements with the same name but different contents.

```
Contents of sellers.el1
[1,] Andrew
Contents of sellers.el1
[1,] 1.000000
Contents of sellers.age
[1,] 29.000000
```

Using the dot operator (**seller.el1**) we will only get the value of the first object with this name (**"Andrew"**). We can, however, access the elements of a list not only by its name but also by its number by using the braces ({}) function. So,

```
sellers{1}
```

gives

```
Contents of el1
[1,] "Andrew"
```

As well as inserting an object we can delete it by using the function **delete**.

```
delete(sellers, 1)
```

deletes the second element (**sellers.el1="Andrew"**) from the list. Furthermore, we can append an object to the end of a list using **append**:

```
sales = list(100, 200, 75)
append(sellers, sales)
sellers
```

appends the list object **sales** as a new element to the list object **seller**:

```
Contents of sellers.seller1.el1
[1,] "Ludwig"
Contents of sellers.seller1.el2
[1,] "Beethoven"
Contents of sellers.seller1.el3
[1,]        20
Contents of sellers.seller2.el1
[1,] "Claudio"
Contents of sellers.seller2.el2
[1,] "Arrau"
Contents of sellers.seller2.age
[1,]        36
Contents of sellers.sales.el1
[1,]       100
Contents of sellers.sales.el2
[1,]       200
Contents of sellers.sales.el3
[1,]        75
```

In contrast to the **insert** function the **append** function assigns the object name of the appended object to the new element name of the list unless it is only a temporary object (expression). If the first parameter is not a list, i.e. not a composed object, then a list is generated by the **append** function, where the first parameter becomes both the name of the resulting list and the name of its first component.

16.7.3   Getting Information on Lists

```
names (x)
      gives the names of all components of a list object x

y = size (x)
      gives the number of elements of x contained in a list

y = comp (obj, com)
      checks whether an object obj has a specific component com or
      not
```

The function **names** gives the names of all components of a list object. The resulting object is a vector, such that one can access single values using the square brackets operator [].

```
names(sellers)[2:3]
```

shows

```
Contents of names(sellers)[2:3]
[1,] "seller2"
[2,] "sales"
```

Note that the output of the third element of the list gives only the name of the list object **sales** itself but not the names of its components.

The **size** function gives the number of elements that are contained in a list. For example,

```
li = list(sellers, matrix(3, 4))
size(li)
```

gives

```
Contents of numcomp
[1,] 2
```

As before, only the elements of the respective list are counted, no matter what type they are or whether they have subobjects.

Finally, comp checks whether a list contains a certain component. For example

```
comp(li,"sellers")
```

returns

```
Contents of comp
[1,]        1
```

whereas

```
comp(li,"seller1")
```

returns

```
Contents of comp
[1,]        0
```

The false value 0 tells us that the list li does not contain a component named seller1.

17 Quantlets and Quantlibs

Wolfgang Härdle, Zdeněk Hlávka and Sigbert Klinke

Quantlets are quantitative procedures that are designed in the XploRe language and run inside the XploRe computing environment. They may be made public on the Internet via an HTML outlet.

If several quantlets serve a common aim and constitute a group of homogeneous code, they may be combined into a set of quantlets to constitute a so-called **quantlib**. XploRe has several such quantlibs, e.g. the collection of time series routines (quantlib **times**) and the smoothing methods (quantlib **smoother**).

17.1   Quantlets

```
proc (y) = pname (x) - endp
      defines a procedure with the input x and output y

func ("file")
      loads the quantlet file.xpl from the standard quantlet path
```

Quantlets are constructed via the XploRe editor. One opens the editor via the New item from the Programs menu. A blank window appears with the name noname.xpl.

Let us construct an example which lets us study the effect of outliers to linear least squares regression. First we generate a regression line of n points by the commands

```
n = 10                      ; number of observations
randomize(17654321)         ; sets random seed
```

```
beta = # (1, 2)              ; defines intercept and slope
x = matrix(n)~sort(uniform(n))
                             ; creates design matrix
m = x*beta                   ; defines regression line
```

quant01.xpl

The vector m now contains the regression line $m(x) = 1 + 2x$ at randomly selected points x. The vector β of regression parameters is given by beta=#(1,2). The points of x are uniformly distributed over the interval $[0, 1]$ by the command uniform(n). Note that we sorted the data with respect to the x component, which is suitable for plotting purposes.

It is now about time to save this file into a user owned file. Let us call this file myquant.xpl We save the file by the command <Ctrl A> or by selecting the option Save as under Programs. We enter the text myquant (the ending ".xpl" is added automatically). The editor window now changes its name. Typically, it is called C:\XploRe\myquant.xpl. We execute the quantlet by clicking the Execute item or by entering <Alt E>. Entering m on the action line in the input window yields the following 10 points:

```
Contents of m
[ 1,]    1.0858
[ 2,]    1.693
[ 3,]    1.8313
[ 4,]    2.1987
[ 5,]    2.2617
[ 6,]    2.3049
[ 7,]    2.3392
[ 8,]    2.6054
[ 9,]    2.6278
[10,]    2.8314
```

The vector m contains the values of the regression line $1 + 2x$. Let's now add some noise to the regression line and produce a plot of the data. We add the following lines and obtain a picture of the 10 data points that are scattered around the line $m(x) = 1 + 2x$. We extracted here the second column of the design matrix x, since the first column of x is a column of ones that models the constant intercept term of the regression line:

```
eps = 0.05*normal(n)         ; create obs error
```

```
y = m + eps                    ; noisy line
d = createdisplay(1,1)
dat = x[,2]~y
```

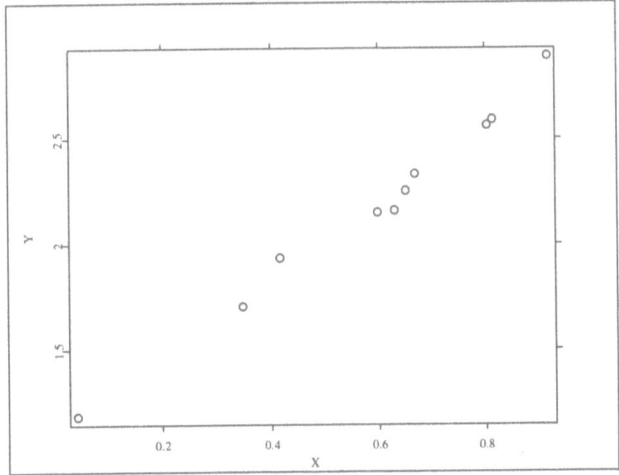 quant02.xpl

If we enter now on the action line the command

```
show(d, 1, 1, dat)
```

we obtain the following plot:

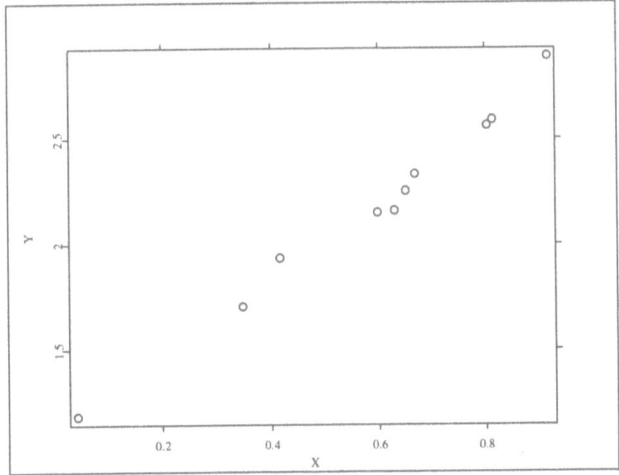

Let's now add the true regression line and the least squares estimated regression line to this plot. We use the command **setmaskl** to define the line mask and the command **setmaskp** to define the point mask. Using the command

```
tdat = x[,2]~m
```

we define the matrix **tdat** containing the true regression line. The command

```
setmaskl(tdat, (1:rows(tdat))', 1, 1, 1)
```

connects all points (2nd parameter) of **tdat** and defines a blue (3rd parameter: colorcode = 1) solid (4th parameter: type code = 1) line with a certain thickness (5th parameter: thickness code = 1). The command

```
setmaskp(tdat, 0, 0, 0)
```

sets the data points to its minimum size 0, i.e. invisible.

```
tdat = x[,2]~m
setmaskl(tdat, (1:rows(tdat))', 1, 1, 1)
                                  ; thin blue line
setmaskp(tdat, 0, 0, 0)           ; reduces point size to min
beta1 = inv(x'*x)*x'*y            ; computes LS estimate
yhat = x*beta1
hdat = x[,2]~yhat
setmaskl(hdat, (1:rows(hdat))', 4, 1, 3)
                                  ; thick red line
setmaskp(hdat, 0, 0, 0)
show(d, 1, 1, dat, tdat, hdat)
```

🔍 quant03.xpl

The result is given in the following picture:

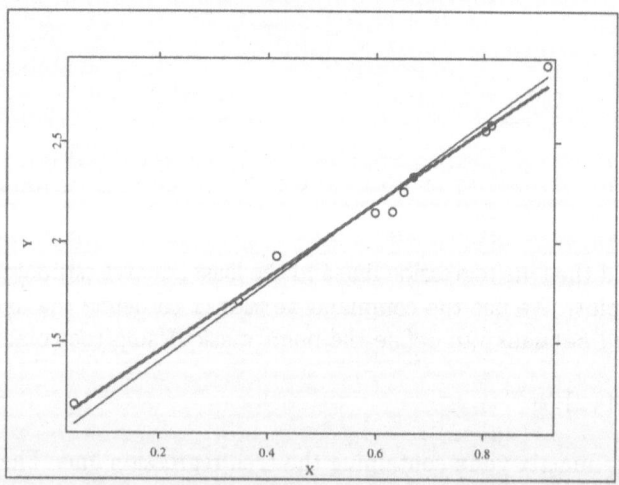

The true regression line is displayed as a thin blue line and the estimated regression line is shown as a thick red line. In order to create a quantlet we encapsulate these commands into a proc – endp bracket. This way we can call the quantlet from the action line once it is loaded into XploRe. We add the line

```
proc() = myquant()
```

as the first line, indent all following commands by the Tab key or by the Format source command in the Tools menu and add as a last line the word

```
endp
```

Altogether we should have the following in the editor window. We save this quantlet by <Ctrl S> or by passing through the Programs menu item:

```
proc() = myquant()
  n = 10                              ; number of observations
  randomize(17654321)                 ; sets random seed
  beta =#(1, 2)                       ; defines intercept and slope
  x = matrix(n)~sort(uniform(n))      ; creates design matrix
  m = x*beta                          ; defines regression line
  eps = 0.05*normal(n)                ; creates obs error
  y = m + eps                         ; noisy line
  d = createdisplay(1,1)
  dat = x[,2]~y
  tdat = x[,2]~m
  setmaskl(tdat, (1:rows(tdat))', 1, 1, 1)
                                      ; thin blue line
  setmaskp(tdat, 0, 0, 0)             ; reduces point size to min
  beta1 = inv(x'*x)*x'*y
  yhat = x*beta1
  hdat = x[,2]~yhat
  setmaskl(hdat, (1:rows(hdat))', 4, 1, 3)
                                      ; thick red line
  setmaskp(hdat, 0, 0, 0)
  show(d, 1, 1, dat, tdat, hdat)
endp
```

quant04.xpl

If we execute this program code via the Execute item, nothing will happen since the code contains only the definition of a quantlet. The quantlet performs the desired action only if it is called. By entering the command

```
myquant()
```

on the action line in the input window, we obtain the same picture as before. The quantlet myquant is now loaded in XploRe, and we can repeat this plot as many times as we want.

Let's now modify the quantlet so that the user of this quantlet may add another observation to the existing 10 observations. This additional observation will be the outlier whose influence on least squares regression we wish to study. We do this by allowing myquant to process an input parameter obs1 containing the coordinates (x, y) of an additional observation. We change the first line of myquant to proc() = myquant(obs1) and add the lines

```
// new x-observation is
  x = x|(1~obs1[1])
```

after the creation of the original design matrix. The first line is a comment, and the second line adds the x-coordinate of the new observation obs1 to the design matrix. Note that we also added a 1 to the first column of the design matrix in order to correctly reflect the intercept term in the least squares regression also for this 11th observation.

The second modification is given by the following two lines:

```
// new y-observation
  y = m[1:n] + eps            ; noisy line
  y = y|obs1[2]
```

The first line is again a comment and the second line adds the normal errors eps to the first n values of the m-observation. The third line adds the y-value of the new observations obs1 to the response values.

We also display the outlying observation in a different way than the other observations.

```
outl = obs1[1]~obs1[2]
setmaskp(outl,4,12,8)
show(d, 1, 1, dat[1:n], outl, tdat, hdat)
```

The second parameter in the **setmaskp** command defines black color, the third parameter defines a star and the fourth parameter sets the size of the star to 8. The **show** command displays the original data as black circles and the outlier as a star. We set a title of the graph with the command

```
setgopt(d,1,1,"title","Least squares regression with outlier")
```

The command **setgopt** can be used to change also other attributes of the graph, e.g. limits, tickmarks, labels, etc.

If we save and execute now the quantlet, we have it ready in XploRe to be executed from the action line:

```
proc() = myquant(obs1)
   n = 10                     ; number of observations
   randomize(17654321)        ; sets random seed
   beta =#(1, 2)              ; defines intercept and slope
   x = matrix(n)~sort(uniform(n))
                              ; creates design matrix
// new x-observation is
   x = x|(1~obs1[1])
   m = x*beta                 ; defines regression line
   eps = 0.05*normal(n)       ; creates obs error
// new y-observation
   y = m[1:n] + eps           ; noisy line
   y = y|obs1[2]
   d = createdisplay(1,1)
   dat = x[,2]~y
   outl = obs1[1]~obs1[2]
   setmaskp(outl,0,12,8)      ; outlier is black star
   tdat = x[,2]~m
   setmaskl(tdat, (1:rows(tdat))', 1, 1, 1)
                              ; thin blue line
   setmaskp(tdat, 0, 0, 0)    ; reduces point size to min
   beta1 = inv(x'*x)*x'*y
   yhat = x*beta1
   hdat = x[,2]~yhat
   setmaskp(hdat, 0, 0, 0)
   setmaskl(hdat, (1:rows(hdat))', 4, 1, 3)
                              ; thick red line
```

```
    show(d, 1, 1, dat[1:n], outl, tdat, hdat)
    title="Least squares regression with outlier"
    setgopt(d,1,1,"title",title)
                              ; sets title
endp
```

⊙ quant05.xpl

By entering

```
myquant(#(0.9,4.5))
```

from the action line, we obtain the following graphic which shows the effects of this outlier on the least squares regression:

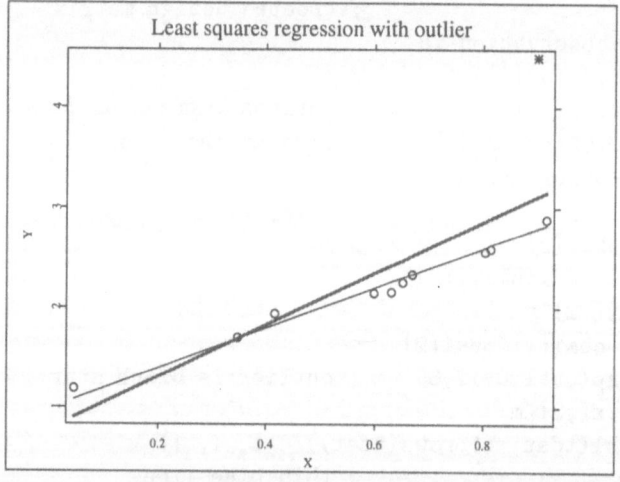

One clearly sees the nonrobustness of the least squares estimator. The additional observation $(0.9, 4.5)$ influences the estimated regression line. The thick red line is different from the true regression line indicated as the thin blue line.

The situation becomes even more extreme when we move the x-observation of the new observation into the leverage zone outside the interval $[0, 1]$. Suppose that we call the quantlet with the new observation $(2.3, 45)$. The x-value of this new observation is clearly outside the range $[0,1]$ of the first uniformly generated 10 design values. The y-value 45 of the new observation is enormous relative to the range of the other 10 values.

```
myquant(#(2.3,45))
```

The effect will be that the thick red line will be even more apart from the blue line. This becomes clear from the following graphic:

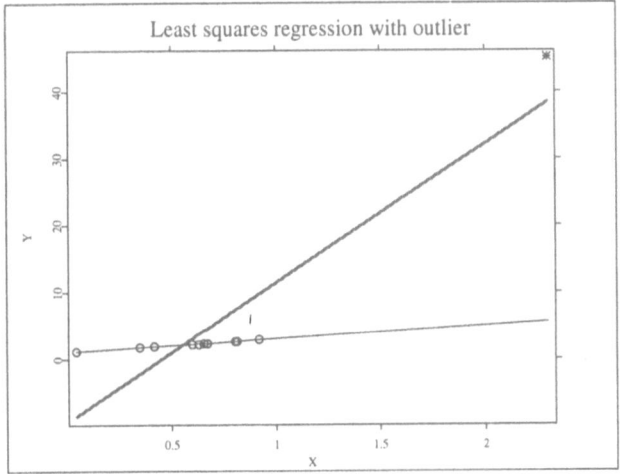

We may now leave XploRe and recall the quantlet when we restart XploRe. Suppose we have done this. How do we call our quantlet again? Assuming that XploRe is installed in the C:\XploRe directory of our computer, we use the command

```
func("C:\XploRe\myquant.xpl")
```

Let us now do this loading of the quantlet automatically by defining a new quantlet myquant2.xpl. This quantlet contains the func command and a call to the quantlet myquant by e.g.

```
myquant(#(0.9, 4.5))
```

If we encapsulate the code into a proc – endp bracket, we have the following quantlet

```
proc()=myquant2()
  myquant(#(0.9, 4.5))
endp
func("C:\XploRe\myquant.xpl")
myquant2()
```

Executing it will reproduce the same picture as before, but note that this time the call to myquant is done from another quantlet. Let us modify this procedure further so that the user may add outlying observations interactively. Suppose we want to see the effect of adding a new observation three times. We realize this by a while – endo construction, which we run exactly three times:

```
proc()=myquant2()
  ValueNames = "x=" | "y="
  defaults = 0.9 | 4.5
  i = 1
  while (i<=3)
    v = readvalue(ValueNames, defaults)
    myquant(v)
    i = i+1
  endo
endp
func("C:\XploRe\myquant.xpl")
myquant2()
```

quant06.xpl

The new quantlet myquant2.xpl first loads the existing quantlet myquant.xpl by the command func("C:\XploRe\myquant.xpl"). The names of the values to be read are defined by ValueNames = "x=" | "y=", the default values are set to $(0.9, 4.5)$ by defaults = 0.9 | 4.5. Then the loop construction with initial value i = 1 and end value 3 guarantees that the commands

```
v = readvalue(ValueNames, defaults)
myquant(v)
```

are executed exactly three times. If we enter for example $(10, 45)$ as the outlier value then we obtain the following graphic:

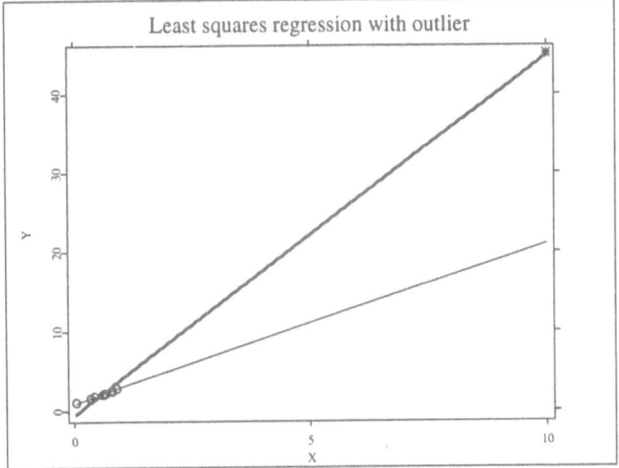

The while – endo construction is explained in more detail in Subsection 17.2.4. For more information on readvalue see Section 17.3.

17.2   Flow Control

XploRe offers a variety of possibilities to implement sophisticated structures
in your programs. With the commands introduced in the following you are
able to make your program react on conditions; you can handle failures while
avoiding or provoking a program's stop, and last but not least, you can give
your program a well-structured, understandable shape.

17.2.1   Local and Global Variables

```
putglobal(x)
     generates a global variable x

getglobal(x)
     reads a global variable x

existglobal("x")
     checks the existence of a global variable x
```

When you are working with procedures, you have to know whether the variables
used in the procedures are accessible in the main program and vice versa. The
answer is: XploRe strictly distinguishes between globally and locally defined
variables.

Variables in procedures can take on the values of global ones. The procedure
then can use the values, change them, and compute various results, but it cannot change the value of the original global variable. However, XploRe helps you
to overcome this restriction by using the `putglobal` and `getglobal` commands.
Note that these as well as the `existglobal` only work within procedures.

XploRe offers you the possibility to easily transfer a variable from the procedure
to the main program. A locally defined variable can then (when the procedure
has been run at least once) be accessed and changed in the main program.

Consider the following example:

```
proc() = test(a, b)
  c = a + b        ; computes c as the sum of a and b,
                   ;    c is now a local variable
```

```
    putglobal("c")  ; declares c to be also GLOBAL variable
  endp

  x = 1
  y = 2

  test(x, y)        ; runs the procedure test taking the
                    ;    values of the globals x and y for
                    ;    the locals a and b

  c                 ; output of c, c is accessible here
                    ;    because it has been transferred by
                    ;    the putglobal command
```
Q quant07.xpl

It produces the following output

```
Contents of c
[1,]    3
```

You can also do this the other way around. Normally in a procedure it is not possible to access global variables with the exception of those transferred in the head of the procedure's definition.

For this purpose, you use the getglobal command, which can transfer the value of a global variable to a local variable. The value can then be used, changed, and results can be computed in the procedure. However, it will not change the global variable in the main program.

```
  proc() = test(a, b)
    z = getglobal("z") ; the local z takes on the value of
                       ;    the global z
    z = a +b +z        ; computes the local z as a sum
    z                  ; output of the "local" z
  endp

  x = 1
  y = 2
  z = 3
```

```
test(x, y)                    ; runs the procedure test with the
                              ;    values of x and y for the locals
                              ;    a and b, respectively, prints the
                              ;    local z

z                             ; prints the global z
```

quant08.xpl

You obtain this result:

```
Contents of z
[1,]    6
Contents of z
[1,]    3
```

We want to remark here that the existence of a global variable can be checked by existglobal.

17.2.2   Conditioning

> if – {else} – endif
>   conditional branching with two branches

In XploRe, you have the possibility to let the execution of one or several commands depend on a logical condition. XploRe is able to check whether a certain condition is fulfilled or not and then executes the relevant commands in either case. Note that the if – else – endif construct only works within procedures.

For example: To compute the square root of a number, you have to make sure that the number is not negative. To this aim it is advisable to use the following if – else – endif construction:

```
proc() = squareroot(a)
  if(a >= 0)                  ; if this condition is true,
                              ;    XploRe will execute the next
                              ;    commandline
```

```
    sqrt(a)             ; computes the squareroot of a
  else                  ;   in case the condition above is not
                        ;   fulfilled XploRe runs the else branch
    "number is negative"
                        ; output of the else branch
  endif                 ; end of the if construction

 endp                   ; end of the procedure
```

Q quant09.xpl

You can check the effect if you call this procedure with a negative argument, e.g.

```
squareroot(-10)      ; runs the procedure with value -10
```

You obtain the following output:

```
Contents of _tmp
[1,]"number is negative"
```

By running the procedure **squareroot** with a positive argument

```
squareroot(9)          ; runs the procedure with value 9
```

you obtain the desired result

```
Contents of sqrt
[1,]    3
```

Note that the **else** branch is optional.

17.2.3   Branching

switch – case(x) – endsw
>   conditional branching with more than two branches

break
>   marks the end of a case block inside the switch environment, the
>   procedure is continued at **endsw**; it can be omitted

default
>   when the program's counter comes inside a **switch** – **endsw** con-
>   struct to this keyword, the following commands are processed in
>   any case; a **default** statement can be finished by the keyword
>   **break**, but does not have to be

If you want to make more than two branching points within your program, you
may use the **switch** – **case** – **endsw** construction.

The following procedure can distinguish whether its argument is positive, negative or zero:

```
proc() = signum(x)       ; defines a procedure named "signum"

  switch                 ; opens the switch branch
    case (x > 0)
      "positive number"  ; output in the case that x > 0
      break
    case (x < 0)
      "negative number"  ; output in the case that x < 0
      break
    default
      "number is zero"   ; output in the case that x = 0
      break
  endsw                  ; end of the switch branch
endp                     ; end of the procedure
```

Q quant10.xpl

By calling it with different arguments

```
signum(10)        ; runs the procedure with value 10
signum(-5)        ; runs the procedure with value -5
signum(0)         ; runs the procedure with value 0
```

you obtain, respectively, the following results

```
Contents of _tmp
[1,]"positive number"
Contents of _tmp
[1,]"negative number"
Contents of _tmp
[1,]"number is zero"
```

17.2.4   While-Loop

```
while - endo
    repeats one or several commands as long as some condition is
    fulfilled
```

You may execute one or several commands repeatedly — as long as a certain condition is fulfilled. For this purpose you have the while – endo loop at your disposal. Note that the while – endo construct only works within procedures.

This kind of loop executes one or several commands as long as a logical condition is fulfilled. The following example explains how the factorial of a natural number can be computed using this loop:

```
proc(j) = factorial(x)
                  ; defines a procedure named "factorial"
    j = 1         ; defines the variable j as 1

    while (x >= 2)  ; as long as this condition is fulfilled,
                  ;     XploRe executes the following commands

        j = j * x   ; computes j as the product of j and x
```

```
    x = x - 1       ; reduces x by 1
  endo              ; end of the while loop
endp                ; end of the procedure
```
🔍 quant11.xpl

After calling the procedure with command

```
factorial(5)       ; runs the procedure with value 5
```

you obtain the factorial 5! calculated with the help of the while – endo loop:

```
Contents of j
[1,]     120
```


17.2.5   Do-Loop

> do – until
>   repeats one or several commands until the condition is fulfilled

Another possibility for looping inside XploRe procedures is provided by the do – until construction. As with the while – endo construct, the do – until works only within procedures.

In contrast to the while – endo loop, the do – until loop executes one or several commands until a certain condition is fulfilled. Since the condition will be checked at the end of the loop, it runs at least once. An example follows:

```
proc(j) = factorial(x) ; defines a procedure named "factorial"
  j = 1                ; defines the variable j as 1
  do                   ; opens the do loop
    j = j *x           ; computes j as the product of j and x
    x = x -1           ; reduces x by 1
  until (x < 2)        ; if the condition is not fulfilled,
                       ;    the loop will be run again
endp                   ; end of the procedure
```
🔍 quant12.xpl

Calling the procedure with argument 5

```
factorial(5)       ; runs the procedure with value 5
```

produces the desired result:

```
Contents of j
[1,]    120
```

17.2.6 Optional Input and Output in Procedures

```
exist("x") or exist(x)
      checks the existence and the type of a variable x
```

Both the input and output parameters of a procedure quantlet are in fact lists. For more information on lists, see Chapter 16.

Due to this list concept, all input and output parameters can be optional. The existence of input parameters can be checked with the `exist` command. Table 17.1 gives all possible return values for `exist`.

Value	Meaning
-1	object does not exist
0	object exists but is empty
1	object exists and is numeric
2	object exists and is text
3	object exists and is of type XPLTIME
4	object exists and is a display
9	object exists and is a composed object (list)
10	object exists and is a quantlet
> 10	object exists as a quantlet and a variable, e.g. if there is a numeric variable cov and the library("xplore") is loaded exist("cov") results in 11

Table 17.1. Return values of `exist`.

Consider the following example:

```
proc(a,b)=myquant3(x,y)
  error(exist("x")<>1, "input x is not numeric!")
  if (exist("y")==0) ; y does not exist
    y=0
  endif

  switch
    case (exist("y")==2) ; if y is a string
      a=x
      break
    case (exist("y")==1) ; if y is numeric
      a=x
      b=y
      break
    default;
      error(1, "y is neither numeric nor string!")
      break
  endsw
endp
```

quant13.xpl

The quantlet myquant3 checks first, whether the input x is numeric. Then the existence of y is checked. If y does not exist, it is set to the numeric value 0.

The switch environment produces different results for the cases that y is string or numeric, respectively. We added an error message for the case that y is neither string nor numeric. You may now test the procedure with different input values:

```
result=myquant3(1,2)
result
```

gives

```
Contents of result.a
[1,]          1
Contents of result.b
[1,]          2
```

This means that the resulting output object is a list comprising the components `result.a` and `result.b`, which contain just the input values.

We can also call `myquant3` without specifying the input y, i.e.

```
result=myquant3(1)
result
```

The result is similar, except that the missing y value is replaced by 0:

```
Contents of result.a
[1,]        1
Contents of result.b
[1,]        0
```

In the case that a string input y is used, the result is again different:

```
result=myquant3(1,"hallo")
result
```

produces only one output

```
Contents of result.a
[1,]        1
```

To be on the safe side, we assign the output always to the same variable `result`. This captures the case of two outputs as well as of one output. The function `exist` can be used again to check whether the output variable `result` contains valid component `result.b`. For our last example

```
exist(result.b)
```

yields

```
Contents of exist
[1,]        0
```

which indicates that the component `result.b` is empty.

17.2.7   Errors and Warnings

```
error (cond, message)
     stops a XploRe program and displays a message

warning (cond, message)
     displays a warning message
```

You have the opportunity to give to the user some hints about problems which
occurred when the program is run. With the error and warning commands,
you can transmit messages to the user of the program.

Both commands check whether a certain condition is true (equal to 1) or not
(equal to 0). If the condition is true, a window will be displayed containing a
message that has been specified within the command. If the condition is false,
the program continues with the next command.

The error command displays the error box and stops immediately if the given
condition is fulfilled. All data and changes to variables that have not been
saved before this moment will be lost.

```
proc() = test(x)    ; defines a procedure
  error(x<0,"Negative argument!")
                    ; displays an error box containing the
                    ;     specified text and stops the program
                    ;     in case x is negative
   sqrt(x)          ; computes the square root in case the
                    ;     preceding command did not lead to
                    ;     a break of the program
  endp              ; end of the procedure
```
                                                     Q quant14.xpl

Calling this procedure with the negative argument

```
  test(-4)              ; runs the procedure with value -4
```

will display only this error message:

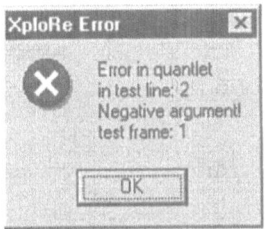

If you do not want to stop the program immediately after checking a problem, you can use the warning command. In this case there is no data loss. If the condition in the warning command is fulfilled, the program continues and shows a warning box after it finishes:

```
proc() = test(x)    ; defines a procedure

  warning(x<0,"Negative argument!")
                    ; displays a warning box containing the
                    ;     specified text in case x is
                    ;     negative at the end of the program

  sqrt(abs(x))      ; computes the squareroot of abs(x),
                    ;     whatever the above command found
                    ;     about x

  endp              ; end of the procedure
```
Q quant15.xpl

Calling this procedure with the negative argument

```
test(-9)            ; runs the procedure with value -9
```

will produce the result

```
Contents of sqrt
[1,]    3
```

accompanied with the warning

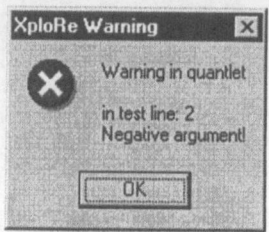

17.3   User Interaction

---

selectitem (headline, items)
    allows the user to choose one or more items before continuing the
    program

values = readvalue (valuenames, defaults)
    allows the user to enter one or more values before continuing the
    program

In the chapter about teachware we have seen a lot of menus which are used
to execute several programs. The command to achieve this is `selectitem`. It
allows us to select one or more items and continues a program as soon as the
user pressed the OK button:

```
selhead = "My headline"; sets the headline
selitem = "item 1"|"item 2"|"item 3"; sets the items
;
; now we open a dialog box with the items and wait
; until  the user presses OK
;
sel = selectitem(selhead, selitem)
sel
```

Q quant16.xpl

The resulting selection box is shown in Figure 17.1.

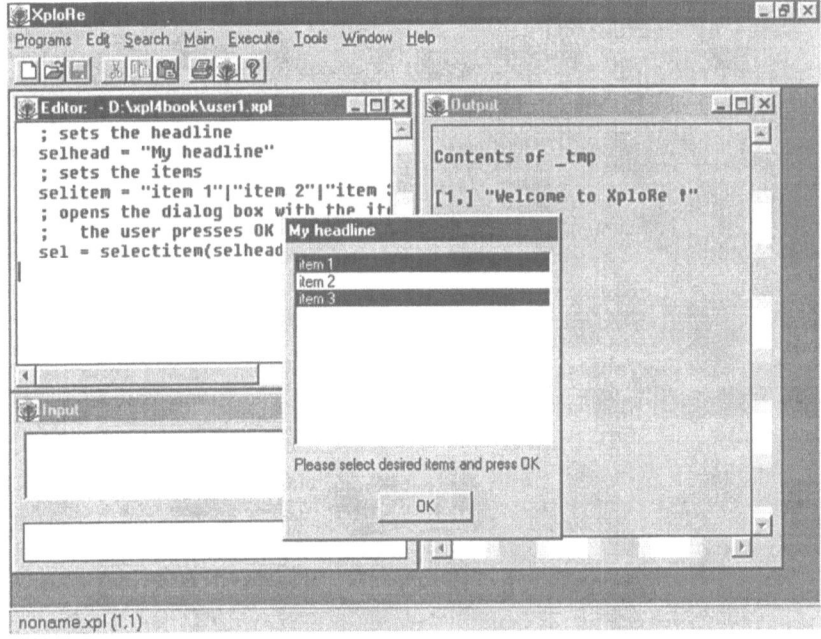

Figure 17.1. The `selectitem` example. ◫ `quant16.xpl`

The return vector `sel` contains zeroes and ones; zero if the user has not selected the item and one if the user has selected the item.

```
Contents of sel
[1,]        1
[2,]        0
[3,]        1
```

Let's now do a complex example: We choose between two univariate regression methods for a given data set. Then we plot the data and the regression estimate.

```
proc()=regression(x,y)
  selhead  = "Choose regression method"
  selitem  = "Linear regression"|"Nadaraya-Watson"
  disp     = createdisplay(1,1)
  continue = 1
```

```
    show (disp, 1, 1, x~y)
    while (continue<>0)
      sel = selectitem (selhead, selitem)
      if (sel[1]==1)
        l = grlinreg(x~y)
        show (disp, 1, 1, x~y, l)
      endif
      if (sel[2]==1)
        h = (max(x)-min(x))/20
        w = sort(x~y)
        mh = regest(w)
        l = setmask(mh, "line")
        show (disp, 1, 1, x~y, l)
      endif
      continue = sum(sel)>0
    endo
  endp

  library("plot")
  library("smoother")
  x = read("bostonh")
  regression(x[,13], x[,14])
```

quant17.xpl

A screenshot of this procedure is shown in Figure 17.2.

Let's now discuss the quantlet regression in detail. First we define our headline and items. Then we plot the data set. Finally we build a loop which will be finished if continue equals zero.

In the loop we open first the selection box and the user selects one of the two methods. Thus we receive in sel either #(1,0) in the case that the user selects Linear regression, or #(0,1) in the case that the user selects Nadaraya-Watson. The two if statements test whether sel[1] equals one or if sel[2] equals one.

In the first case we compute a linear regression; in the second case we compute the Nadaraya–Watson regression estimator (see Chapter 6)

$$\widehat{m}_h(x) = \frac{\sum_{i=1}^{n} K\left(\frac{x_i - x}{h}\right) y_i}{\sum_{i=1}^{n} K\left(\frac{x_i - x}{h}\right)}$$

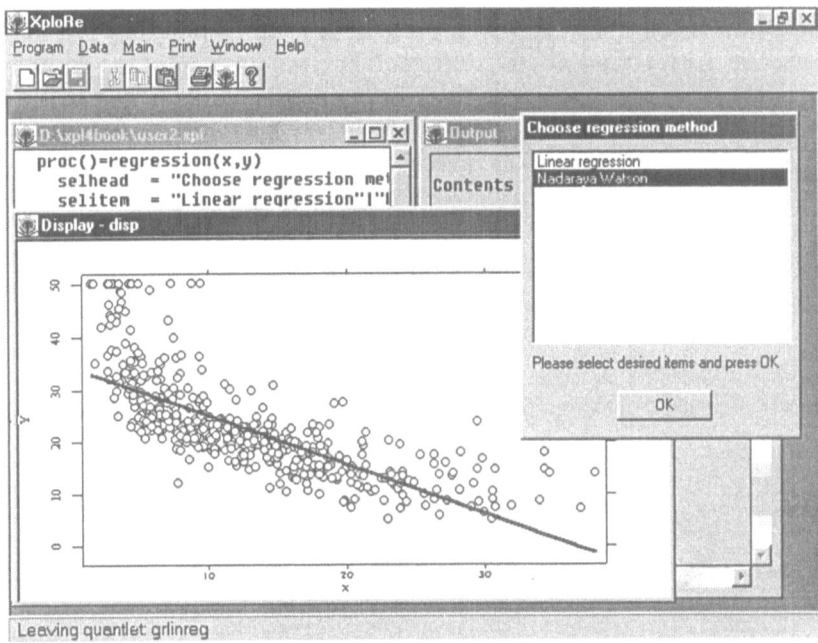

Figure 17.2. A `selectitem` in action. Ⓠ `quant17.xpl`

and plot it.

However, two possibilities are not handled well: If the user selects nothing (`sel = #(0,0)`) or if the user selects both (`sel = #(1,1)`). In the latter case, we would always get the Nadaraya–Watson estimator.

The case `sel = #(0,0)` can be handled by `continue = sum(sel)` which simply finishes our procedure. We add up all elements and compare if the sum is larger than zero. If we have selected one or more items, then the sum will be positive and `continue` will be true (=1). Otherwise `continue` will be false (=0).

Our last regression program is modified such that we can also handle the case `sel = #(1,1)` properly.

We know that the bandwidth h is important for the nonparametric Nadaraya–Watson estimator of the true regression function m. In our example we have always used $h = 0.05(\max x_i - \min x_i)$. Many algorithms have been created to determine the "right" bandwidth for a given data set. (This issue is addressed in detail in Chapter 6.) However, sometimes it turns out that the human eye chooses better than any algorithm. Thus we use `readvalue` to get the bandwidth interactively from the user. Let's do a simple example first:

```
item = "item 1"|"item 2"|"item 3"   ; set the items
def  = 0|0|0                         ; sets the default values
;
; now we open a dialog box with the items and
; wait until the user presses OK
;
val  = readvalue(item, def)
```

◌quant18.xpl

The resulting `readvalue` box is shown in Figure 17.3

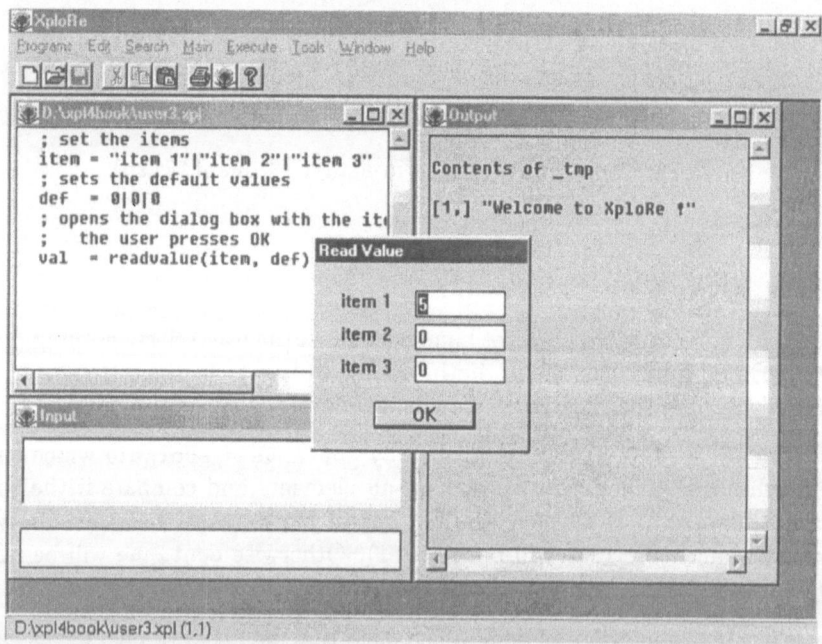

Figure 17.3. The `readvalue` example. ◌`quant18.xpl`

`readvalue` has two input parameters: A text string `item` such that the user knows what we are asking him and a default value `def`. The output of `readvalue` is a vector of the same length as **def**. Note that **def** can also be a text vector rather than a numerical vector. In this case the output of **readvalue** is a text vector too. Note that you cannot mix numerical and text items in **def** !

Let's now add **readvalue** to our regression program to allow the user to choose the bandwidth interactively:

```
proc()=regression(x,y)
  selhead  = "Choose regression method"
  selitem  = "Linear regression"|"Nadaraya-Watson"
  disp     = createdisplay(1,1)
  continue = 1
  show (disp, 1, 1, x~y)
  h = (max(x)-min(x))/20
  while (continue<>0)
    sel = selectitem (selhead, selitem)
    n   = sum(sel)
    if (n>0)
      if (sel[2]==1)
    .     h = readvalue ("Bandwidth", h)
      endif
      disp = createdisplay(1, n)
      i    = 1
      if (sel[1]==1)
        l = grlinreg(x~y)
        show (disp, 1, i, x~y, l)
        i = i+1
      endif
      if (sel[2]==1)
        w  = sort(x~y)
        mh = regest(w, h)
        l  = setmask(mh, "line")
        show (disp, 1, i, x~y, l)
        i = i+1
      endif
    endif
    continue = (n>0)
  endo
endp
```

```
library("plot")
library("smoother")
x = read("bostonh")
regression(x[,13], x[,14])
```

Q quant19.xpl

Figure 17.4 shows this quantlet in action.

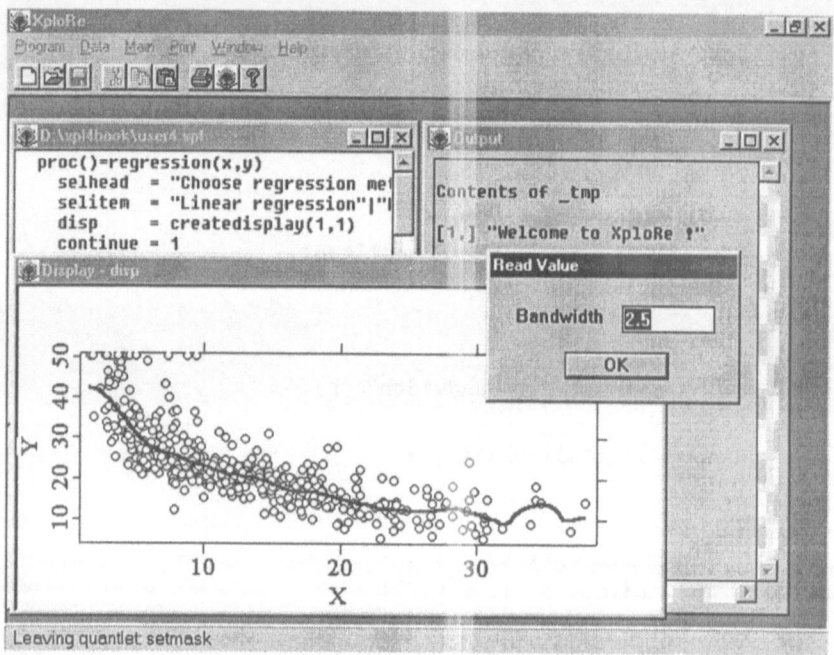

Figure 17.4. readvalue in action.   Q quant19.xpl

17.4 APSS

```
makehelp(options,file)
      invokes the automatic translator for generating help pages

setenv("xpl4outhelp",path)
      sets the APSS output stream
```

Once we have constructed a fine quantlet, we are interested in other people using it. This is done by publishing the quantlet technology via an HTML file. Other students and scientists may use the quantlet then inside the XploRe Auto Pilot Support System (APSS), see Figure 17.5.

We are now ready to publish our first quantlet with the tools that are available in XploRe. First we define where the HTML file should be written. We do this via the `setenv` command by setting the APSS output stream to go into the directory `C:\XploRe`

```
setenv("xpl4outhelp","C:\XploRe")
```

We may check the effect of this command by clicking the Main item and then the Info item. We see that the output stream is set to the desired directory.

In the next step it is necessary to provide all information which will be contained in the resulting HTML file. This is done very easily by including the following header directly in the source code of our quantlet. A template of the header is inserted automatically via the Insert help item in the Tools menu and you can fill in all the desired information. You should do this very carefully in order to allow other users to run your quantlet comfortably. Typically, it starts immediately after the `proc` statement.

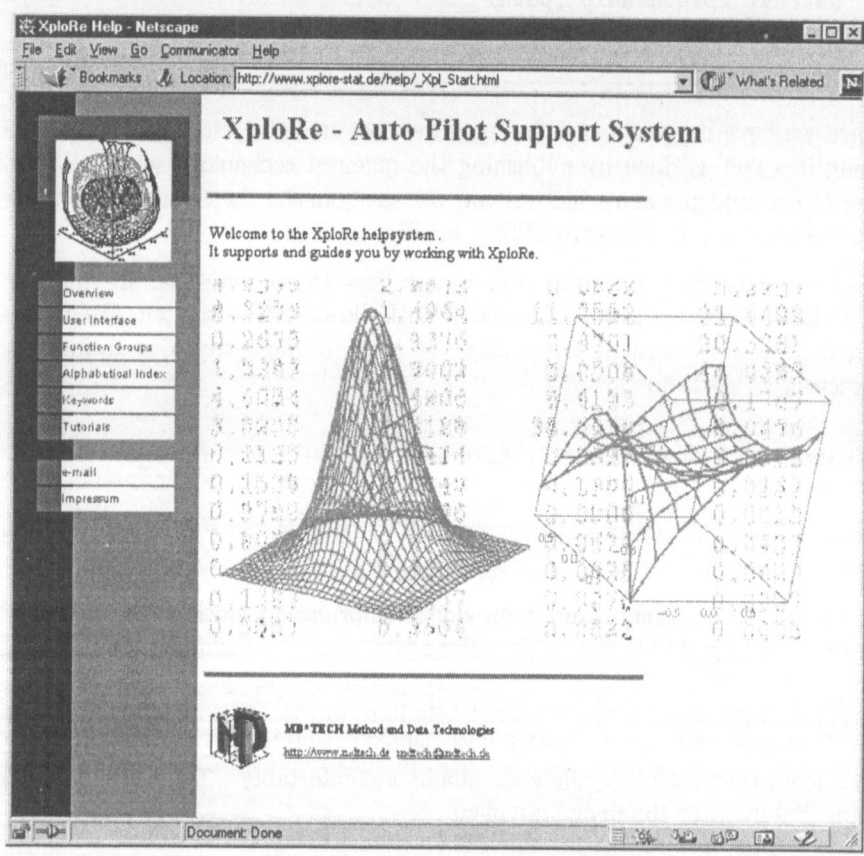

Figure 17.5. APSS start page.

Recall our first quantlet myquant from Section 17.1. The beginning of the
quantlet myquant.xpl should look like this:

```
proc() = myquant(obs1)
; -----------------------------------------------------------
;  Library      stats
; -----------------------------------------------------------
;  See_also     linreg
; -----------------------------------------------------------
;  Macro        myquant
; -----------------------------------------------------------
;  Description  plot of the effect of outliers in linear
;               regression model
; -----------------------------------------------------------
;  Usage        myquant(obs1)
;  Input
;    Parameter  obs1
;    Definition  2x1 vector - coordinates of an outlier
;  Output
;    Parameter
;    Definition
; -----------------------------------------------------------
;  Notes    works fine
; -----------------------------------------------------------
;  Example  myquant(#(10,45))
; -----------------------------------------------------------
;  Result   produces a plot of true regression line and
;               the regression line
; -----------------------------------------------------------
;  Keywords     myquant
; -----------------------------------------------------------
;  Reference    lecture notes, XploRe manual
; -----------------------------------------------------------
;  Link
; -----------------------------------------------------------
;  Author       anonymous
;-----------------------------------------------------------
    n = 10
```

quant20.xpl

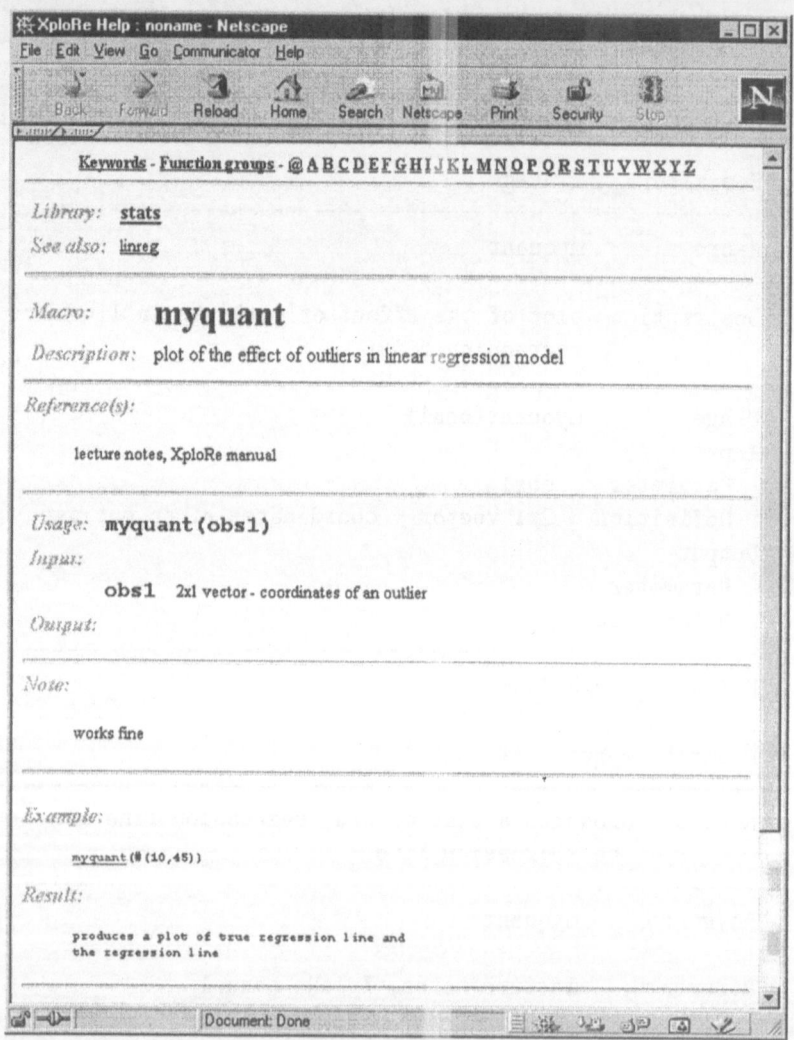

Figure 17.6. Helpfile for myquant.

Now you are ready to create the HTML help file with the `makehelp` command

```
makehelp("html","C:\XploRe\myquant.xpl")
```

This command automatically creates an HTML help file `myquant.xpl` in directory `C:\XploRe` which was beforehand specified as the output directory for the help file by the command `setenv`. You can have a look at the resulting file with Netscape Navigator (Figure 17.6), Internet Explorer or any other World Wide Web browser of your choice.

The first argument of `makehelp` specifies which type of help file will be created. Possible choices are `"html"` for HTML help file, `"ascii"` for ascii help file and `"latex"` for manual page. The second argument specifies the XploRe program for which the help file will be created.

You do not have to restrict yourself to generating only one help file. You can easily create the complete APSS (Auto Pilot Support System). The command

```
makehelp("html","C:\XploRe")
```

generates the complete help system from the files in directories `C:\XploRe\hlp2` and `C:\XploRe\lib` and writes the resulting files in `C:\XploRe\help` and `C:\XploRe\help\text` directories. Before you call this command, you should check whether these directories exist and whether you have permission to write into them.

17.5   Quantlibs

```
library("lib")
      loads XploRe quantlib lib, a library of quantlets
```

If you have several quantlets which serve the same purpose, you can combine them easily into a **quantlib**, i.e. a library of quantlets. If you want to use quantlets stored in some library, you have to load the library with the command

```
library("myqlib")
```

This command will load all quantlets which are listed in file `myqlib.lib` in your `XploRe\lib` directory. You can very easily create your own quantlib. Create file which looks like this:

```
;*********************************************************
;************* The XploRe Myqlib library **************
;*********************************************************
;****************** date : 991204 *********************
;*********************************************************
myquant            ; My first quantlet
myquant2           ; My second quantlet
myquant3           ; My third quantlet
regression         ; My interactive regression quantlet
```

and save it under the name `myqlib.lib` to the directory `XploRe\lib` together with quantlets `myquant.xpl`, `myquant2.xpl`, `myquant3.xpl`, `regression.xpl`. Now you are ready to use your first quantlib. Entering

```
library("myqlib")
```

on the command line will load quantlets listed in the file `myqlib.lib`, i.e. `myquant`, `myquant2`, `myquant3`, `regression` and you can start to use them.

It is important that the quantlib is saved in the directory `XploRe\lib` together with all quantlets it refers to.

It is also possible to edit existing quantlibs and you are free to modify them in any way which you find convenient. You can very easily create a system of quantlibs which best suits your needs.

The current list of existing quantlibs is shown in the following Table.

Library Name	Library
eiv*	Error in variables models
finance	Financial library
gam*	Generalized additive models
glm	Generalized linear models
gplm*	Generalized partially linear models
graphic	Graphics library
ista*	Interactive statistics
kernel	Kernel library
math	Mathematical functions
metrics	Econometrics library
multi*	Multiple time series
nn	Neural networks
plm*	Partially linear models
plot	Plotting library
pp*	Projection pursuit
rpclib*	Remote procedure calls
salsa*	Seasonal adjustment
smoother	Smoother methods
stats	Statistics
times	Time series analysis
tware	Teachware statistics
twave	Teaching wavelets
wavelet	Quantlets for wavelet analysis
xclust*	Cluster analysis
xplore	Basic stuff

Note that the quantlibs marked by * are not part of the Academic Edition of XploRe.

Appendix

Appendix

A Customizing XploRe

There exist two possibilities to customize XploRe:

- the file XploRe.ini and

- the file startup.xpl

Whereas the entries in the XploRe.ini influence the behavior of XploRe by setting environment variables, startup.xpl contains executable XploRe code.

A.1   XploRe.ini

A.1.1   The ini File

Since XploRe does not make any entries in the Windows registry, we need another mechanism to get the basic settings into XploRe. Thus, if you double click on XploRe, then the file XploRe.ini will be read as the first file. If you start XploRe from a DOS box or create a link, then you might use the following syntax to force XploRe to read a specific ini-file:

```
xplore -ini C:\xplore\myini.ini
```

Let's analyze the standard XploRe.ini file which contains a set of lines like

```
# The following variables are influencing the behavior of \xplore{}:
#
#   os                      operating system (cannot be overwritten)
...
#
```

```
# should point to the next XploRe WWW site,
#
#   Germany (Berlin),      http://www.xplore-stat.de
#   USA (Fairfax, VA),     http://www.galaxy.gmu.edu/~xplore/
#
xpl4nethome=http://www.xplore-stat.de
#
# basic directory given by the system administrator
#
xpl4syshome=C:\XploRe
```

The lines starting with # will be considered as comments and ignored by
XploRe. The beginning of the file lists all possible environment variables which
can be modified. The only exception is the environment variable os, which is set
internally by XploRe. Most of the environment variables can also be changed
within XploRe by the command setenv.

At the begin of XploRe.ini, we find the environment variables outheadline,
outlayerline, outlineno and outmaxdata. The effects of these variables and
their default values are described in Chapter 15. The following variables from
browser to rpclink are of minor interest for Windows, since they are intended
for the XploRe server and for the Unix version of XploRe. The exception is
startup, which tells XploRe which startup program it should read (see next
section, the default is startup.xpl).

Then we will find four sets of variables named xpl4...home, xpl4...lib,
xpl4...data, xpl4...help, xpl4...dll, xpl4...prog, xpl4...temp. These
variables tells XploRe where to read and write data, libraries, programs, help
and temporary files.

xpl4...home is for the convenience of the user. E.g. if xpl4syshome is
set to C:\XploRe, then the variable xpl4syslib will be set automatically to
C:\XploRe\lib, xpl4sysdata will be set to C:\XploRe\data and so on. None
of the xpl4...home variables can be modified from within XploRe.

The xpl4net... variables are used to get information over the Internet. Only
xpl4nethelp is used to find a help system, if no local help system is found.
To use this possibility, you need to have connection to the Internet. These
variables cannot be modified from within XploRe. The xpl4sys... variables
are to find the locally installed libraries, data and programs which come with
XploRe. Normally the system administrator will set these variables on the
directory where XploRe is installed. These variables cannot be modified from
within XploRe.

In contrast to all other variables, the xpl4out... variables should contain only one directory. This is the place where XploRe writes the data, e.g. written by the command write, and all other files. A variable that should be modified with care is xpl4outtemp, which tells XploRe where it can write its temporary files. If this points to a directory which does not exist or where the user has no write permission, then XploRe will not start at all.

Finally there is a set of variables xpl4home, ..., xpl4temp which can be modified by the user.

The following XploRe.ini file is a minimal setup for XploRe, assumed XploRe is installed in C:\XploRe:

```
xpl4nethome=http://www.xplore-stat.de
xpl4syshome=C:\XploRe
```

We point to the XploRe home URL and to the local installation directory of XploRe. Assume now you move the whole installation from C: to D:. The only thing you will have to change is the entry xpl4syshome:

```
xpl4nethome=http://www.xplore-stat.de
xpl4syshome=D:\XploRe
```

A.1.2 Composing Paths

Whenever XploRe tries to read data (e.g. if you called the command **read** with no absolute path), then it will try to read the data from

1. the directories given by xpl4outdata,

2. the directories given by xpl4data,

3. the directories given by xpl4sysdata and

4. the directories given by xpl4netdata (planned for the future).

This applies to all other operations, like loading libraries, starting the help system, and so on.

A.2   startup.xpl

The startup.xpl is analogous to the well-known autoexec.bat under DOS or Windows. After reading the ini-file XploRe tries to read the startup file and to execute the XploRe code inside it before appearing on the screen.

Note that you cannot define quantlets within the startup file and therefore cannot use control flow elements like looping, branching, etc.

The standard startup file startup.xpl contains only one line

```
"Welcome to XploRe !"
```

which appears immediately in the output window if you start XploRe. You might use the startup file to load some often used data sets or libraries.

B Data Sets

B.1 Netincome–Food Expenditures

The data file `nicfoo.dat` contains a discretized subsample of 140 observations from the 1973 U.K. Family Expenditure Survey, see Härdle (1990).

Column	Type	Description
1	metric	netincome, standardized and outliers removed
2	metric	food expenditures, same scale as netincome

B.2 U.S. Companies

The data file `uscomp2.dat` contains measurements for 79 U.S. companies out of the top 500.

Column	Type	Description
1	text	company
2	metric	assets
3	metric	sales
4	metric	market value
5	metric	profits
6	metric	cash flow
7	metric	employees
8	text	branch

B.3   CPS 1985

The data file cps85.dat, see Berndt (1991), contains an extract of the May 1985 U.S. Current Population Survey (CPS) with 534 observations.

Column	Type	Description
1	metric	years of education
2	binary	1 if lives in south
3	binary	1 if nonwhite
4	binary	1 if Hispanic
5	binary	1 if female
6	binary	1 if married with spouse present (in household)
7	binary	1 if married female with spouse present
8	metric	years of labor market experience (= age-education-6, minimum = 0 imposed ex post)
9	metric	years of labor market experience squared
10	binary	1 if working on a union job
11	metric	natural logarithm of average hourly earnings
12	metric	age in years
13	binary	1 if working in manufacturing industry
14	binary	1 if working in construction industry
15	binary	1 if occupation is managerial or administrative
16	binary	1 if occupation is sales worker
17	binary	1 if occupation is clerical worker
18	binary	1 if occupation is service worker
19	binary	1 if occupation is professional/technical worker

B.4   Boston Housing

The data in bostonh.dat were collected by Harrison and Rubinfeld (1978). They comprise 506 observations for each census district of the Boston metropolitan area. The data set was analyzed in Belsley, Kuh and Welsch (1988, pp. 244–261) which introduced a number of transformations.

Column	Variable	Type	Description
1	CRIM	metric	per capita crime rate by town
2	ZN	metric	proportion of residential land zoned for lots over 25,000 sq.ft.
3	INDUS	metric	proportion of nonretail business acres per town
4	CHAS	binary	Charles River dummy variable (= 1 if tract bounds river, 0 otherwise)
5	NOXSQ	metric	nitric oxides concentration (parts per 10 million)
6	RM	metric	average number of rooms per dwelling
7	AGE	metric	proportion of owner-occupied units built prior to 1940
8	DIS	metric	weighted distances to five Boston employment centers
9	RAD	metric	index of accessibility to radial highways
10	TAX	metric	full-value property tax rate per $10,000
11	PTRATIO	metric	pupil–teacher ratio by town
12	B	metric	$1000(Bk - 0.63)^2$ where Bk is the proportion of blacks by town
13	LSTAT	metric	% lower status of the population
14	MEDV	metric	median value of owner-occupied homes in $1000's

B.5   Lizard Data

The data in `lizard.dat` are from McCullagh and Nelder (1989) and concern the daytime habits of two species of lizard, grahami and opalinus. They were collected by observing occupied sites or perches and recording the appropriate description, namely species involved, time of day, height and diameter of perch and whether the site was sunny or shaded.

Column	Type	Description
1	binary	indicator for the height of the perch (in feet): 0 if < 5, 1 if ≥ 5
2	binary	indicator for the diameter of the perch (in inch): 0 if ≤ 2, 1 if > 2
3	binary	indicator whether the site was sunny or shaded: 0 if sun, 1 if shade
4	categorical	indicator for the time of day: 1 if early, 2 if midday, 3 if late
5	metric	number of grahami observed
6	metric	total number of lizards observed

B.6 Kyphosis Data

The data in kyphosis.dat are from Hastie and Tibshirani (1990, p. 301) and were collected on 83 patients undergoing corrective spinal surgery. The objective was to determine important risk factors for kyphosis, or the forward flexion of the spine at least 40 degrees from vertical, following surgery. The risk factors are age in month, the starting vertebrae level of the surgery and the number of vertebrae involved.

Column	Type	Description
1	metric	the age (in month)
2	metric	starting vertebrae level of the surgery
3	metric	number of vertebrae involved
4	binary	indicator for the presence of kyphosis: 0 if absent and 1 if present

B.7 Swiss Bank Notes

The data file bank2.dat contains measurements on 100 genuine and 100 forged old Swiss 1000 franc bills from Flury and Riedwyl (1988). The first 100 rows correspond to the genuine bills and the second 100 rows to the forged data.

Column	Type	Description
1	metric	length of the bill
2	metric	height of the bill, measured on the left
3	metric	height of the bill, measured on the right
4	metric	distance of inner frame to the lower border
5	metric	distance of inner frame to the upper border
6	metric	length of the diagonal

B.8 Earnings Data

The data file earnings.dat contains 500 observations, collected from 100 individuals that have been observed over 5 time periods.

Column	Type	Description
1	index	individual
2	index	time period
3	metric	monthly earnings
4	metric	working experience in years
5	metric	years of schooling
6	binary	1 if male, 2 if female

B.9 Westwood Data

The file westwood.dat contains 10 observations on lot sizes and man hours for
10 production runs, see Neter, Wasserman and Kutner (1989).

Column	Type	Description
1	index	individual
2	metric	lot size
3	metric	man hours

B.10 Pullover Data

The data file pullover.dat contains data on "classic blue" pullovers sales in
10 periods, see Härdle and Simar (1999).

Column	Type	Description
1	metric	sales (in pullovers)
2	metric	price (in DM)
3	metric	advertisement costs in local newspapers (in DM)
4	metric	presence of shop assistant (in hours per period)

B.11 Geyser Data

The file geyser.dat contains 267 observations from the Old Faithful Geyser
in the Yellowstone National Park.

Column	Type	Description
1	metric	duration (in minutes) of an eruption
3	metric	waiting time (in minutes) to the next blowout

Bibliography

Belsley, D.A., Kuh, E. and Welsch, R.E. (1980). *Regression Diagnostics*, Wiley.

Berndt, E. (1991). *The Practice of Econometrics*, Addison–Wesley.

Flury, B. and Riedwyl, H. (1988). *Multivariate Statistics, A Practical Approach*, Cambridge University Press.

Härdle, W. (1990). *Applied Nonparametric Regression*, Econometric Society Monographs No. 19, Cambridge University Press.

Härdle, W. (1991). *Smoothing Techniques, With Implementations in S*, Springer, New York.

Härdle, W. and L. Simar (1999). *Applied Multivariate Statistical Analysis*, Course Script.

Harrison, D. and Rubinfeld, D.L. (1978). Hedonic prices and the demand for clean air, *J. Environ. Economics & Management*, **5**: 81–102.

Hastie, T. J. and Tibshirani, R. J. (1990). *Generalized Additive Models*, Vol. 43 of *Monographs on Statistics and Applied Probability*, Chapman and Hall, London.

McCullagh, P. and Nelder, J. A. (1989). *Generalized Linear Models*, Vol. 37 of *Monographs on Statistics and Applied Probability*, 2 edn, Chapman and Hall, London.

Neter, J., Wasserman, W. and Kutner, M. H. (1989). *Applied Linear Regression Models*, Irwin.

Index